future arquitecturas s.l. 编

未来建筑竞标 中国 第2辑

10欧罗潘复兴

FUTURE ARQUITECTURAS COMPETITIONS CHINA II
EUROPAN 10 REVITALIZATION

图书在版编目 (CIP) 数据

未来建筑竞标. 中国. 第2辑：10欧罗潘复兴：汉英对照 / 西班牙未来建筑出版社编. —杭州：浙江大学出版社，2011.2
ISBN 978-7-308-08414-7

Ⅰ. ①未… Ⅱ. ①西… Ⅲ. ①建筑设计—世界—现代—图集 Ⅳ. ①TU206

中国版本图书馆CIP数据核字（2011）第020055号

浙江省版权局著作权合同登记图字：11-2011-14号

未来建筑竞标 中国 第2辑 10欧罗潘复兴

FUTURE ARQUITECTURAS COMPETITIONS CHINA II EUROPAN 10 REVITALIZATION

future arquitecturas s.l. 编

责任编辑：黄娟琴
文字编辑：李峰伟 张凌静
封面设计：future arquitecturas s.l. 未来建筑
出版发行：浙江大学出版社
（杭州天目山路148号 邮政编码：310007）
（网址：http://www.zjupress.com）
印 刷：杭州多丽彩印有限公司
印 张：15
开 本：889mm×1194mm 1/8
字 数：192千
版印次：2011年2月第1版 2011年2月第1次印刷
书 号：ISBN 978-7-308-08414-7
定 价：150.00元

浙江大学出版社发行部邮购电话：88925591

典型案例分析

在过去的十年中，中国与拉丁美洲之间就科学技术这样的重要文化科技类课题引发了爆发式的贸易增长，并促进了双方的共同发展。继普通话之后，西班牙语已成为世界第二大母语。如果当今时代的开端从根本上减少了世界对技术与数学的探求，那么科学的发展就俨然拓宽了具体的生活空间和原始的思维空间。很少有时代会像现在这个时代一样经历过如此深入且飞速的转型以及全球化进程，使得社会和文化结构亦随之而改变。

中国以及印度和印度尼西亚等东南亚国家通过强劲的增长来抵御国际金融危机。由于高度强调人力资源并将其作为差动因素，这些国家在科学、研究与培训方面的进展尤为显著。高品质创新技术学术项目的开发很好地体现了高校与商业界的紧密联系，该联系也极有可能促成科技类大学开发具有商业价值的商业项目。

"2010年第17届泛美基多建筑双年展"延续了他们的事业， 32年来一直努力为那些旨在研究和转换建筑行业内不同标准和知识并将其作为一种文化表达的定期组织分析能见利益。Terence riley被委任为2011深圳•香港城市\建筑双城双年展（SZHK）总策展人，他将成为自2005年该双年展开展以来的第一个国际策展人。此次双年展将把目光对准快速发展的中国城市化和建筑类活动，尤其是深圳和香港地区。预计来自30个国家的约6万名观众将参加此次双年展并一同探讨关于现代化大都市崛起的有关问题。中国蓬勃发展中的经济的知识和创造性基础是什么？是她的快速城市化进程还是其建筑和艺术发展？所需的代价是什么？这一增长的盲点又是什么？

作为先行者的深圳和香港及基多对于中国和拉丁美洲的发展究竟会起到什么特别作用？

法国画家Pierre Auguste Renoir采用印象主义手法，通过简单的形体描绘来传达当前与历史的存在感。

MODEL AND ANALYSIS

In the last decade there has been a trade explosion between China and Latin America in important topics culturally and intellectually as sciences and technology that have promoted their growth. After Mandarin language, Spanish is the second native language in the world. If the roots of the beginning of the Modern Age had reduced the world to the exploration of the technique and mathematics, the development of sciences leads the horizon to the specific world of life, to the original space of thinking. Few epochs like the present have been subject to such profound and quick processes of transformation and globalization which have affected the social and cultural structures.

China and Southeast Asia, e.g. India, Indonesia, resist the international economic crisis with important growths, where the progress of science, research and training stand out because of its high values emphasizing human capital as a differential factor. Thus, the close link between university and business worlds which develop high quality academic projects of innovation and technology, will hold options to become business projects of commercial value of the scientific university.

The XVII Pan-American Architecture Biennale of Quito, BAQ 2010 has successfully continued its career after 32 years analysing the visible benefits of its periodical organization aimed at the research and the transfer of different criteria and knowledge within architecture as a cultural expression. Terence riley has been appointed chief curator for the 2011 Shenzhen & Hong Kong Bi-City Biennale of Urbanism\Architecture (SZHK). He will be the first international curator for the biennale since it started in 2005. The Biennale will focus on the fast progress of urbanization and architectural activities in China, particularly Shenzhen and Hong Kong. The approximately 60,000 visitors, from 30 countries, expected to attend the meeting will be exploring issues regarding the emergence of the modern metropolis. What is the intellectual and creative underpinning of China's burgeoning economy, its rapid urbanization, its architectural and artistic development? What are the costs, and the blindspots, of this growth?

What is the special role played by Shenzhen & Hong Kong and Quito as laboratories for China's and Latin America's development?

The French painter Pierre Auguste Renoir, spread through his impressionism the presence and the history of a glance setting its identity with its shape.

浙江大学建筑工程学院现有规划、建筑、土木、水利四大学科，基本涵盖了国家基本建设领域的全部学科，涉及到建筑、市政、交通、水利、铁道、港口与海洋工程等主要产业领域。在学科上具有良好的互补性和交叉性，为学科发展奠定了良好的基础。

浙江大学建筑设计研究院始建于1953年，是国家重点高校中最早成立的甲级设计研究院之一。业务范围有高层、超高层的大型办公、宾馆、商业综合体、行政办公楼；学校校园规划与设计；影剧院、图书馆、博物馆等文化建筑；居住区规划与设计；体育建筑；医院类建筑；城市设计；智能建筑设计、室内设计；风景园林与景观设计；市政公用工程；岩土工程；幕墙设计；古建筑和近现代建筑的维修保护、文物保护规划等。

The College of Civil Engineering and Architecture of Zhejiang University has a wide coverage of the fields of national capital construction including: building design and construction, municipal engineering, transportation, water conservancy, railway and harbor and offshore engineering. And it forms a construction with Regional and Urban Planning, Architecture, Civil Engineering and Hydraulic Engineering to have the integration of production, learning and research.

Architectural Design and Research Institute of Zhejiang University was founded more than half a century ago in 1953, it has been one of the earliest Grade-A design institutes established among state key universities. The business covers high-rise or super high-rise large office, hotel, business complex and administrative office building; planning and design of campus; cinema, library, museum and other cultural buildings; residential community planning and design; construction of sports facilities; construction of hospital; urban design; intelligence architectural design, indoor design; landscape architecture and landscape design; municipal public engineering; geotechnical engineering; curtain walling design; maintenance and protection of ancient building and modern building, planning of cultural relics protection, etc.

海明厂区改造·德清

Refurbishment of Haiming Factory · China

陈翔 Chen Xiang · 朱培栋 Zhu Peidong

委托项目 commission

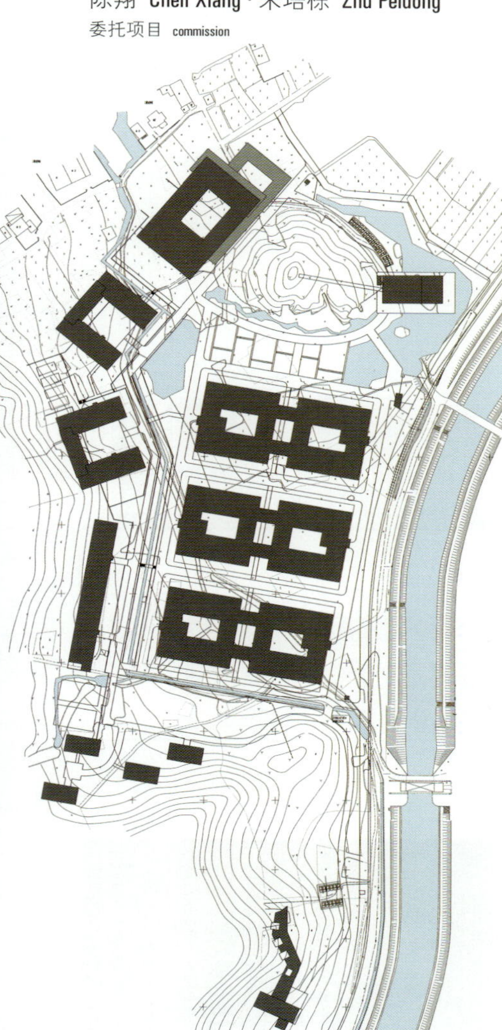

地块位置 SITE PLAN

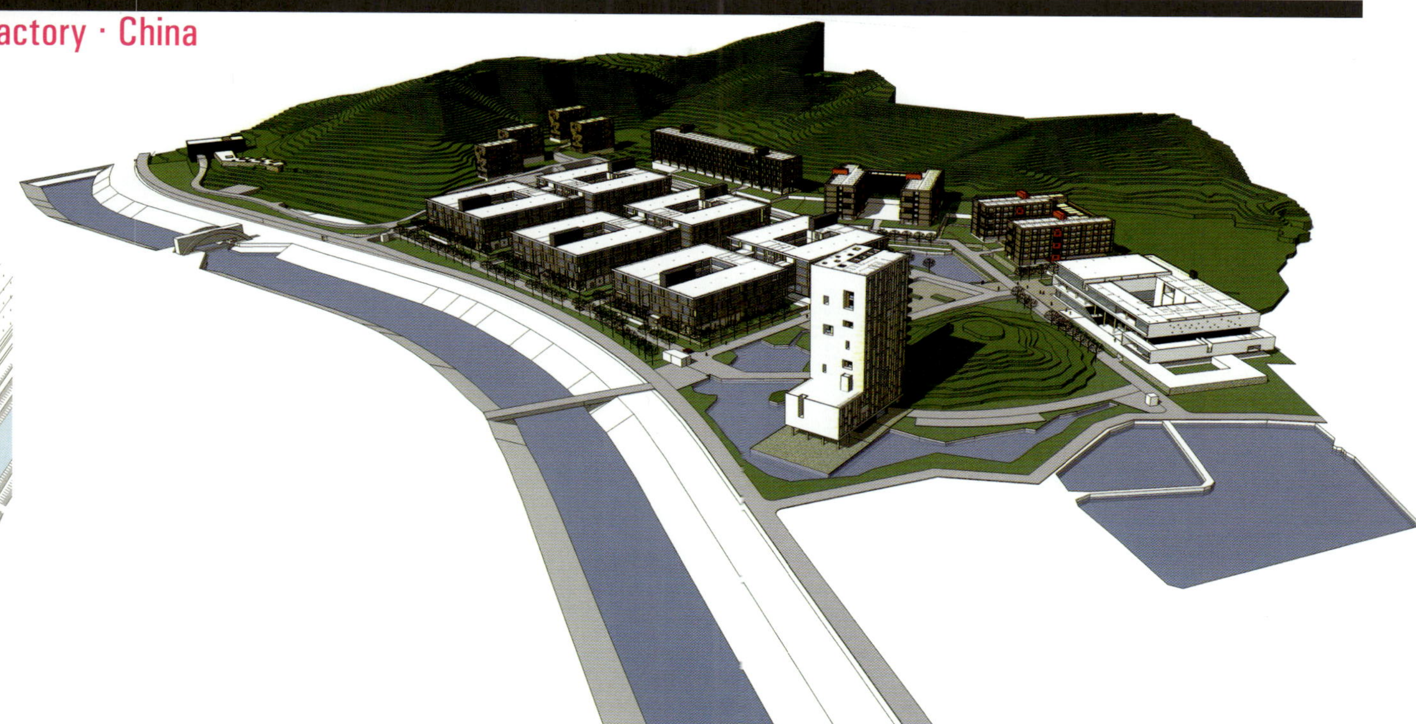

设计以场所为依据，因循山势，治理水系，以使建筑与基地建立和谐的关系。在严格的造价限制下，通过地方材料的选择、利用和加工，形成了具有地域特征的建构方法。常规建材和建造方式的解构，重塑了建筑的日常表情，建筑表皮与自然风景则成为了适宜技术下的共生元素。合适的地点、适合的设计、恰当的材料、相宜的方法，共同来完成适宜的建构。

Taking account of the site in terms of mountain features and river systems, the design aims for the harmony between building and foundation. Limited by the strict construction cost, it has built the local characteristics into the building through selection, use and processing of local materials. Artful use of the regular construction materials and methods will enrich the common expression of the building, forming a symbiosis relationship between the building appearance and natural scenery with applicable techniques. By combing the proper locality, design, materials and methods, it will bring about the most suitable building.

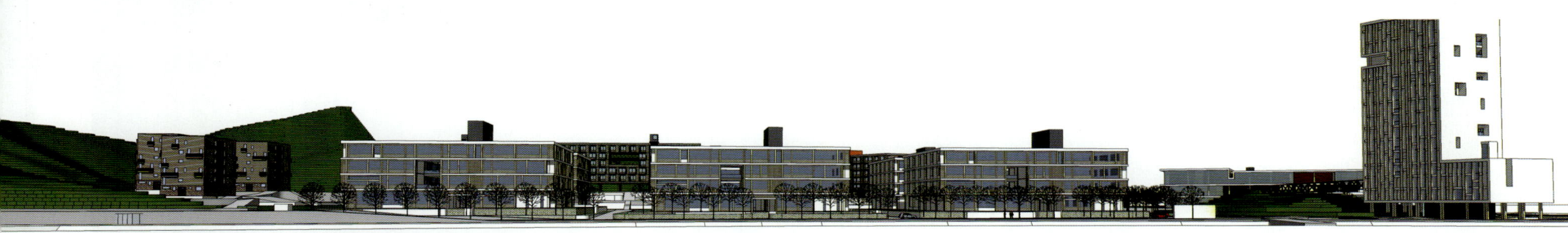

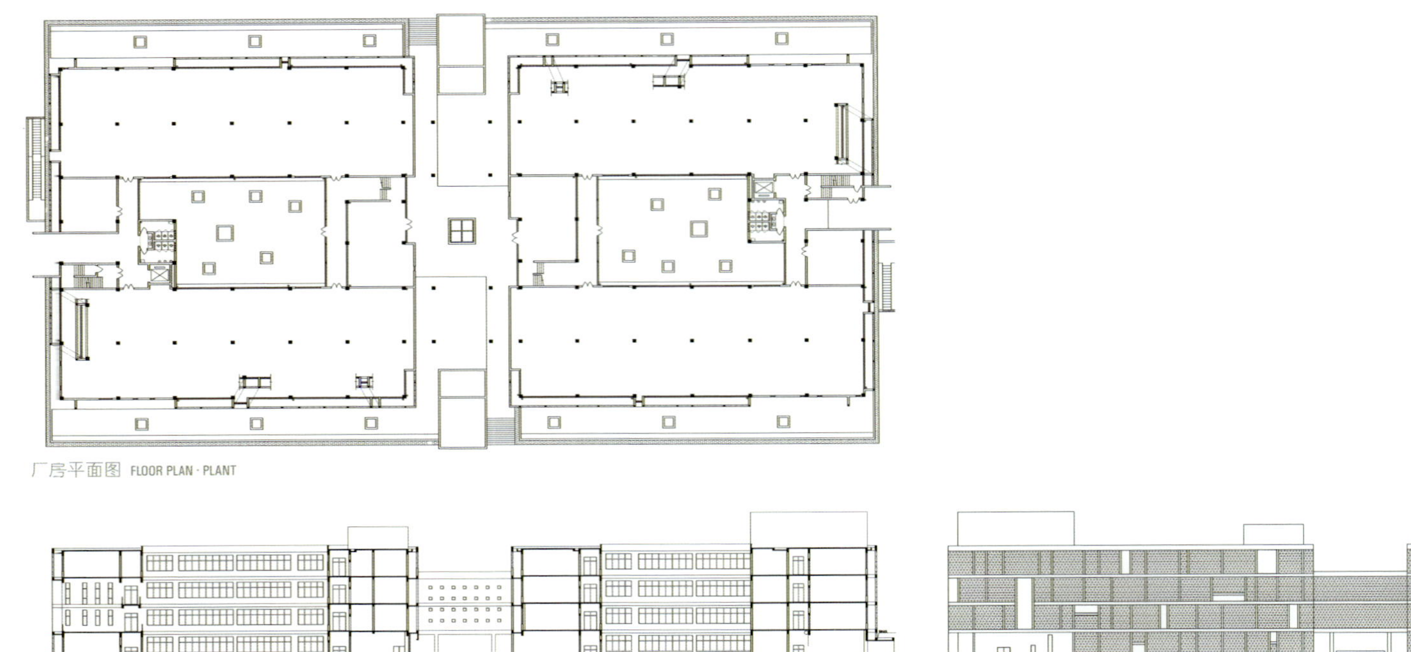

厂房平面图 FLOOR PLAN · PLANT

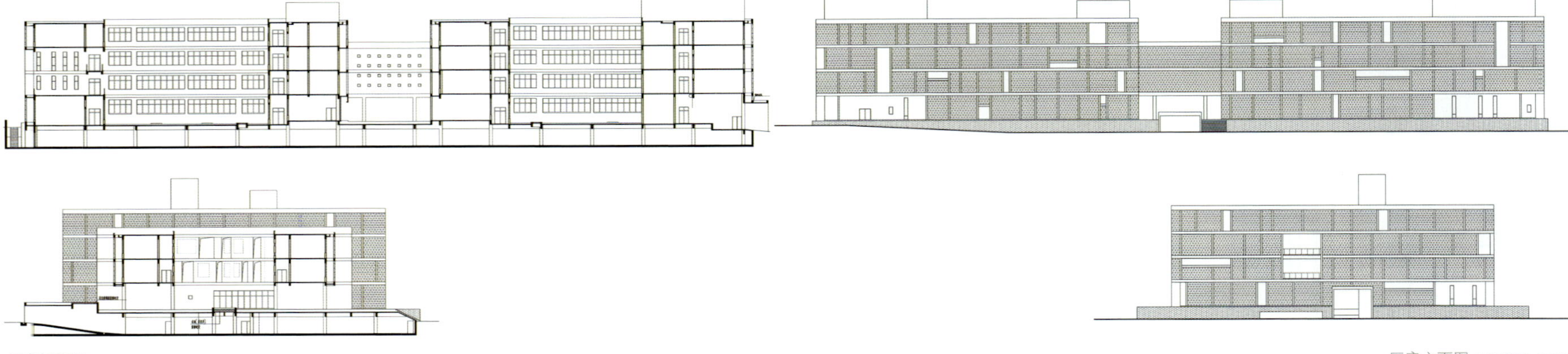

厂房剖面图 SECTIONS · PLANT

厂房立面图 ELEVATIONS · PLANT

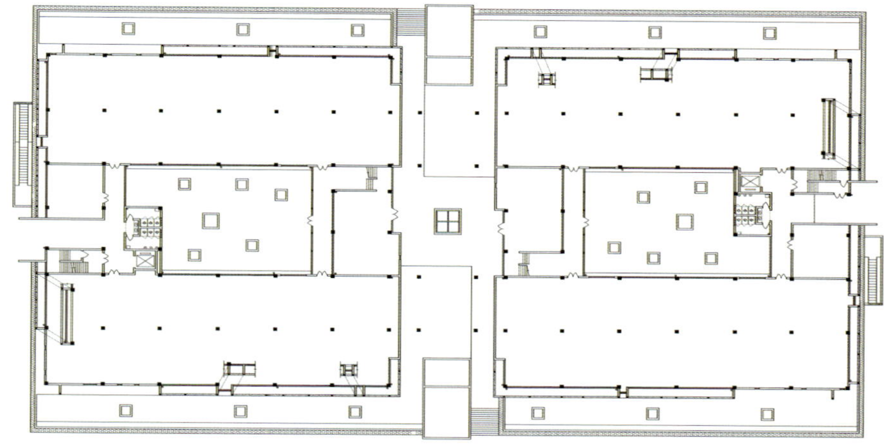

食堂平面图 FLOOR PLAN · CANTEEN

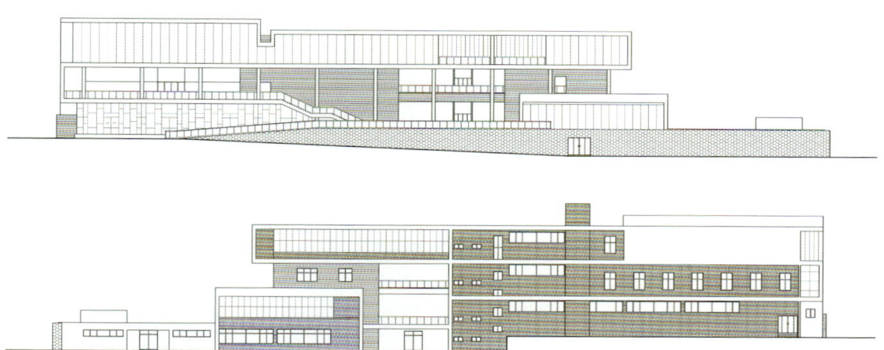

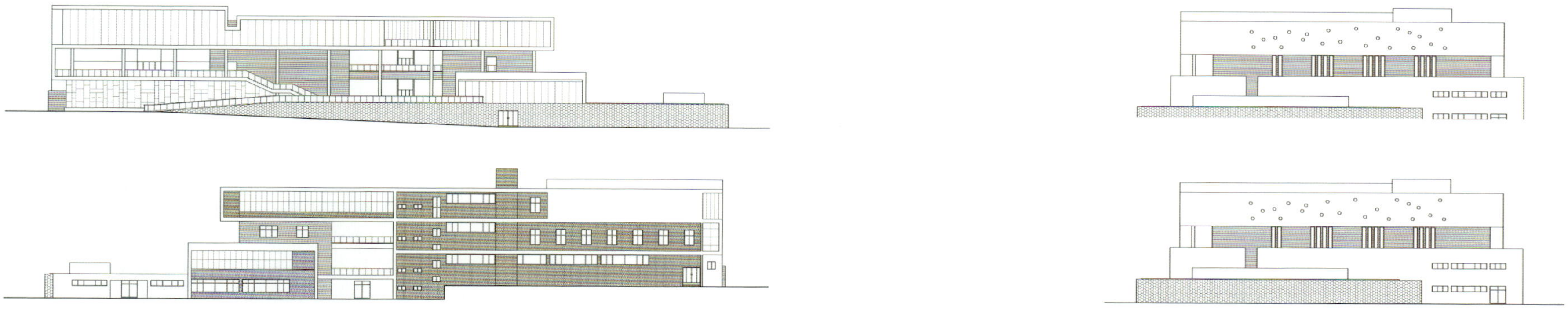

食堂立面图 ELEVATIONS · CANTEEN

常德大剧院 · 湖南

Changde Grand Theatre · China

陈帆 Chen Fan · 吴璟 Wu Jing

邀标参标方案 invited competition proposal

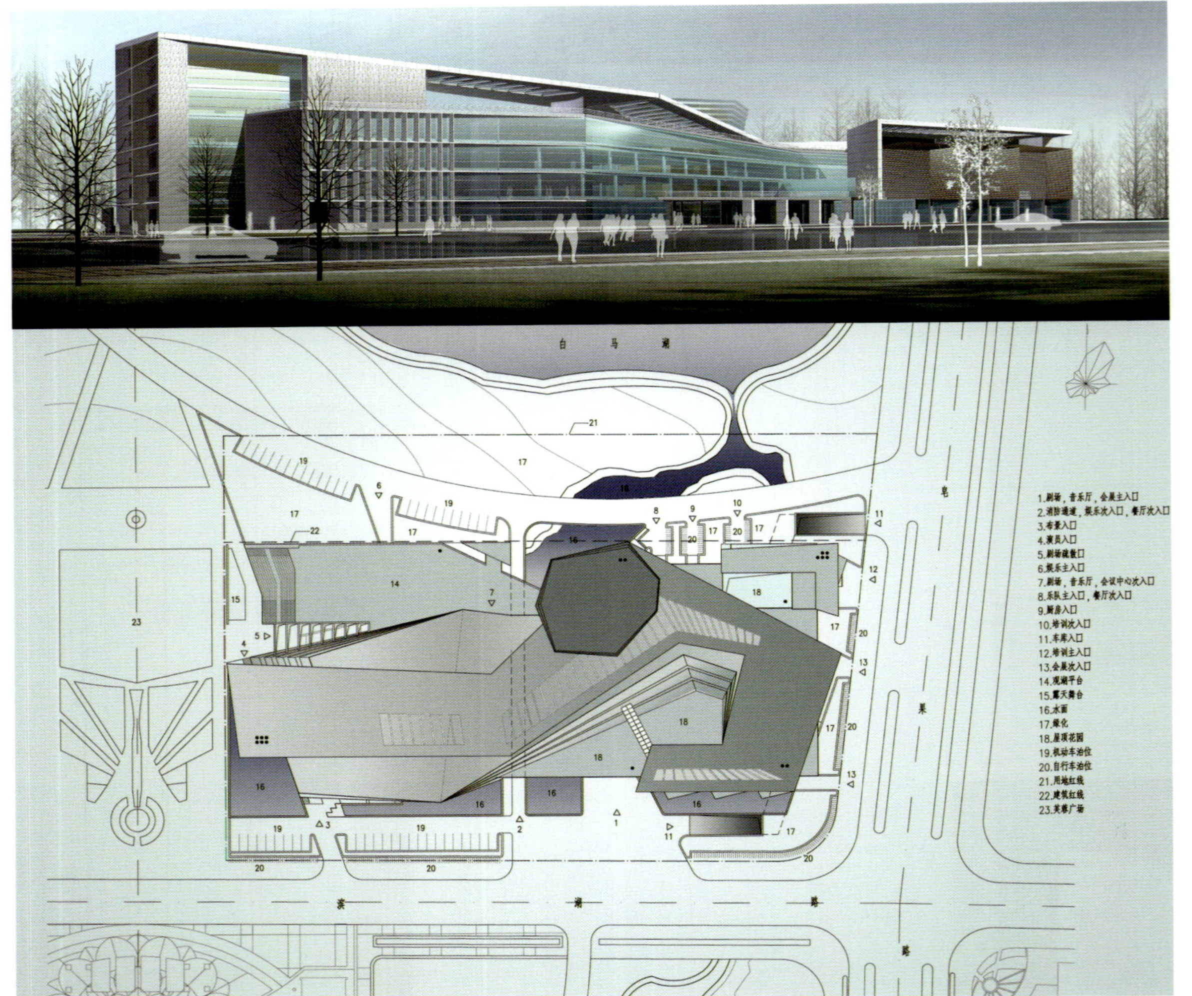

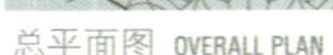

总平面图 OVERALL PLAN

1. 剧场、音乐厅、会展主入口 MAIN ENTRANCE OF THEATRE, AUDITORIUM, EXHIBITION
2. 消防通道、娱乐次入口、餐厅次入口 FIRE EXITS, SECONDARY ENTRANCE OF AMUSEMENT, SECONDARY ENTRANCE OF CANTEEN
3. 布景入口 ENTRANCE OF SCENERY
4. 演员入口 ENTRANCE OF ARTISTS
5. 剧场疏散口 EVACUATION PORT OF THEATRE
6. 娱乐主入口 MAIN ENTRANCE OF AMUSEMENT SPACE
7. 剧场、音乐厅、会议中心次入口 SECONDARY ENTRANCE OF THEATRE, AUDITORIUM, CONFERENCE
8. 乐队主入口、餐厅次入口 MAIN ENTRANCE OF BAND, SECONDARY ENTRANCE OF CANTEEN
9. 厨房入口 ENTRANCE OF KITCHEN
10. 培训次入口 SECONDARY ENTRANCE OF TRAINING
11. 车库入口 ENTRANCE OF PARKING
12. 培训主入口 MAIN ENTRANCE OF TRAINING
13. 会展次入口 SECONDARY ENTRANCE OF EXHIBITION
14. 观湖平台 LAKE VIEW DECK
15. 露天舞台 AIR STAGE
16. 水面 WATER SHEETS
17. 绿化 VEGETATION
18. 屋顶花园 ROOF GARDEN
19. 机动车泊位 PARKING SPACES FOR CARS
20. 自行车泊位 PARKING SPACES FOR BICYCLES
21. 用地红线 SITE BOUNDARY
22. 建筑红线 BUILDING BOUNDARY
23. 芙蓉广场 HIBISCUS SQUARE

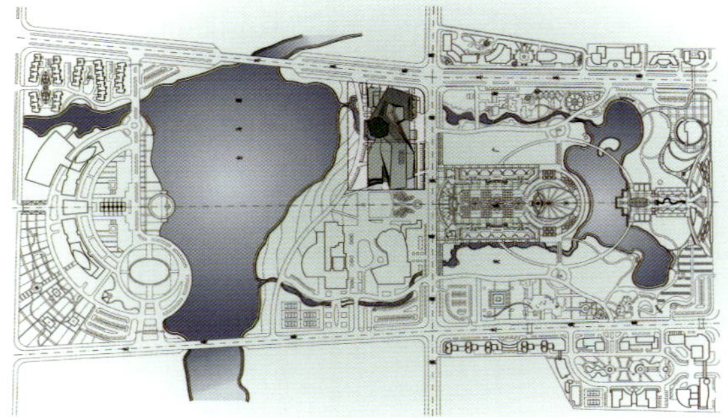

地块位置 SITE PLAN

常德市大剧院用地32,184 m2，建筑面积31,253 m2。地上六层，地下一层。其主要功能组成：(1) 1,200座剧场；(2) 600座音乐厅；(3) 会展中心；(4) 餐饮娱乐中心；(5) 培训中心。设计构形的特点是通过折线形展开的空间形态，将复杂的功能和繁复的空间关系凝结成为一个整体，既满足多功能综合性要求，又构成完整的城市空间形态。

Changde Grand Theatre covers an area of 32,184 m2, with a building area of 31,253 m2 in 6 floors and one ground floor. Its main functions include: (1) Theatre with a seating capacity of 1,200; (2) Music Hall with a seating capacity of 600; (3) Exhibition Center; (4) Catering and Recreational Center; (5) Training Center. By integrating the complex functions and its spatial relations into one entirety, the design features the shape of broken lines, which can not only satisfy the comprehensive function requirements, but also form a complete urban spatial pattern.

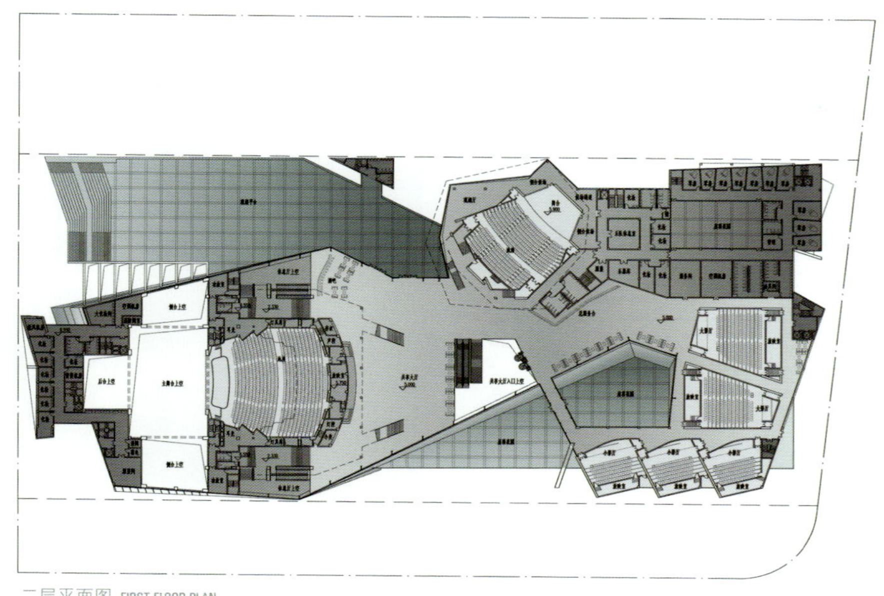
二层平面图 FIRST FLOOR PLAN

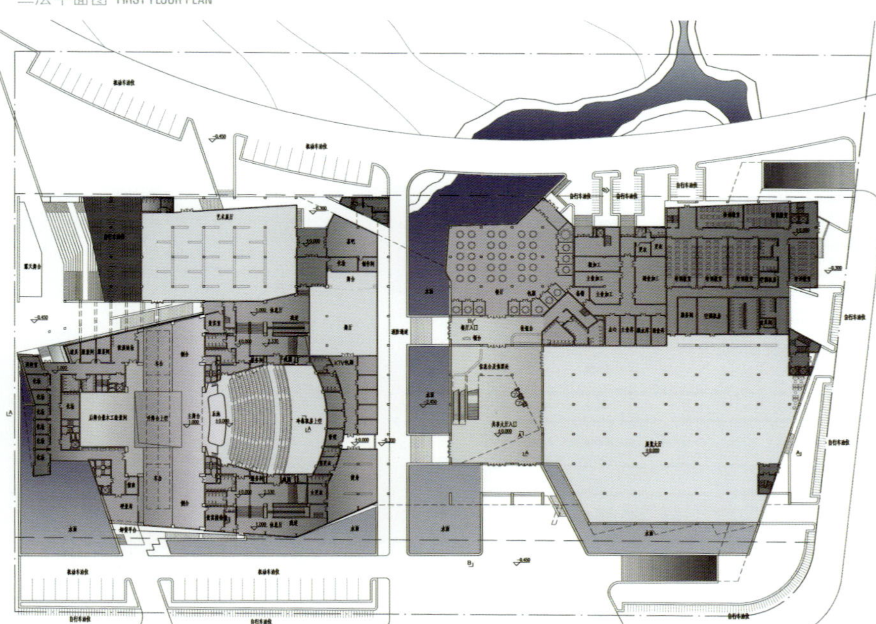
底层平面图 GROUND FLOOR PLAN

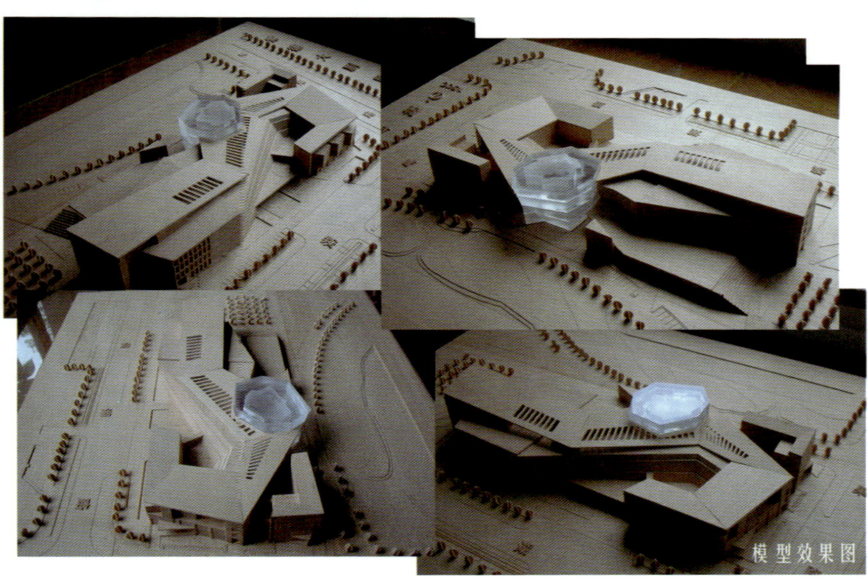
模型效果图

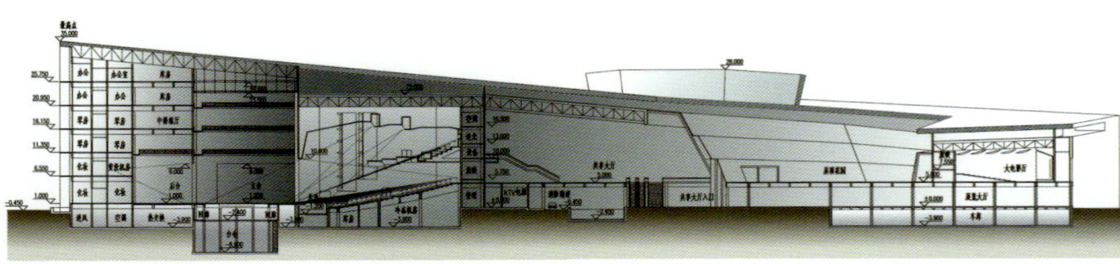
剖面图1 SECTION 1

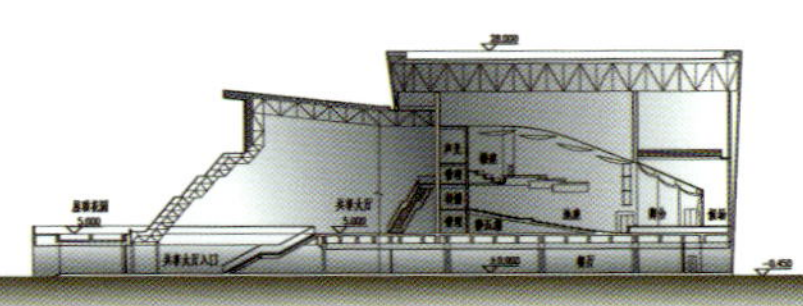
剖面图2 SECTION 2

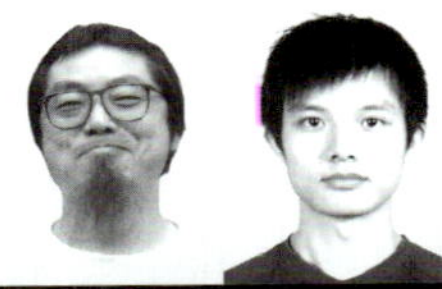

萧山国际机场高速收费站·杭州

Highway Toll Station for Hangzhou Xiaoshan International Airport

陈帆 Chen Fan · 章鹏峰 Zhang Pengfeng

参标方案 competition proposal

本方案将云这一实体通过归纳、抽象进而引入到收费站的设计中，建筑整体由两排“云”组成，每排15朵，期间镶以玻璃，以便底部获得较好的采光。每朵“云”的形态各自不同，结构上互为连接组成一个整体。因此，既独立又不失联系。整个屋面采用太阳能电池方阵，为收费窗口提供部分电力。

方案整体色调以白色为主。其中整个顶部高低错落，形态变化丰富。与之前的厚重、演示相反，给人以轻盈、通透之感。

This proposal introduces the “cloud” into the design through the methods of induction and abstraction. The building takes the shape of two rows of 15 clouds each, inlaid with glass for great day-lighting. Each “cloud” is unique in shape and yet forms a harmonious entirety together, each one being independent yet connected. The whole rooftop is equipped with solar batteries to provide electricity for toll booths.

The overall color for the proposal is white. The top of the building is well-proportioned with rich changes in structure. Compared with the dignified, tight one, this one gives a slim, graceful and transparent impression.

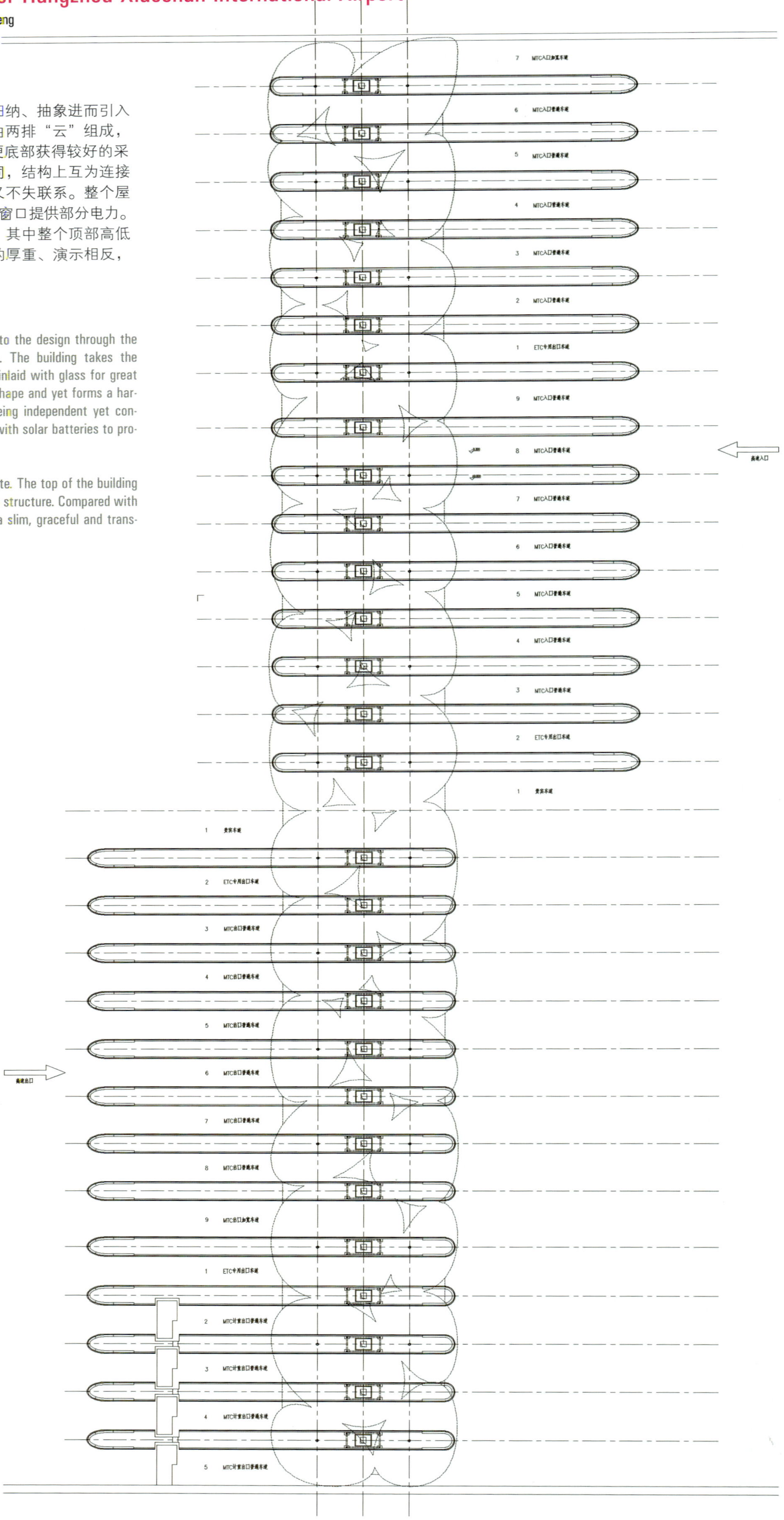

平面图 FLOOR PLAN

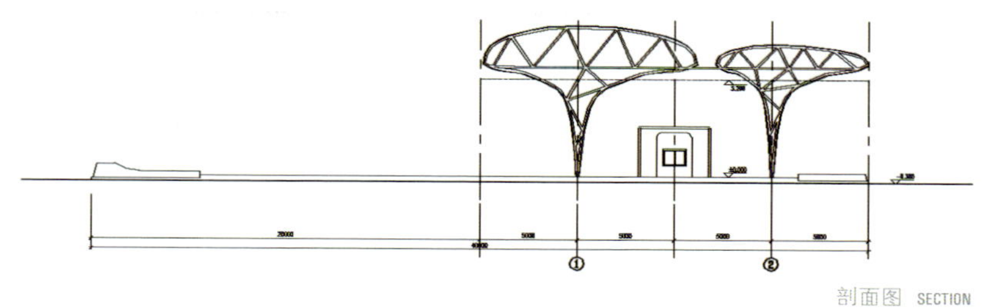
剖面图 SECTION

中国禄丰侏罗纪世界遗址馆 · 云南

China Lufeng Ruins Museum of Jurassic World

杨易栋 Yang Yidong · 王玉平 Wang Yuping · 彭怡芬 Peng Yifen

国内邀标中标 national invitation competition winner

ZUCCEA (浙江大学建筑工程学院)

项目负责 project manager · 杨易栋 Yang Yidong

建筑设计 design architect · 王玉平 Wang Yuping · 彭怡芬 Peng Yifen

结构工程 structural engineering · 干钢 Gan Gang · 邵剑文 Shao Jianwen

给排水工程 plumbing installation · 张钧 Zhang Jun

电气工程 electrical installation · 郑国兴 Zheng Guoxing

暖通工程 heating installation and ventilation · 余俊祥 Yu Junxiang

建筑经济 architectural economy · 孙文通 Sun Wentong

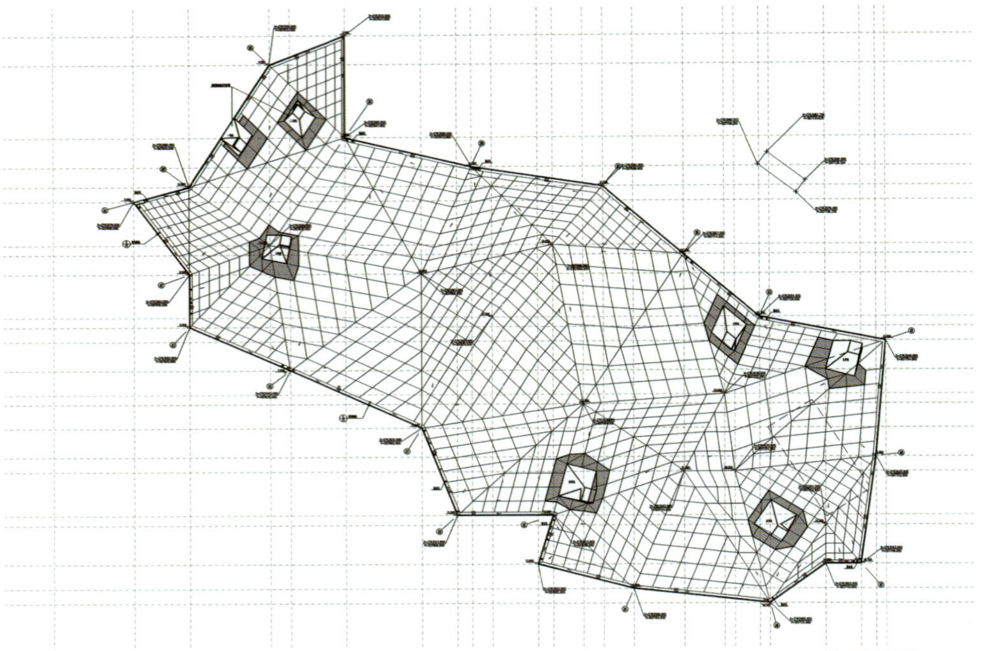

总平面图 OVERALL PLAN

屋顶平面图 ROOF PLAN

平面图标高4.50-7.50 FLOOR PLAN LEVEL 4.50-7.50

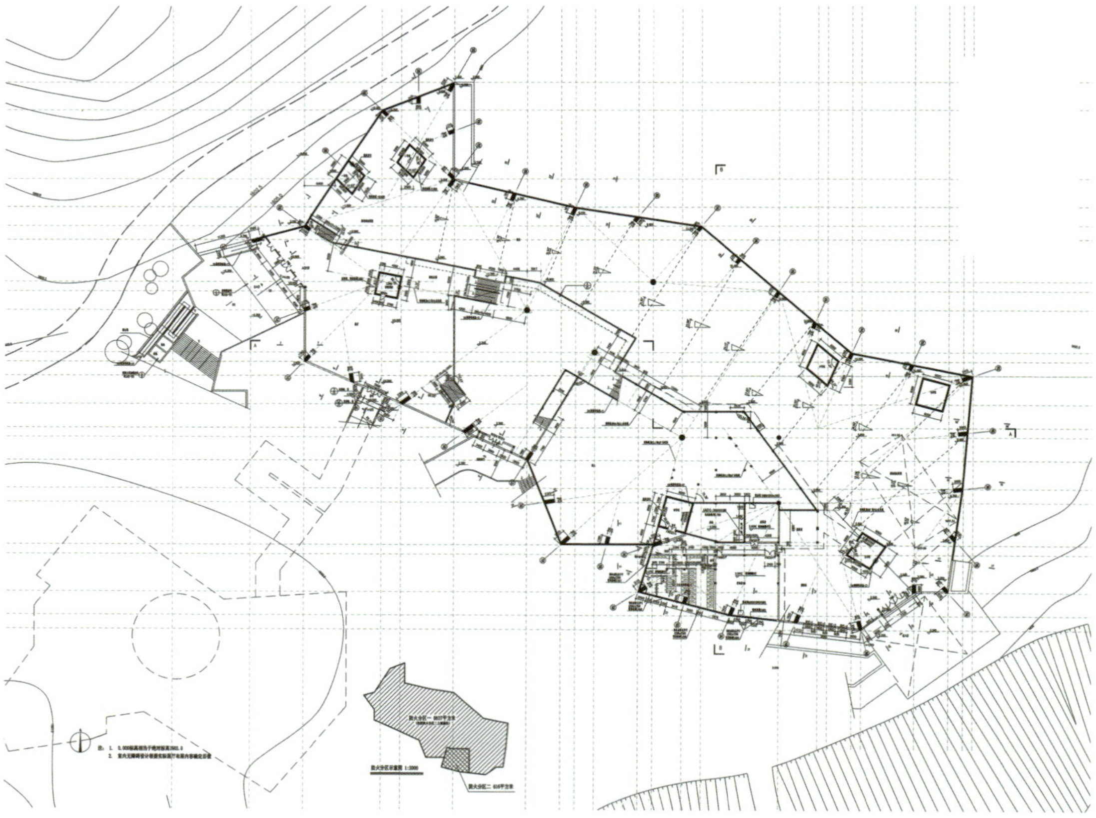

平面图标高0.00-2.50 FLOOR PLAN LEVEL 0.00-2.50

该项目直接建于化石发掘现场之上，建筑整体依山就势，盘旋于山谷之间，如同地貌景观一样，无论从形体构成还是立意上都很好地表达了建筑的主题，并与自然地貌浑然一体。

设计通过倾斜、扭转、切削等建筑手法，模拟山崩地裂的山体滑坡，表达出了恐龙灭绝的震撼场景。

Perching on the mountainside in its entirety, this project is located in the fossil excavation site. Winding among the mountain valley as its geomorphologic landscape, it conveys a vivid expression of the building theme both in structure and in concept, and blends perfectly with the natural features.

By dint of such architectural methods as sloping, twisting and cutting, it imitates the conditions when landslides take place, giving the impression of the sensational scene of the extinction of the dinosaurs.

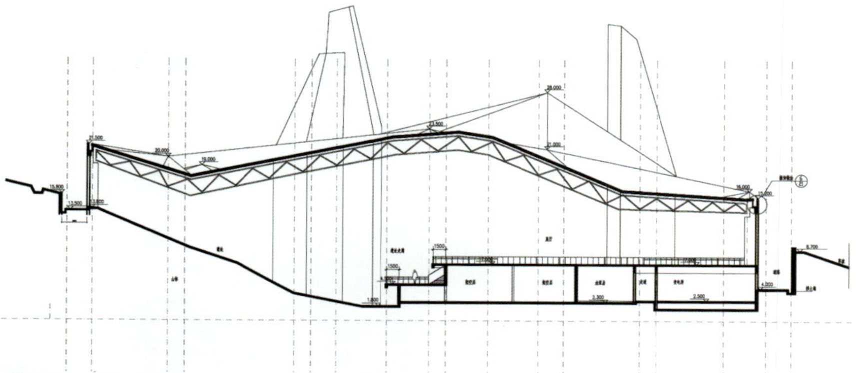

剖面图B-B SECTION B-B

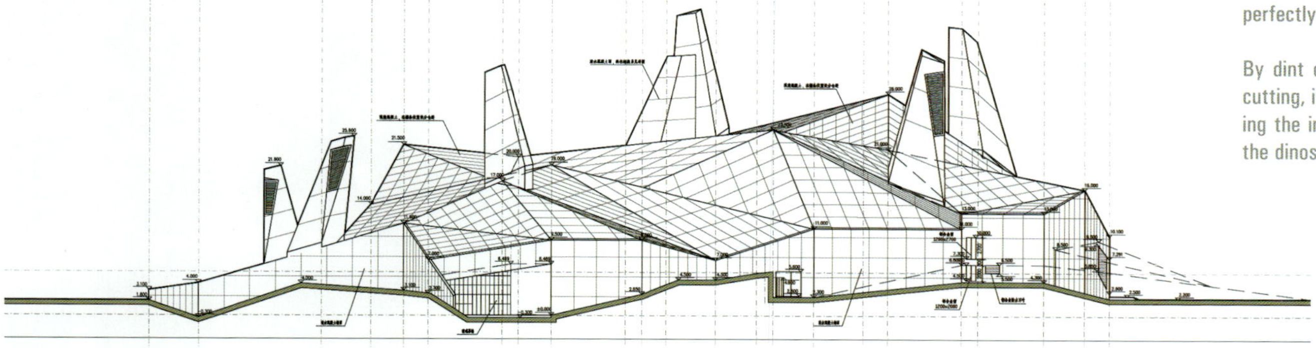

西立面图 WEST ELEVATION

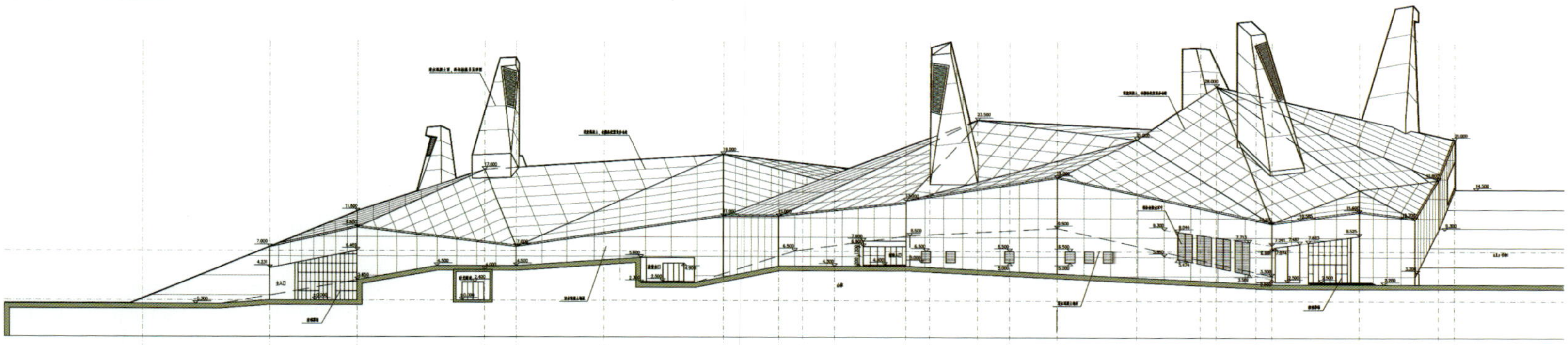

南立面图 SOUTH ELEVATION

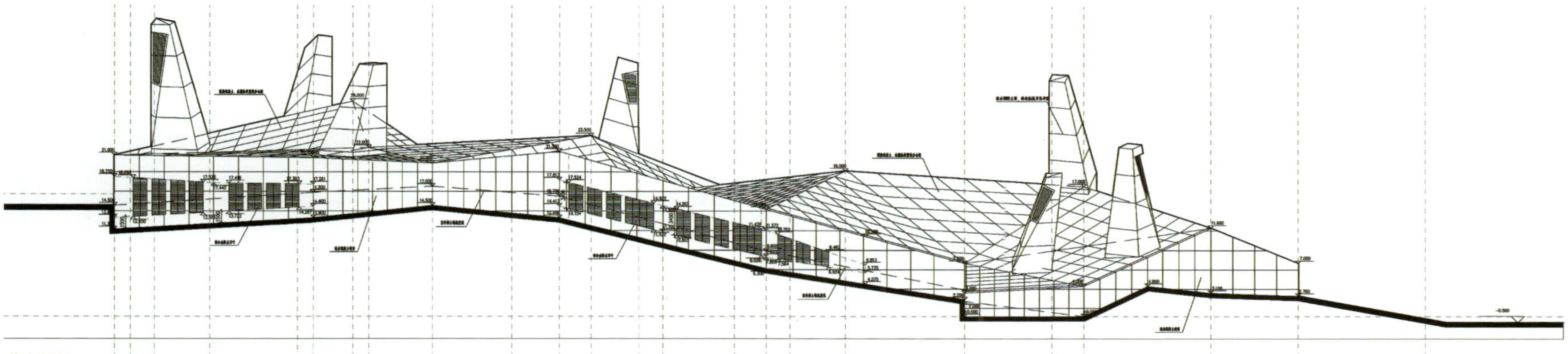

北立面图 NORTH ELEVATION

天一阁古籍库房·宁波

Tianyi Pavilion Stack Room for Ancient Books

胡慧峰 Hu Huifeng · 姚冬晖 Yao Donghui · 钱晨 Qian Chen

国内竞标中标 national competition winner

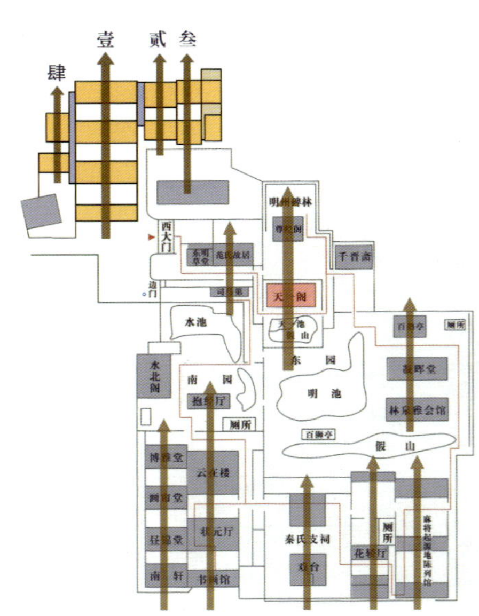

ZUCCEA (浙江大学建筑工程学院)
项目负责 project manager · 胡慧峰 Hu Huifeng
建筑设计 design architect · 姚冬晖 Yao Donghui · 钱晨 Qian Chen
结构工程 structural engineering · 阚建忠 Kan Jianzhong
给排水工程 plumbing installation · 王大伟 Wang Dawei
电气工程 electrical installation · 丰建华 Feng Jianhua
暖通工程 heating installation and ventilation · 刁岳峰 Diao Yuefeng

2008年第四届中国威海国际建筑设计大奖赛佳作奖
First Prize of 2008 Assessment and Examination of the 4th International Grand Prix for Architectural Design in Weihai, China.

宁波天一阁是我国现存历史最久的私家藏书楼，也是亚洲最古老的藏书楼。值天一阁建阁440周年之际，天一阁博物馆在天一阁西北侧用地扩建古籍书库及辅助用房，解原有库房不能妥善保护珍贵古籍之忧，为古老藏书搭建一个现代化、数字化、高科技化的居所，以尽传统文化珍宝“薪火相传”的责任。

我们的设计是对天一阁现存建筑的发展和延续，是天一阁建筑群的一部分。这不但体现在所处基地的位置上，也体现在格局和风貌上。中国文章常讲立意的“起、承、转、合”，我们的设计也因此结构逻辑，起自基地现状，承接聚落肌理，转换功能布局，合于园林空间。

Ningbo Tianyi Pavilion, now on its 440th anniversary of its foundation, is the oldest existing private library in China, even in Asia as a whole. Ancient books and auxiliary rooms of Tianyi Pavilion Museum have been added to the northern wing of the Tianyi Pavilion, relieving the burden of the original stack room by housing the previous ancient books. The museum aims to honor the responsibility of carrying forward the precious traditional cultural treasures by providing a modern, digitalized and technologically advanced place for the ancient books.

Our design is the development and continuance of the Tianyi Pavilion's original buildings, forming a part of the Tianyi Pavilion cluster. This will be reflected not only in its location, but in its layout and style and features as well. Chinese articles have always laid stress on the four steps in the composition of an essay—opening, developing, changing and concluding; and our design has put these features into the building to reflect locality conditions, site texture, functional layout transformation, which all conclude in the garden space.

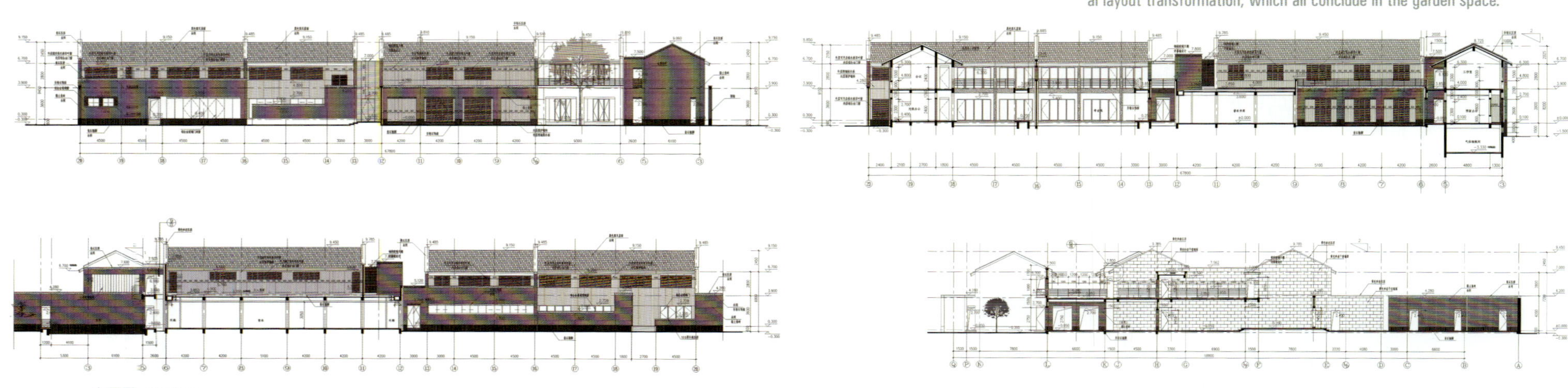

立面图 ELEVATIONS

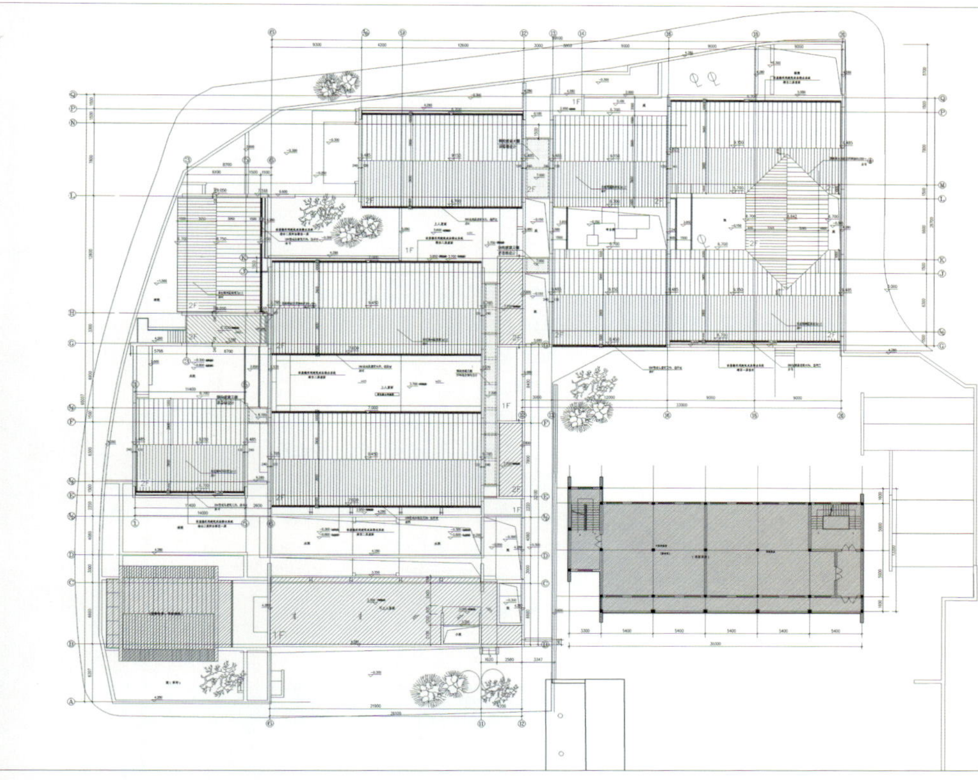

屋顶平面图 ROOF PLAN

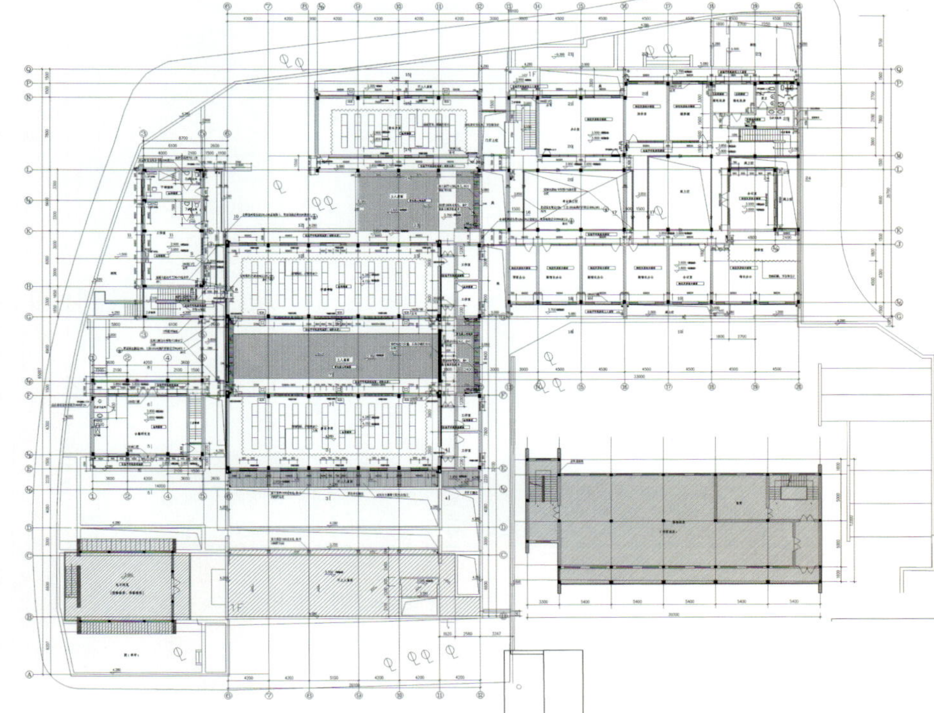

二层平面图 FIRST FLOOR PLAN

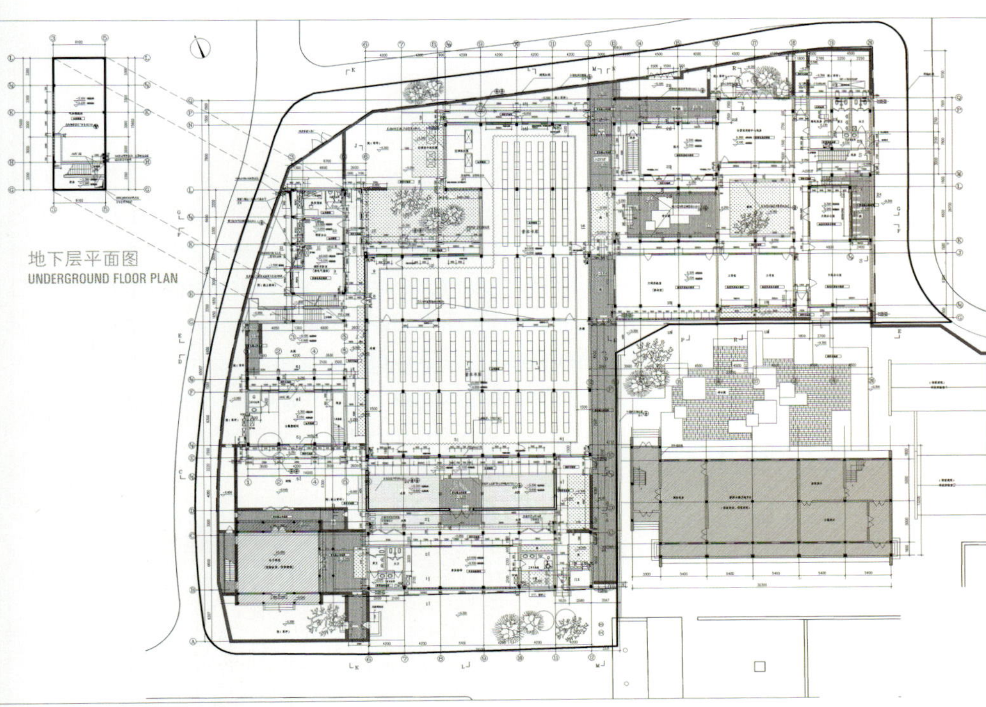

地下层平面图
UNDERGROUND FLOOR PLAN

底层平面图 GROUND FLOOR PLAN

德赫蒙斯克大图书馆·奥斯陆 Deichmanske main library · Norway

竞标 · competition
德赫蒙斯克大图书馆
Deichmanske main library in Norway

竞标类型 · competition type
限制性邀标
restricted competition by invitation

项目地点 · site area
奥斯陆 · 挪威 Oslo · Norway

主办方 · promoter
奥斯陆政府文化事务局 HAV Eiendom AS · Cultural Affairs of Oslo Municipality

获奖者 · awards

一等奖 · first prize
LUND HAGEM ARKITEKTER AS · ATELIER OSLO AS (建筑师事务所)
Svein Lund · Einar Hagem · Per Suul · Vegar Voraa · Anders Melsom · Quentin le Guen-Geffroy
Camille Bonneu (建筑师)
Jonas Norsted · Marius Mowe · Nils Ole Bae Brandtzæg · Thomas Liu
landscape architects: Agence Ter from France (建筑师)

二等奖 · second prize
SCHMIDT HAMMER LASSEN (建筑师事务所)

四等奖 · fourth prize
DAVID CHIPPERFIELD ARCHITETCS (建筑师事务所)

四等奖 · fourth prize
SNØHETTA (建筑师事务所)

荣誉提名奖 · honourable mention
KISTER SCHEITHAUER GROSS ARCHITEKTEN (建筑师事务所)

荣誉提名奖 · honourable mention
A-LAB (建筑师事务所)
Adnan Harambasic · Odd Klev · Geir Haaversen · Katrine Holm · Joana Sa Lima · Andreas Tingulstad · Anette Apold · Jarand Midtgaard · Tonje Løvdahl · Caroline Weill (建筑师)

入围 · finalist
XAVEER DE GEYTER ARCHITECTEN (建筑师事务所)
Xaveer De Geyter · Michael Smith · Emilie Augeard · Christophe Antipas · Wouter Van Daele Rui Zenha · Michel Sikorski (建筑师)

入围 · finalist
WIEL ARETS ARCHITECTS & ASSOCIATES BV (建筑师事务所)
Wiel Arets · Bettina Kraus · Jos Beekhuijzen · Jochem Homminga (建筑师)

合作 (c) Tobias Gehrke · Julius Klatte · Miguel Valerio · Cindy Wouters

入围 · finalist
ALLMANN SATTLER WAPPNER ARCHITEKTEN (建筑师事务所)
Markus Allmann · Amandus Sattler · Ludwig Wappner (建筑师)

团队 team: Carola Dietrich · Charbel Ali Azar · Katharina Brunn · Uwe Ernst · Pedro Ferreira Fabian Getto · Leila Hussein · Eva Faulhaber · Ekin Özdil · Severin Oswald · Janine Sander Daniel Sommer
结构工程 structural engineer: Knippers Helbig, Advanced Engineering
能源顾问 energy consultant: energydesign asia · Ingenieursgesellschaft mbH
标识顾问 signage consultant: Integral Ruedi Baur

入围 · finalist
SeARCH I.C.W. CHRISTIAN MÜLLER ARCHITECTS (建筑师事务所)

入围 · finalist
PLASMA STUDIO (建筑师事务所)
合伙人 partners: Eva Castro Iraola · Holger Kehne
联合合伙人 associated partner: Ulla Hell

入围 · finalist
70°N ARKITEKTUR + DAHL & UHRE ARKITEKTER(建筑师事务所)
合作 (c) Art: Åsa Sonjasdotter · Tromsø Art Academy
water and vegetation: Ellen Braae · Forest & Landscape

入围 · finalist
LUNDGAARD & TRANBERG ARCHITECTS
团队 team: Kenneth Warnke · Henrik Schmidt · Lene Tranberg · Susanne Skov · Malene Hjortsø Kyndesen · Jarl Engelbrecht Vindnæs · David Sherman · Erik Frandsen · Merete Harder Rikke Jørgensen · Mikkel Nielsen

日程安排 · schedule

招标 · Announcement 11.2008
评审结果 · Jury's results 02.2009

评审团 · jury

主席 Chairman:
Berit Kjøll
Mikko Heikkinen · SAFA建筑师 Architect SAFA, 芬兰 Finland
Henrik Lundberg · MNAL建筑师 Architect MNAL, 斯塔万格 Stavanger
Inge Hareide · MNAL建筑师 Architect MNAL, 奥斯陆 Oslo
Birgit Cold · MNAL建筑师 Architect MNAL, 奥斯陆 Oslo
Espen Dag Rydland · 项目经理 Project Manager, 奥斯陆市政府 Oslo Municipality
Eivind Hartmann · 建筑师 Architect, 奥斯陆市政府 Oslo Municipality
Rolf Hapel · 负责人 Director, 奥尔胡斯公共图书馆 Århus Public Library, 丹麦 Denmark
Maija Berndtson · 负责人 Director, 赫尔辛基公共图书馆 Helsinki Public Library, 芬兰 Finland
Eva Hagen · 执行负责人 Managing Director, 奥斯陆 Oslo
Bjørn Agner · MSc结构工程 Civil Engineer MSc, 奥斯陆 Oslo
Odd Einar Dørum · 议员 Member of Parliament
Britt Hildeng · 议员 Member of Parliament

十家建筑事务所受邀直接参与了此次竞赛，另外十家建筑事务所或团队在一系列的审核评比中脱颖而出。该竞赛的目标之一是如何在过去的港口区打造一个紧凑的、具有人文尺度和多重城市职能的城市。**另一个目标在于将Deichman Axis打造成为一个环保的、有利于可持续发展的先锋建筑作品。**

Ten architect offices were invited directly to participate in the competition and an additional ten architects offices/ teams where chosen through a process of request for qualification. The objective was to demonstrate how a versatile and overlapping programming of urban functions can assist in creating a compact city on a human scale in the former harbour areas of the city. **A further objective was that the Deichman Axis could be a pioneer project for an environment-friendly and sustainable development process.**

德赫蒙斯克大图书馆 · 奥斯陆

Deichmanske main library · Norway

一等奖 · First Prize

diagonale (竞标代码)

Lund Hagem Arkitekter and Atelier Oslo (建筑师事务所)

尝试多重入口的设计

该项目设计理念，把场所分为三幢建筑：给每幢建筑设定一个人文尺度，并使该项目与城市风格融为一体；使图书馆面向Operaallmenningen，以便游客能以最佳视角欣赏城市风貌、峡湾风景以及周边的奥斯陆群山；将图书馆建在公众的视野范围内，其顶部以悬臂梁的方式向外伸出，向游客们展示其风采；图书馆的外立面切口处作为其入口，分别位于建筑物的东面、西面和南面，邀请来自四面八方的游客；图书馆还与城市风貌交相辉映，阳光自其外立面反射开去，打造了一个平静安宁的内部空间。夜晚，该建筑灯火璀璨，灯光勾勒出的外观也会随着图书馆内部举行的活动不同而发生变化。

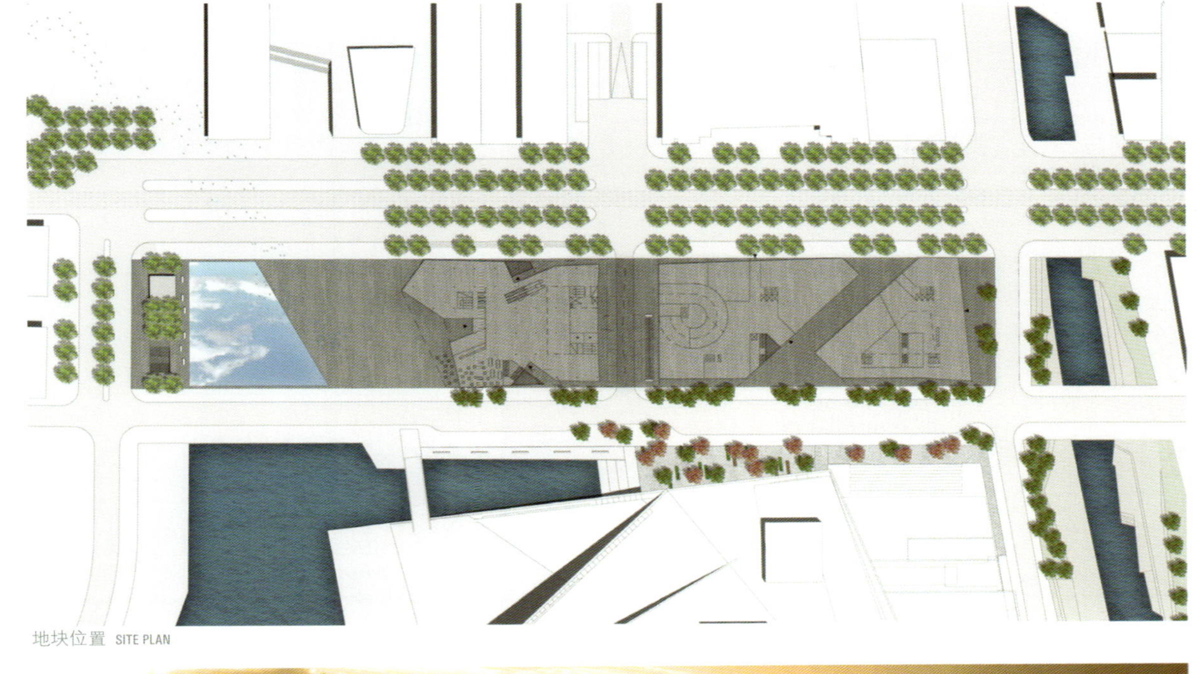

地块位置 SITE PLAN

MULTIPLE ENTRANCE POSSIBILITIES

The project proposes: To divide the site into three buildings: providing each building with a human scale and integrating the project into the city; to place the library on the site towards Operaallmenningen: the visitors are offered the best views towards the city, the fjord and the surrounding green hills of Oslo; to make the Library visible to the public: the top of the library cantilevers out to announce its presence to the visitors arriving; to create entrances to the east, west, and south: big cuts in the facade mark the entrances on three sides of the building, inviting the public coming from all parts of the city; to communicate with the city: the façade diffuses the sunlight, giving a calm feeling to the interior. At night, the building will glow and change appearance as a reflection of all the different activities and events inside the library.

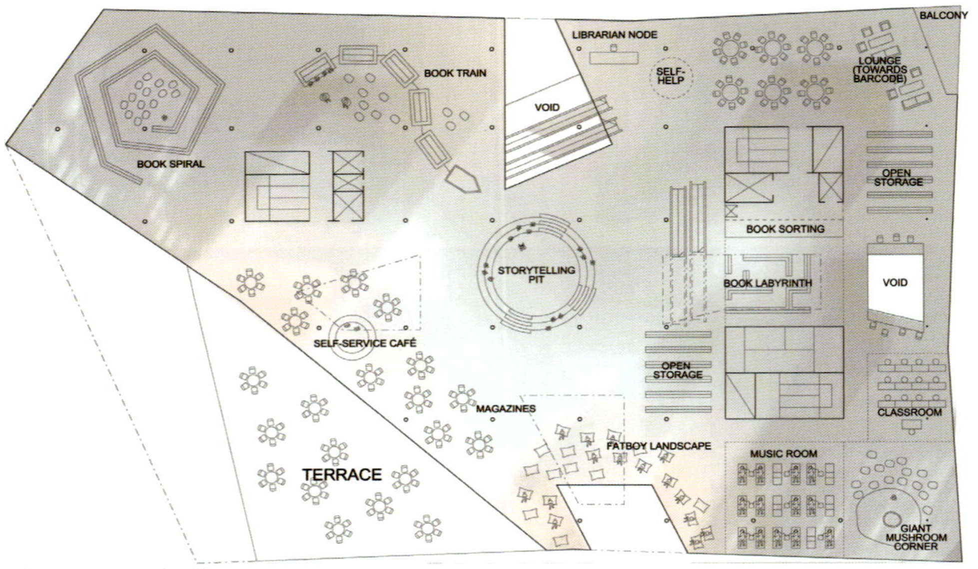

二层多功能层平面图（多媒体视听）FIRST FLOOR PLAN · MULTIPURPOSES (MEDIA-SOUND)

1. 公共服务 GENERAL PUBLIC SERVICES
2. 咨询中心 INFORMATION CENTRE
3. 24小时图书馆 24-HOUR LIBRARY
4. 礼堂与影院 AUDITORIUM AND CINEMA
5. 行政中心 ADMINISTRATION
6. 特殊部门 SPECIALIZED DEPARTMENTS
7. 客服工作人员 PERSONNEL SERVING THE PUBLIC
8. 机房管理 SORTING ROOM
9. 内部保障 INTERNAL SUPPORT
10. 仓库 STORAGE
11. 货物接收 RECEPTION OF GOODS
12. 技术基建 TECHNICAL INFRASTRUCTURE

商业区 COMMERCIAL AREA

A. 酒店 HOTEL
B. 会议中心 CONFERENCE CENTER
C. 办公室 OFFICE
D. 商铺 SHOP
E. 电影院 CINEMA
F. 技术基建 TECHNICAL INFRASTRUCTURE
G. 多功能厅 MULTI-PURPOSE

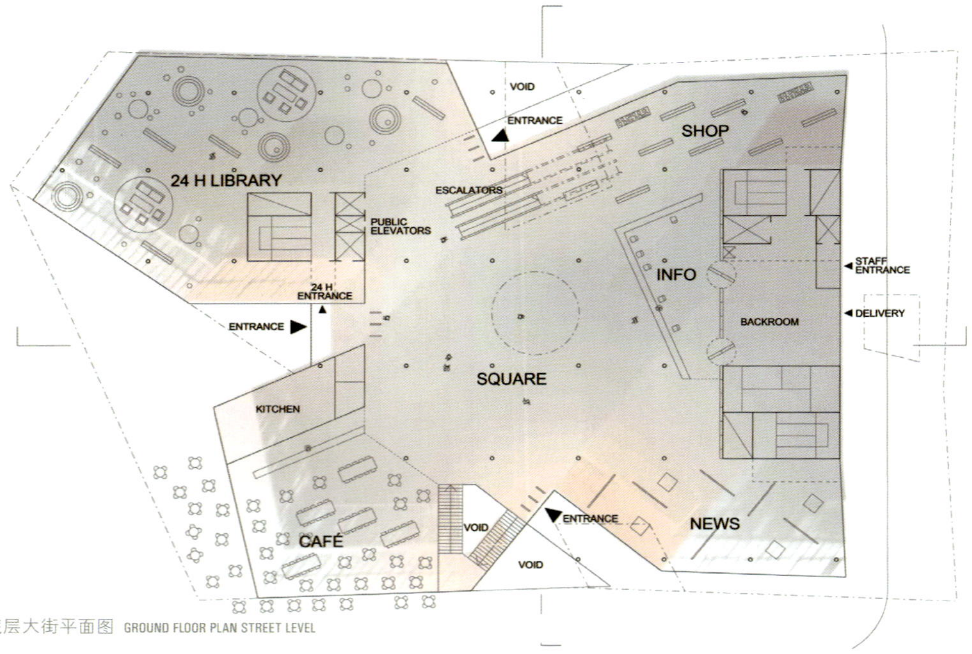

底层大街平面图 GROUND FLOOR PLAN STREET LEVEL

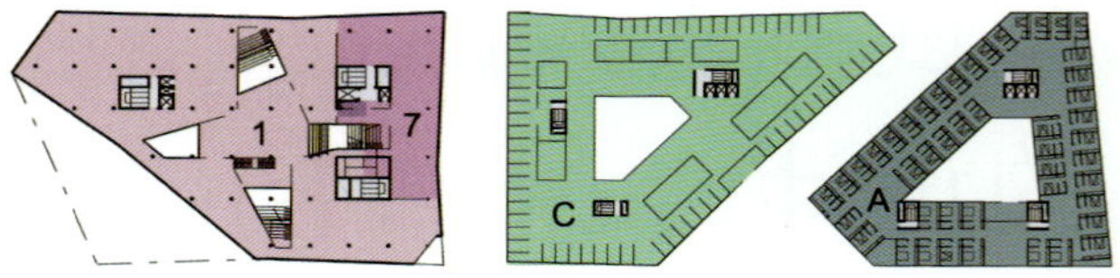

顶层 ROOF PLAN

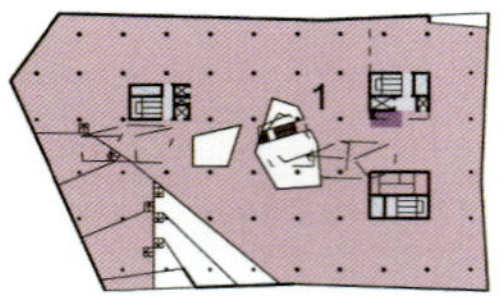

平面 4 FLOOR PLAN 4

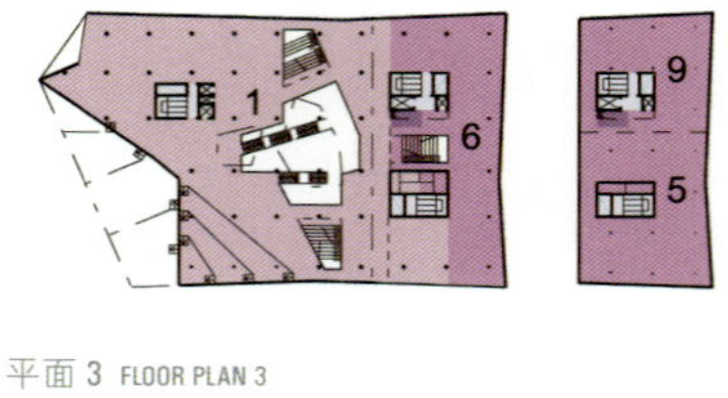

平面 3 FLOOR PLAN 3

平面 2 FLOOR PLAN 2

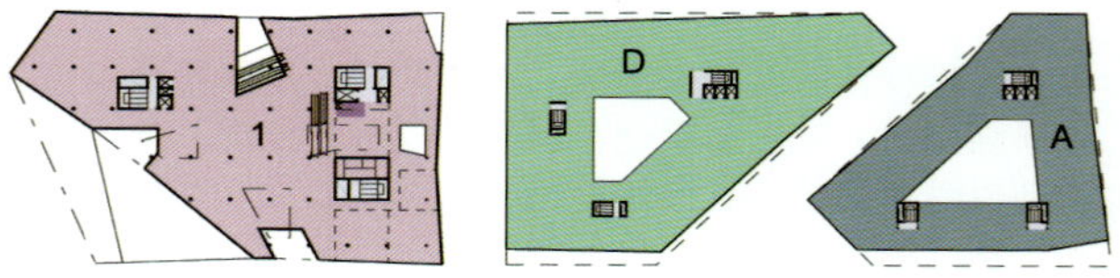

平面 1 FLOOR PLAN 1

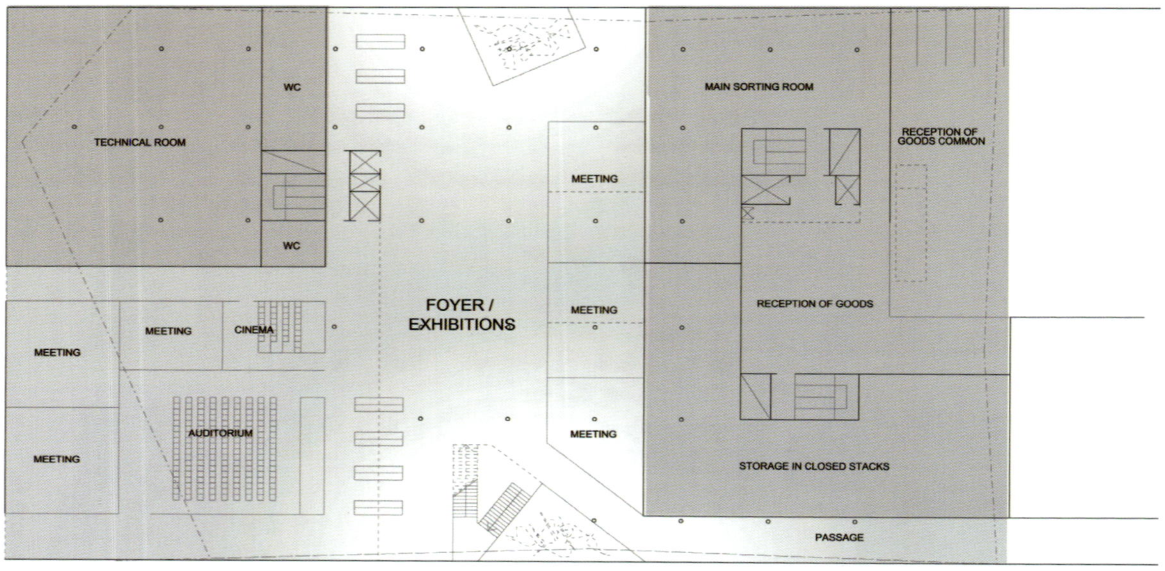

底层 GROUND FLOOR PLAN

地下层平面图 · 活动·会议 UNDERGROUND FLOOR PLAN · EVENTS-MEETINGS

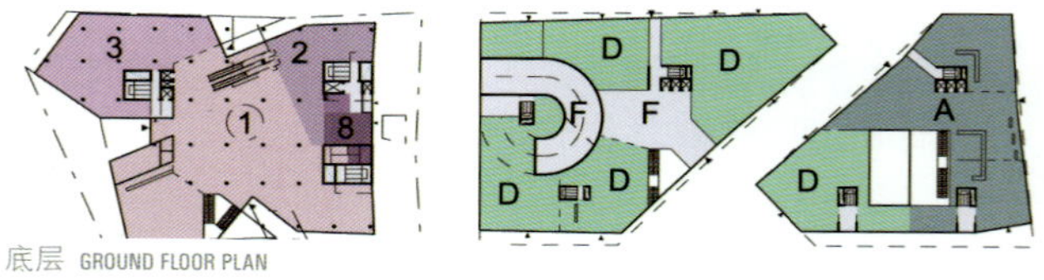

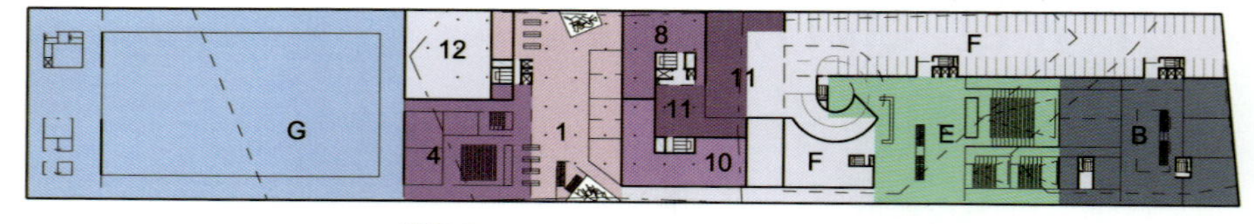

平面 -1 FLOOR PLAN -1

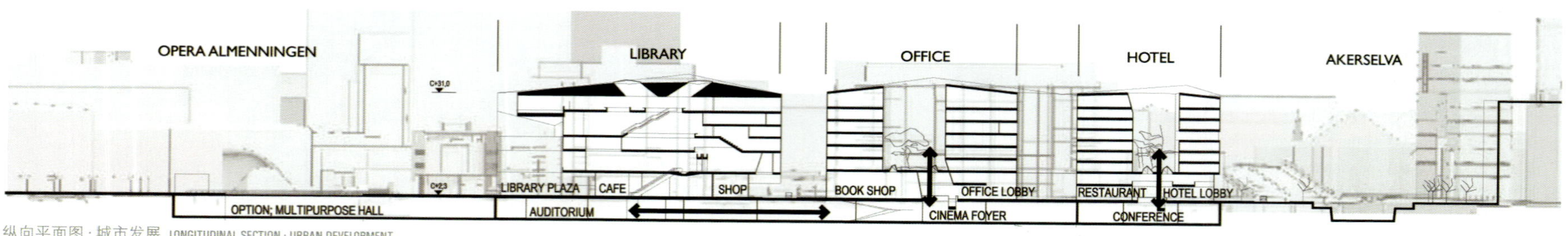

纵向平面图 · 城市发展 LONGITUDINAL SECTION · URBAN DEVELOPMENT

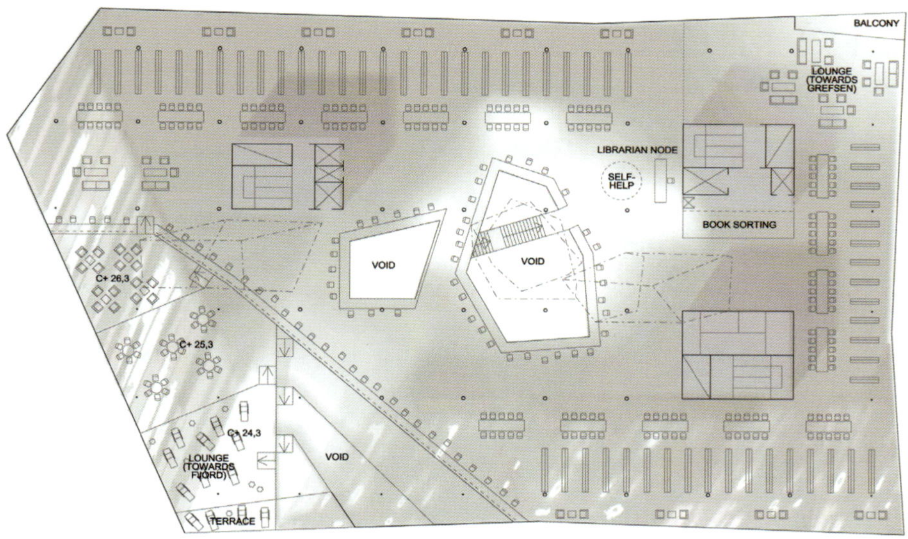

五层平面图 FOURTH FLOOR PLAN

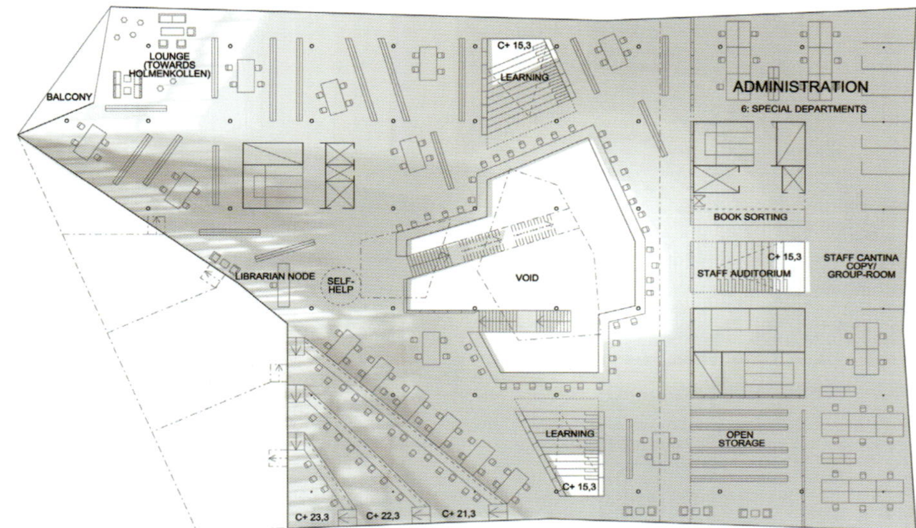

四层平面图 THIRD FLOOR PLAN

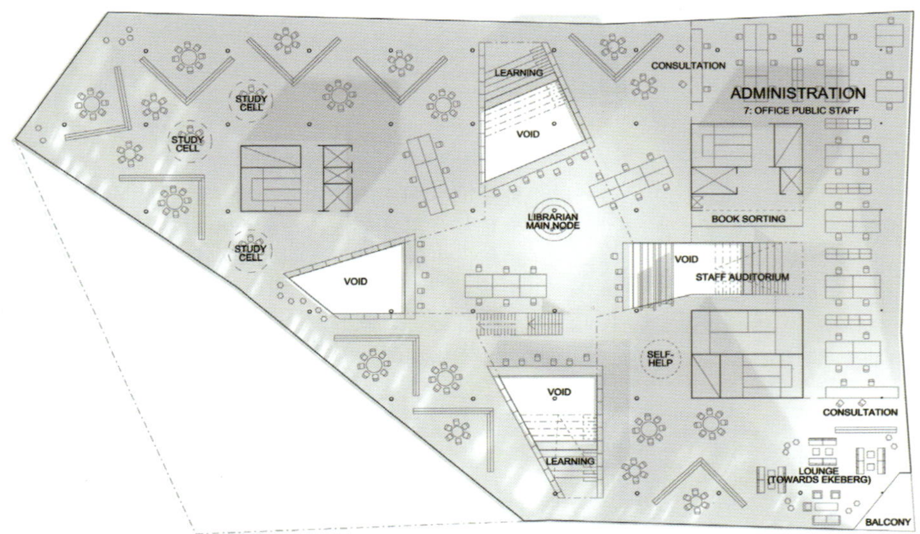

三层平面图 SECOND FLOOR PLAN

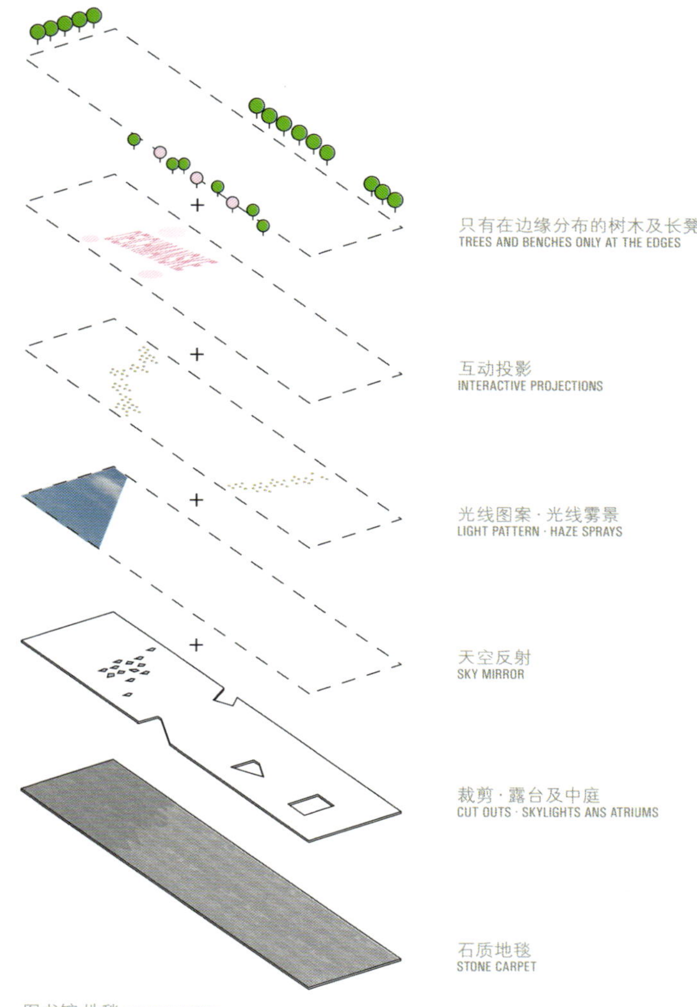

图书馆-地毯 LIBRARY CARPET

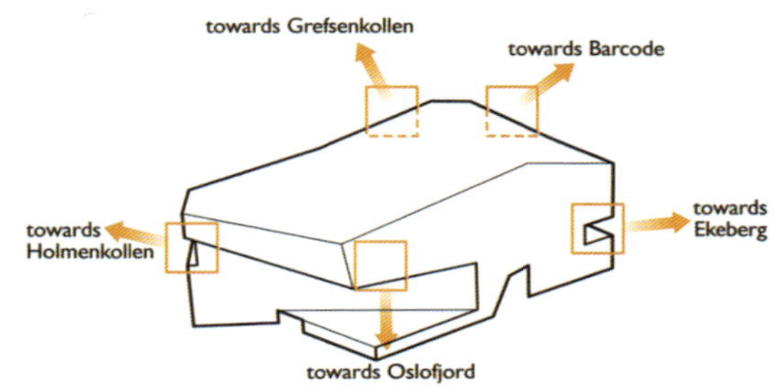

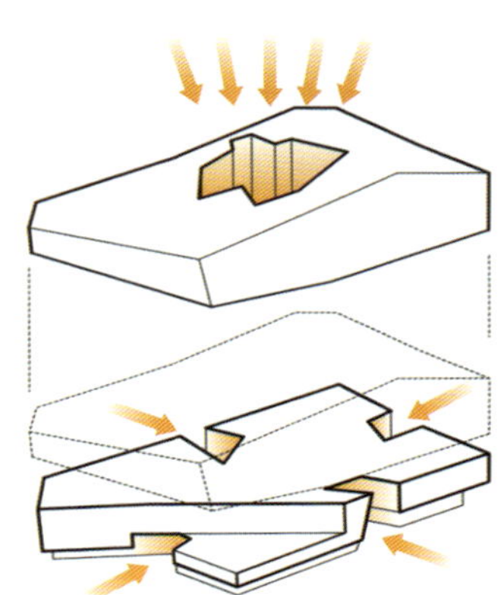

德赫蒙斯克大图书馆 · 奥斯陆

Deichmanske main library · Norway

荣誉提名奖 · Honourable Mention

openminds (竞标代码)

A-lab (建筑师事务所)

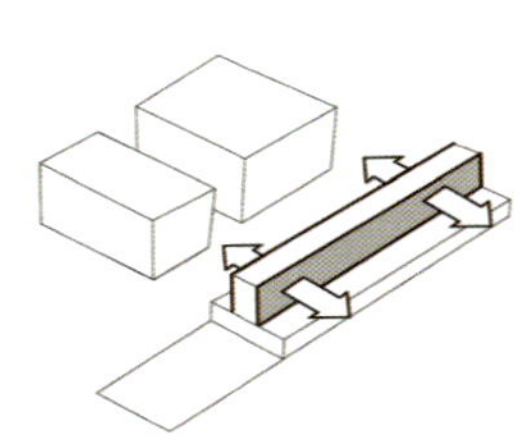

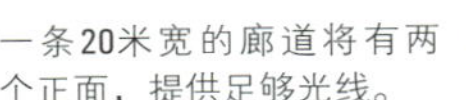

一条20米宽的廊道将有两个正面，提供足够光线。

One strip of 20m width will allow two fronts for program above and enough sky light for library.

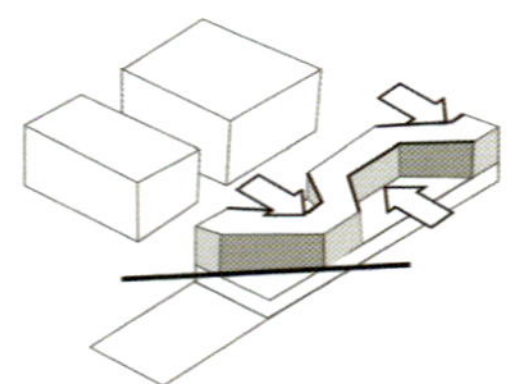

如果我们引入一条视线(按照规律)，使廊道外形不规则，从而增加空间以及形成一个使面向大街的建筑长度断裂的变换的剖面。

If we introduce site line (regulation), we deform the strip to increase the area and promote a variable section that creates a breakdown of the length of building seen from street.

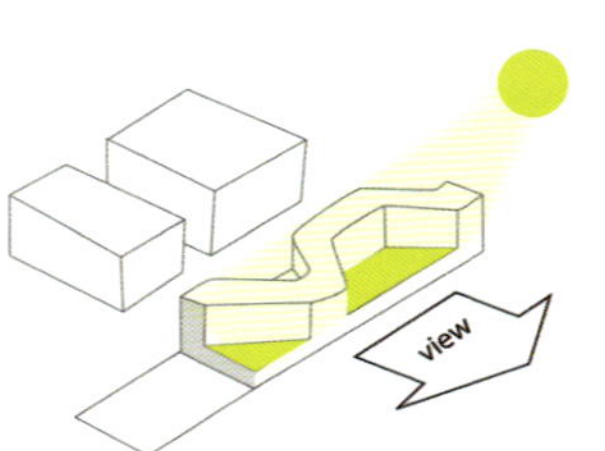

更多朝南的展示空间以及更多面海的外立面。

Bigger exposition to south and more facade facing sea.

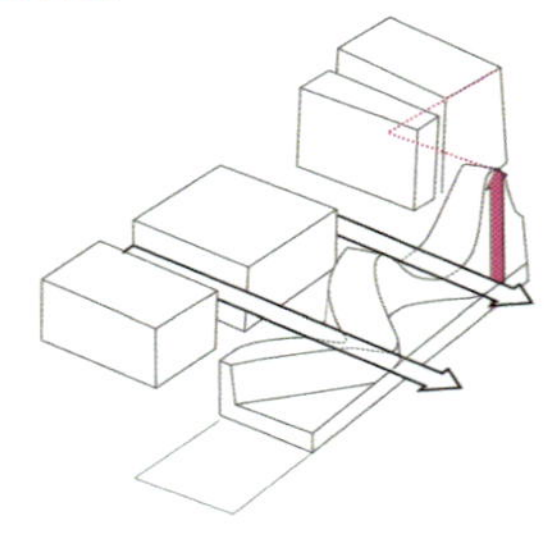

降低建筑的高度从而增加空气的流通以及后排建筑的视线。为了保持设计面积，使面向Barcode的东立面增长至66米(Barcode的设计限高)。

To lower down building to increase air circulation and providing sights to streets behind. In order to maintain area, in east façade, facing Barcode, the height raises to 66m (regulated height in Barcode).

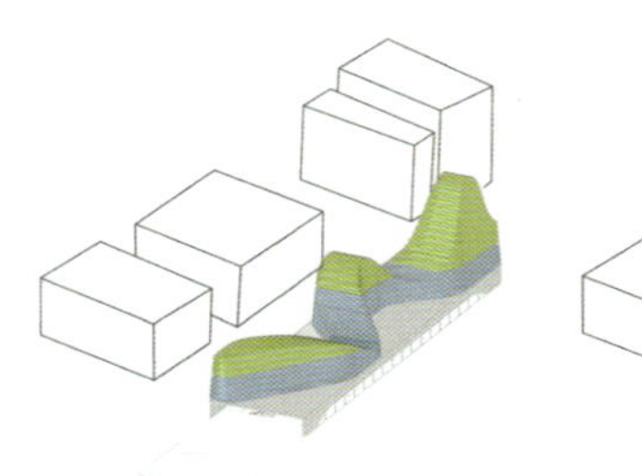

图书馆+办公室/住宅 = 可能设置

library + office/housing = possible configuration

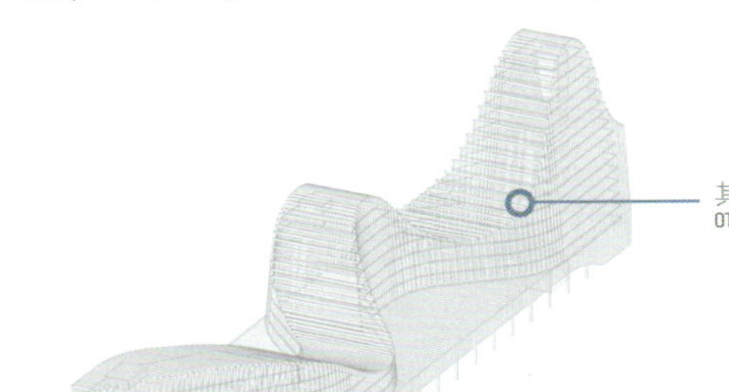

其他功能 OTHER PROGRAMS

自由场所 FLEXI-FIELD

图书馆24小时街道 24H LIBRARY STREET

重物装卸 HARD-DRIVE

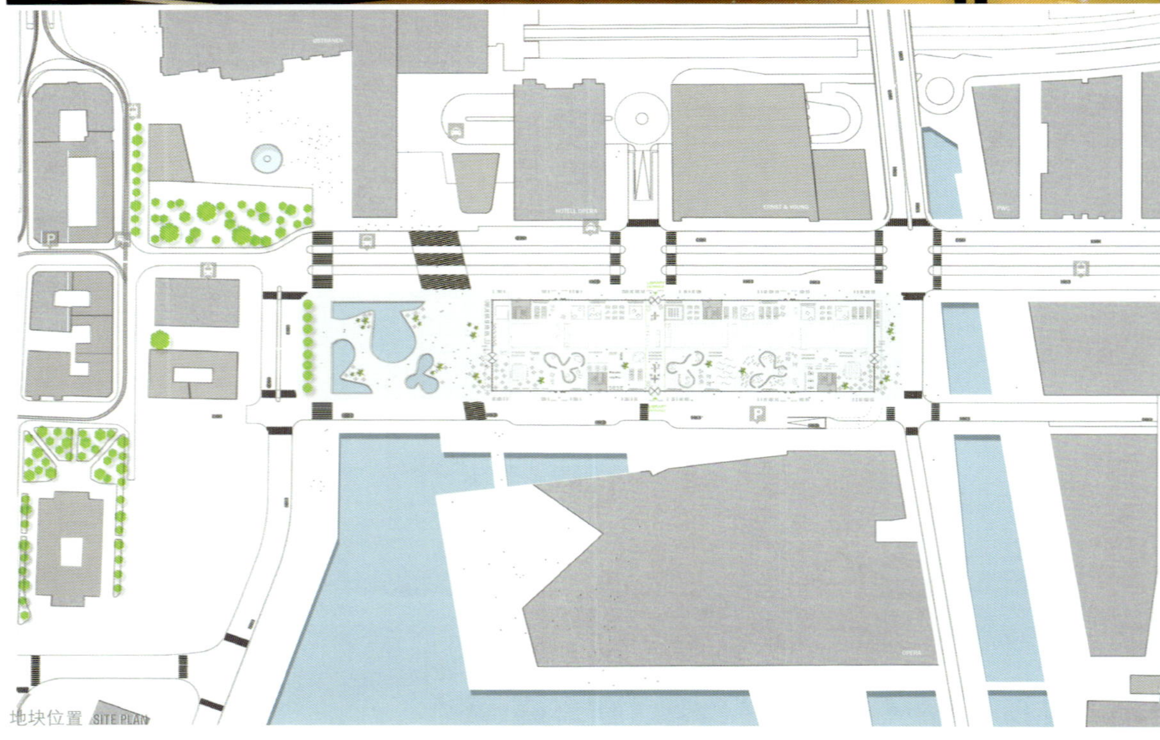

地块位置 SITE PLAN

灵活度

该图书馆横跨东西而建，打造了一个流畅的室内公共空间，毗邻位于剧院屋顶之上的大型都市室外空间。图书馆作为一个城市的公共场所，为了极大地加强其开放性，该项目拟为其设计一个倒转的传统图书馆的造型。图书馆首层通过书籍的成行排列形成了多条通道，构建了图书馆“活动区”和“安静区”之间的过渡地带。办公室和其他房间位于图书馆顶层，形似一条跃于建筑物之上的绸带。

FLEXIBLE

The library occupies the site entirely from east to west, creating a continuous, indoor public space next to the large urban outdoor space above the opera roof. In order to maximize the openness of the library as an urban public space, the project is designed as an inverted traditional library hall. The line of books forms multiple passages on the ground floor and becomes the filter between the "active zone" and the "quieter" zone of the library. Offices and housing are situated on top of the library in the shape of a continuous band.

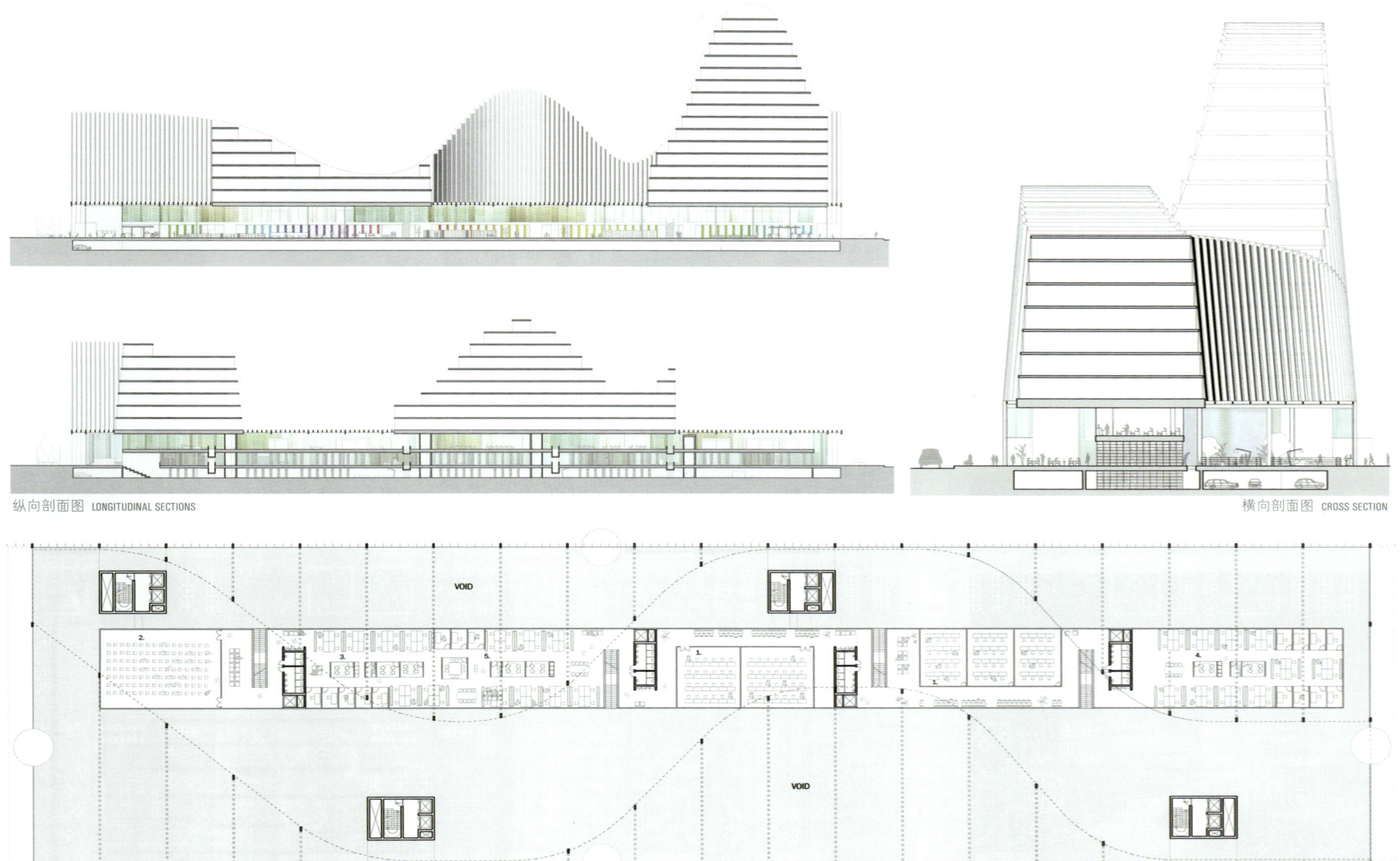

典型层平面 TYPICAL FLOOR PLAN

1. 工作室·会议中心 QUIET STUDY · CONFERENCE CENTER
2. 礼堂 AUDITORIUM
3. 办公室·行政中心 OFFICES · ADMINISTRATION
4. 办公室·特殊部门 OFFICES · SPECIALIZED DEPARTMENTS
5. 图书馆工作站 WORKSTATIONS LIBRARIANS

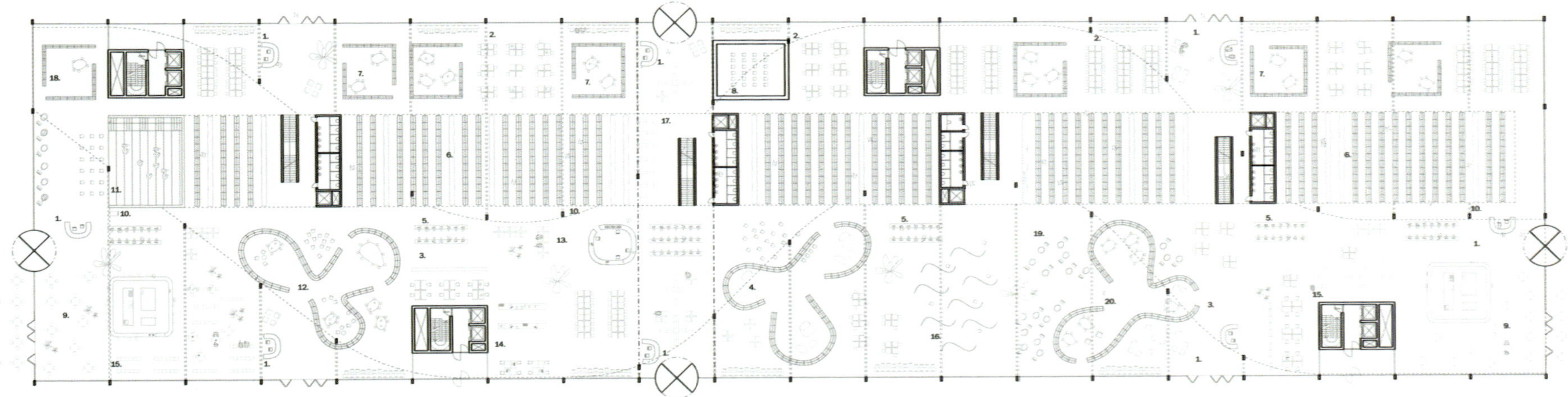

入口层平面 ACCESS FLOOR PLAN

1. 咨询处 INFO POINT
2. 阅读学习室 READING AND STUDY
3. 图书馆工作站 WORK STATIONS LIBRARIANS
4. 每周主题 THEME OF THE WEEK
5. 互联网 INTERNET
6. 图书手机备份 BOOK COLLECTION HARD-DRIVE
7. 主题 THEME
8. 电影院 CINEMA
9. 咖啡 CAFE
10. 图书归还处 BOOKDROP
11. 公开礼堂 OPEN AUDITORIUM
12. 儿童图书 CHILDREN BOOKS
13. 咨询处·商铺 INFOCENTER · SHOP
14. 复印及装订处 SUPPORT FUNCTIONS COPY/BINDING
15. 杂志+報纸 MAGAZINES+NEWSPAPER
16. 展厅 EXHIBITION
17. 24小时图书馆 24-HOUR LIBRARY
18. 特别收藏 SPECIAL COLLECTION
19. 视听室 AUDIO-VIDEO
20. 青年漫画 COMICS FOR THE YOUTH

德赫蒙斯克大图书馆 · 奥斯陆

Deichmanske main library · Norway

入围 · Finalist

三幢主要的建筑体

我们在图书馆前为公众设计了一个水上公园，其由不同形式的水景组成。公园是城市风貌与峡湾风光之间的过渡地带。三幢主要的建筑体特征鲜明，相邻而立。水面上浮有一个透明的球体，与一个封闭直立的建筑体交织在一起，共同构成了图书馆。球体和平板均与一个水下建筑体相连，从这里可以直通商务楼底下的停车场和运载电梯。

THREE BASIC VOLUMES

We propose a water park for the public space in front of the library; a public park, made out of water in all its different forms. The park is the ultimate mediator between the city and the fjord. Three basic, contrasting volumes are set next to the other. A transparent sphere, sitting on the water surface, is intersected by a standing and hermetic volume. Together they form the library. Both the sphere and the slab are connected to an underwater volume, which is linked to the parking under the commercial building and to a delivery elevator.

232511 (竞标代码)

Xaveer De Geyter architecten (建筑师事务所)

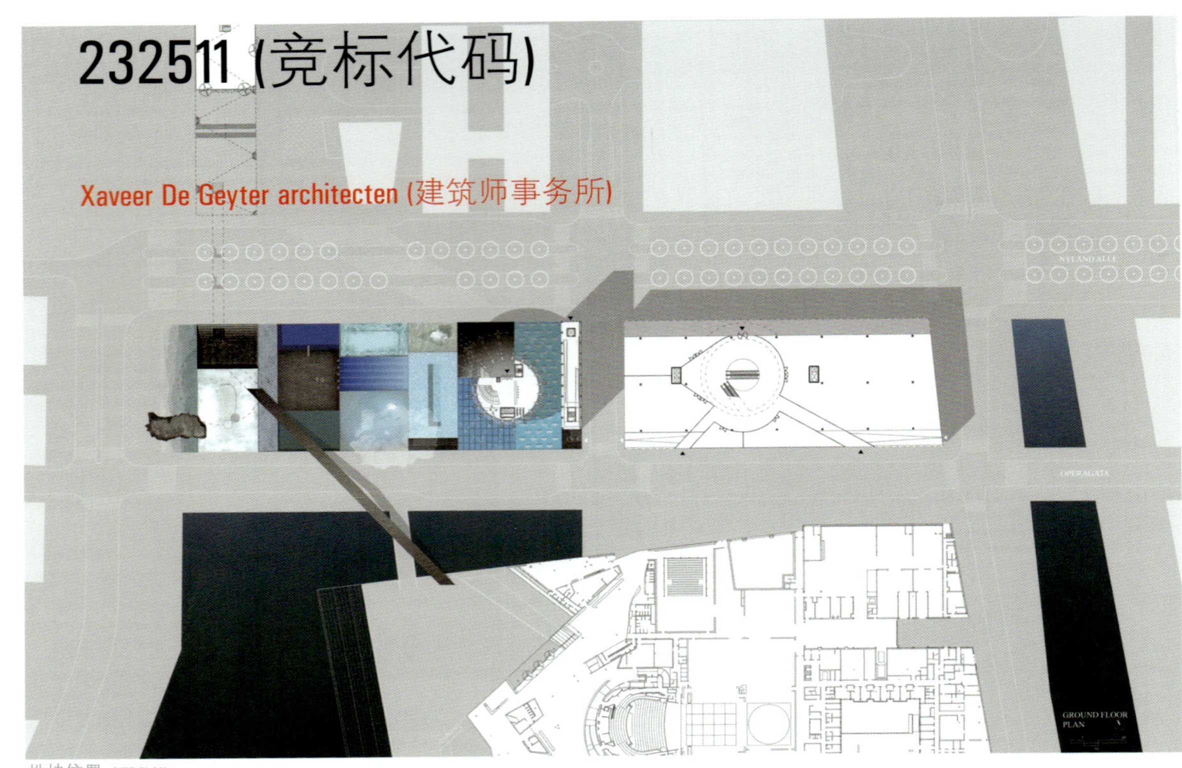

地块位置 SITE PLAN

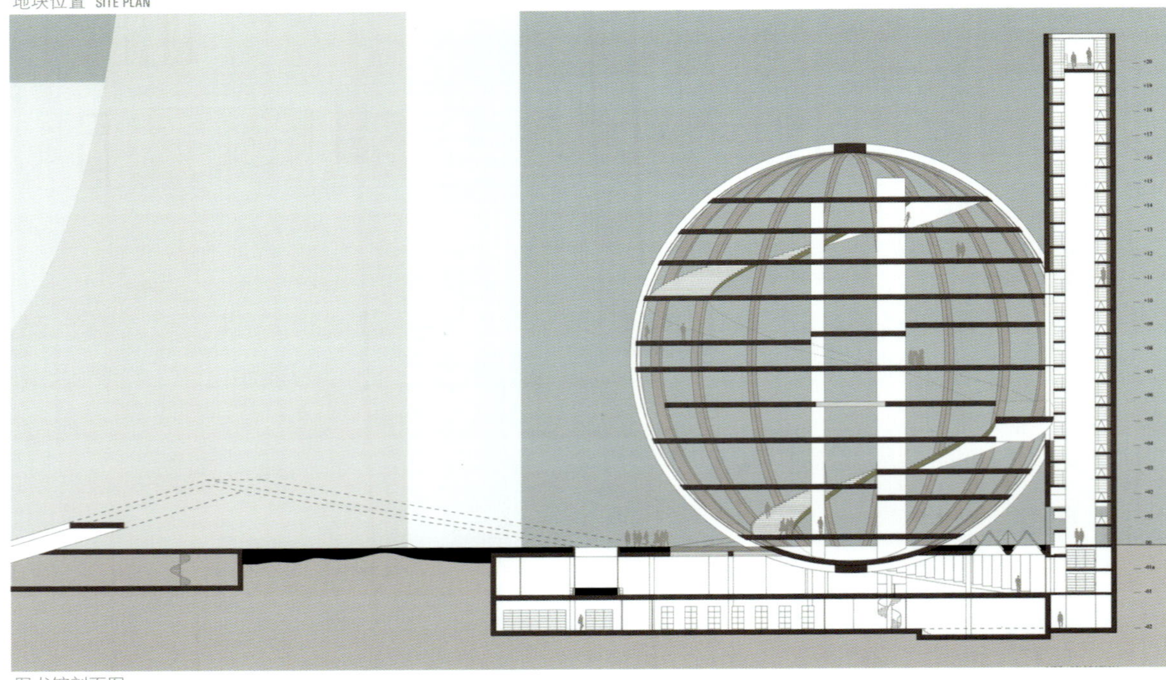

图书馆剖面图 LIBRARY SECTION

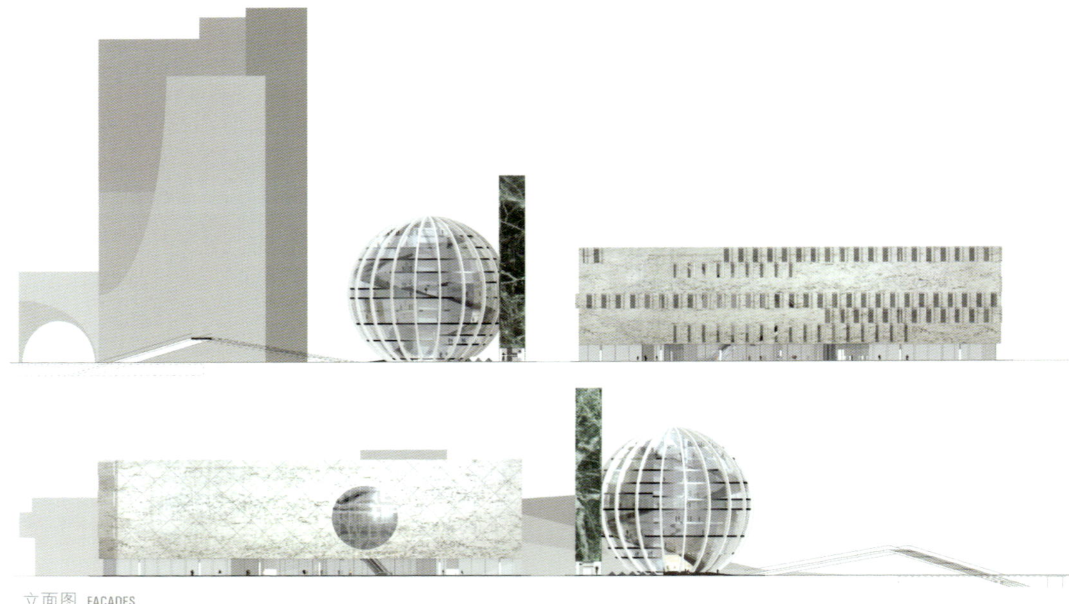

立面图 FAÇADES

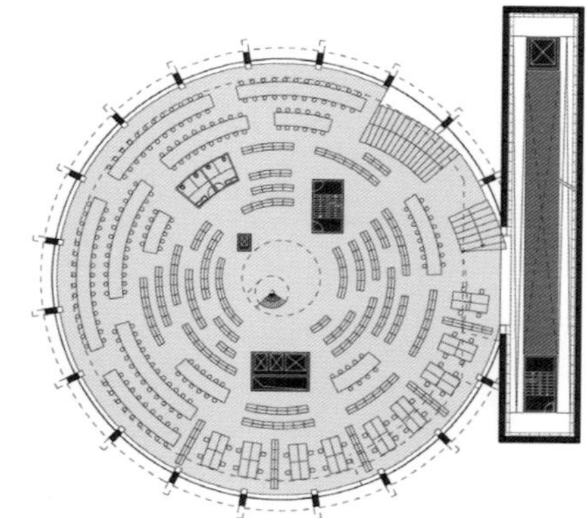

三层平面图 +11.25米 FLOOR PLAN 3 +11.25M

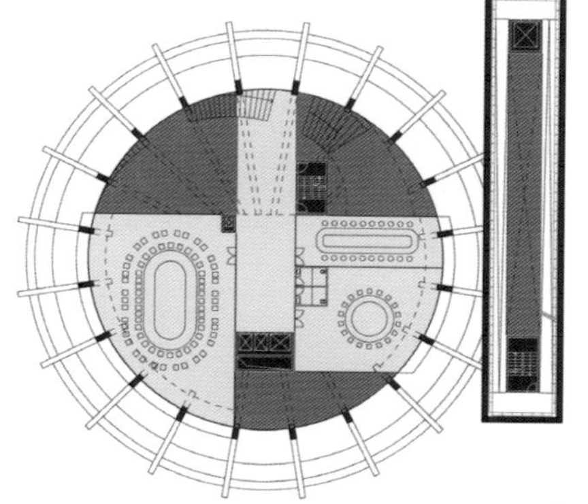

九层平面图 会议厅 +33.25米 FLOOR PLAN 9 CONFERENCE +33.25M

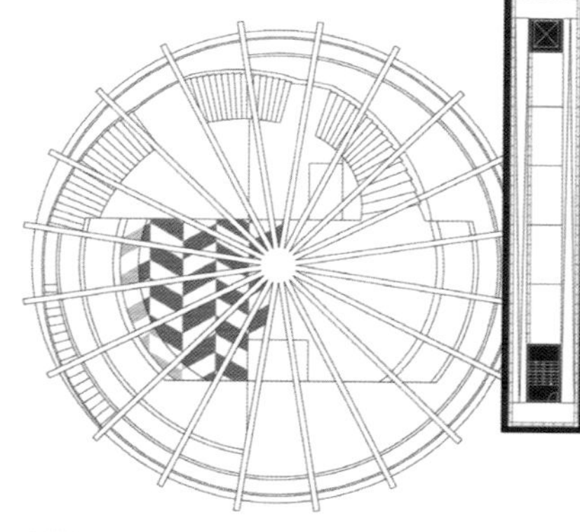

顶层平面图 ROOF PLAN

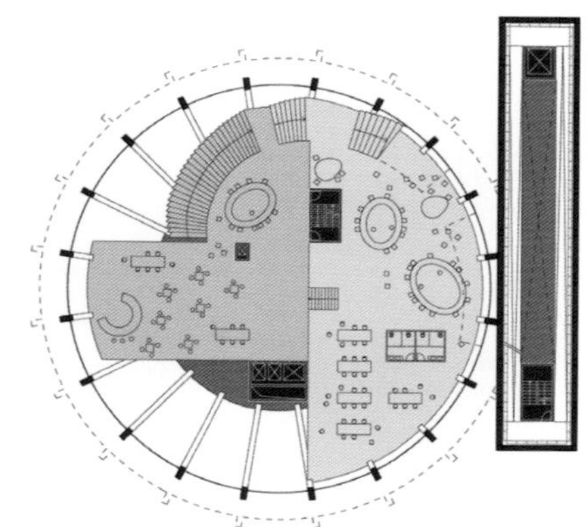

二层平面图 咖啡·新到达 +7.95米 FLOOR PLAN 2 CAFE · NEW ARRIVALS +7.95M

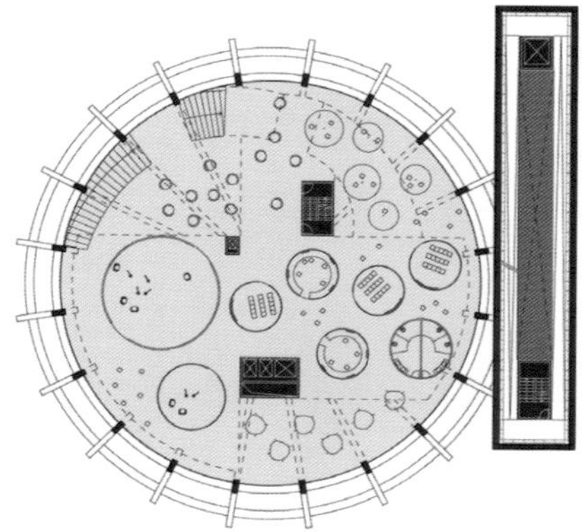

八层平面图 视听室 +30.00米 FLOOR PLAN 8 AUDIO-VISUAL +30.00M

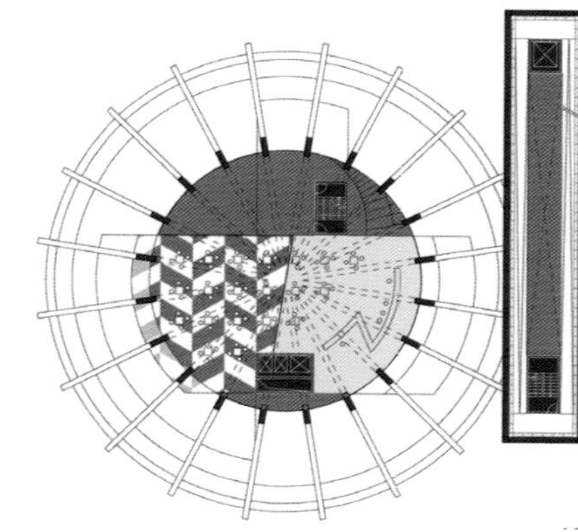

十层平面图 食堂 +36.50米 FLOOR PLAN 10 CANTEEN +36.50M

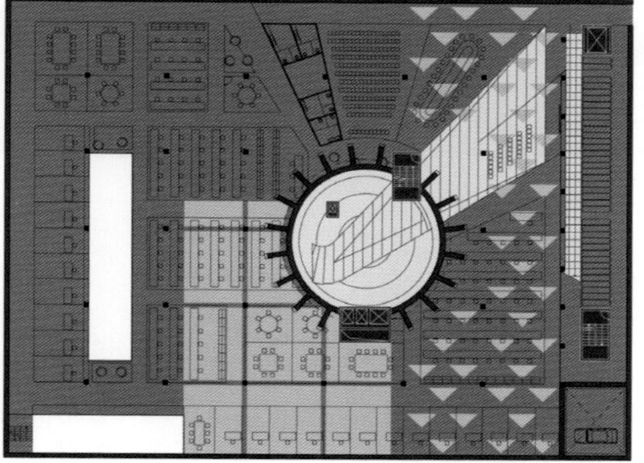

地下夹层平面图 -1.5米 FLOOR PLAN -0.5 -1.5M

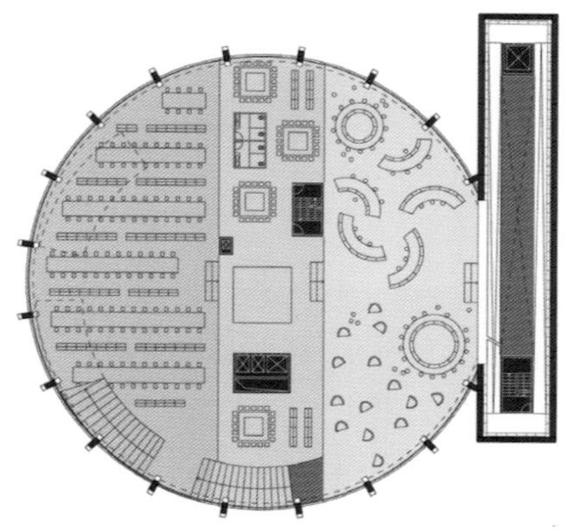

六层平面图 +22.75米 FLOOR PLAN 6 +22.75M

空地剖面图 SECCION VOID

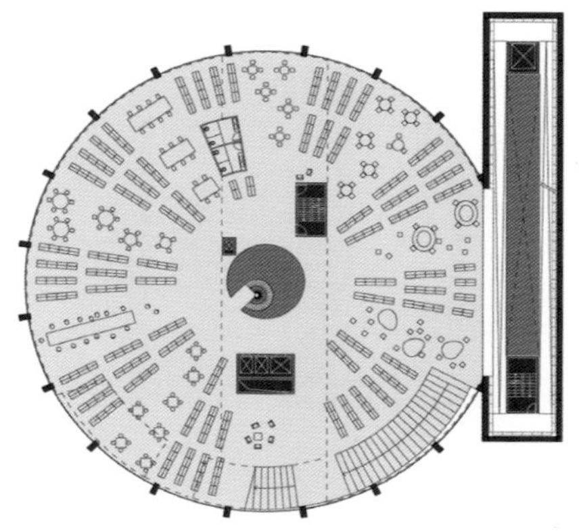

地下一层平面图 -5.0米 FLOOR PLAN -1 -5.0M

五层平面图 +18.75米 FLOOR PLAN 5 +18.75M

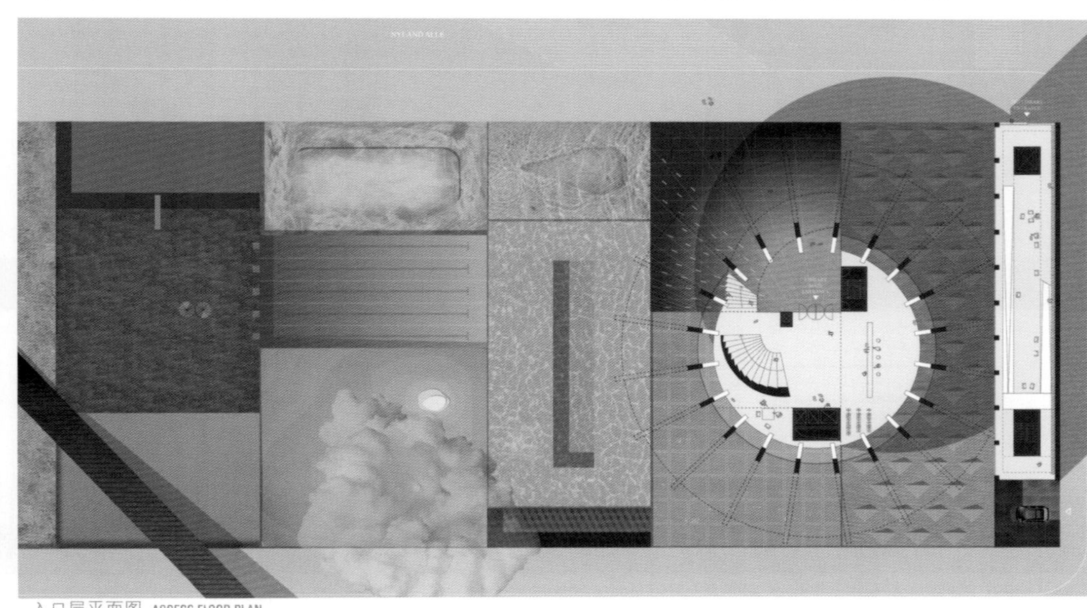

入口层平面图 ACCESS FLOOR PLAN

坡面剖面图 SECCION RAMPS

德赫蒙斯克大图书馆 · 奥斯陆

Deichmanske main library · Norway

入围 · Finalist

sugar (竞标代码)

Wiel Arets Architects & Associates bv (建筑师事务所)

独立体量

新图书馆布局紧凑，分成许多个均匀分布的独立体量。通过旋转，每个独立结构均彼此相对，从而构成了一个公共空间，打造出不同的建筑外观和标新立异的视角。除了紧凑的布局之外，图书馆的设计还在目前的环境基础上增加了城市足迹的构思，优化了亲水环境；同时，设计师还精心构思，融入了老城元素、城市特点以及人文尺度。

VOLUMENES SOLITARIOS

The new library's compact program has been divided into a number of volumes – "solitaries" - that have been evenly dispersed throughout the site. By rotating each volume towards one another, a public space is defined, promoting to diverse exteriors, as well as unexpected perspectives. Despite its concentration, the proposed urban footprint introduces a finer scale to the waterfront than the existing situation; cultivating aspects associated with old town, as well as domestic and human scales.

城市剖面 URBAN SECTIONS

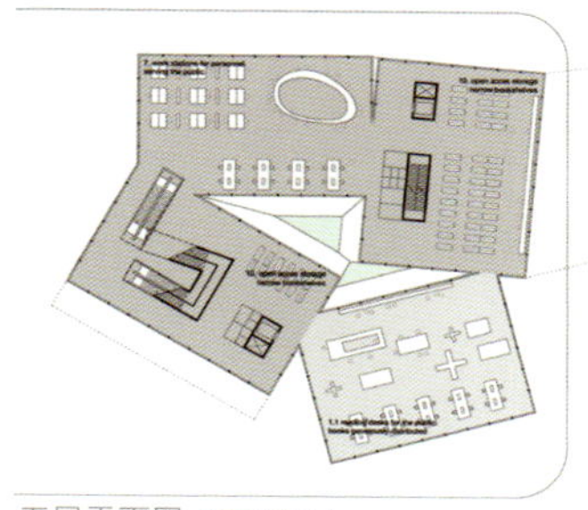

五层平面图 FLOOR PLAN 4

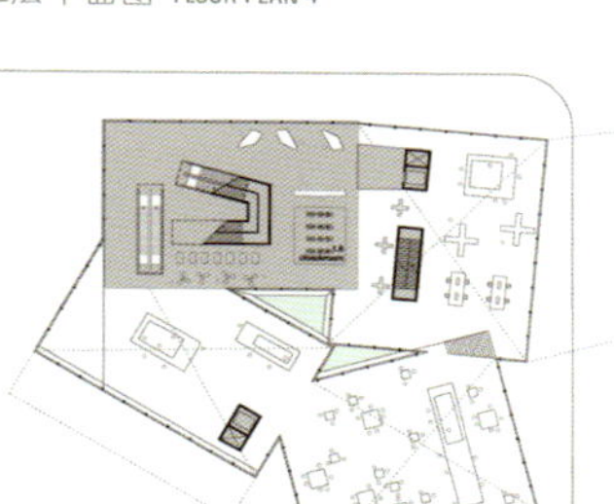
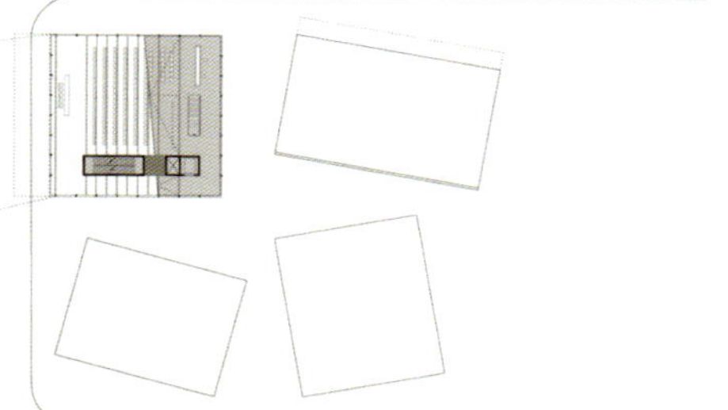
二层平面图 FLOOR PLAN 1

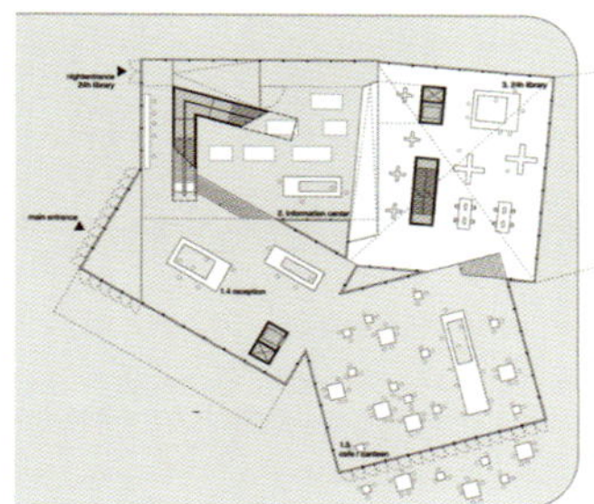
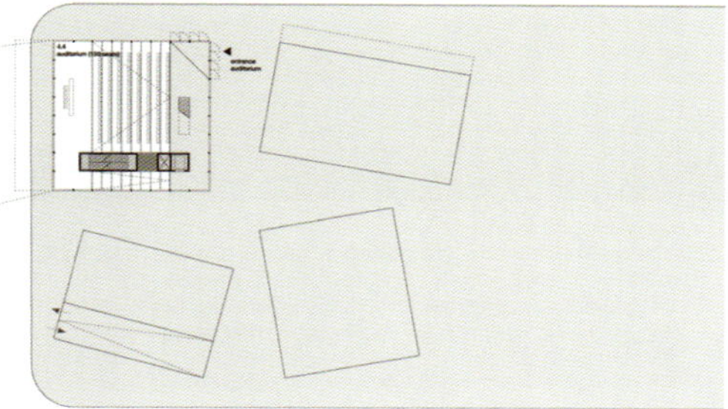
底层平面图 FLOOR PLAN 0

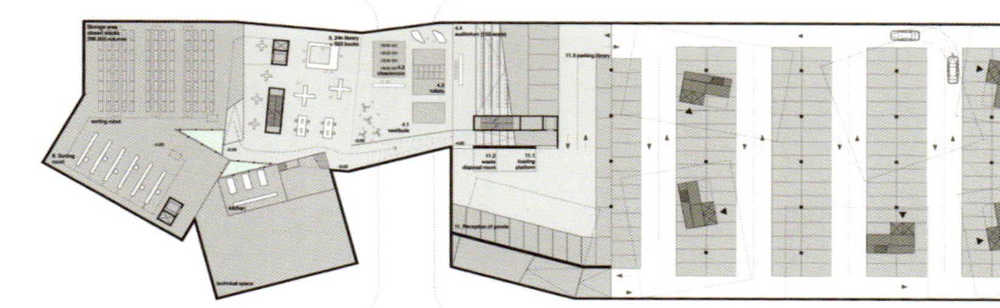

地下城平面图 BASEMENT FLOOR PLAN

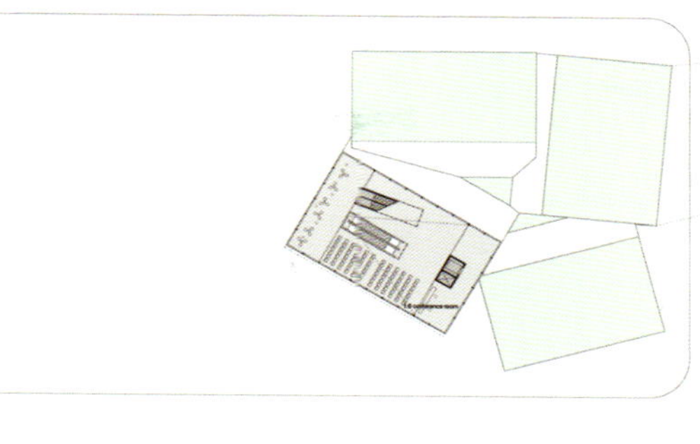

十层平面图 FLOOR PLAN 9

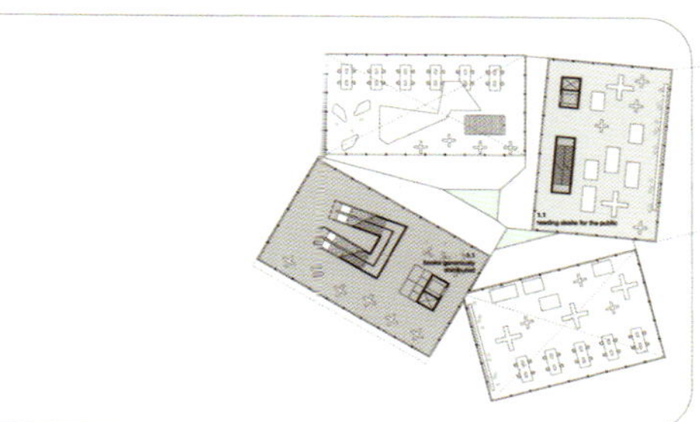
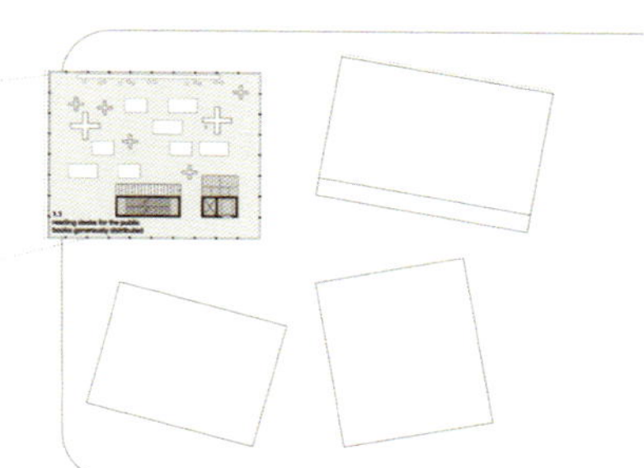
九层平面图 FLOOR PLAN 8

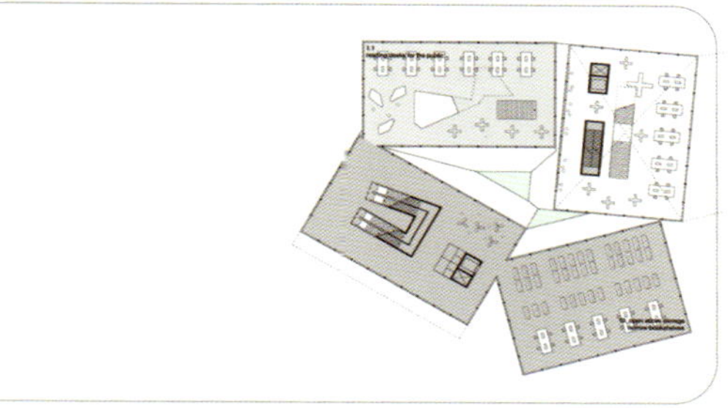
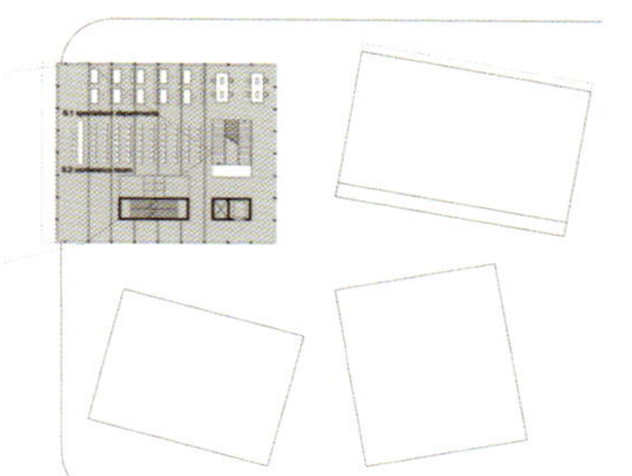
七层平面图 FLOOR PLAN 6

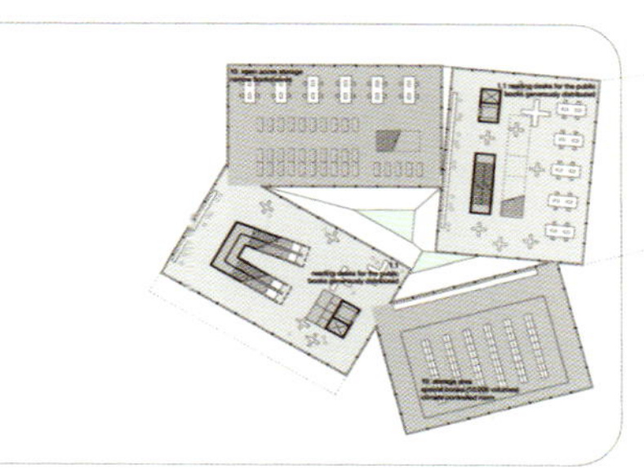
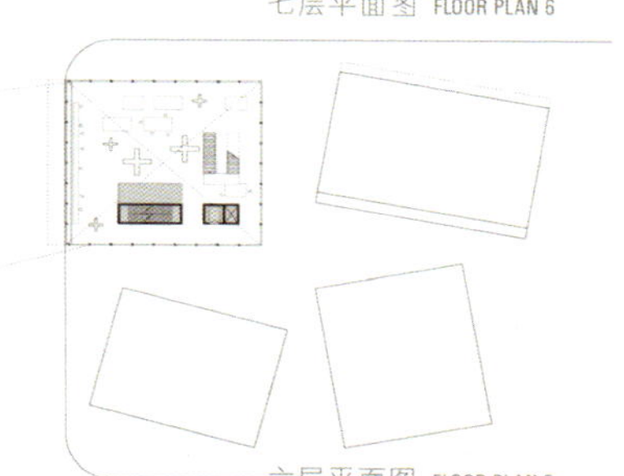
六层平面图 FLOOR PLAN 5

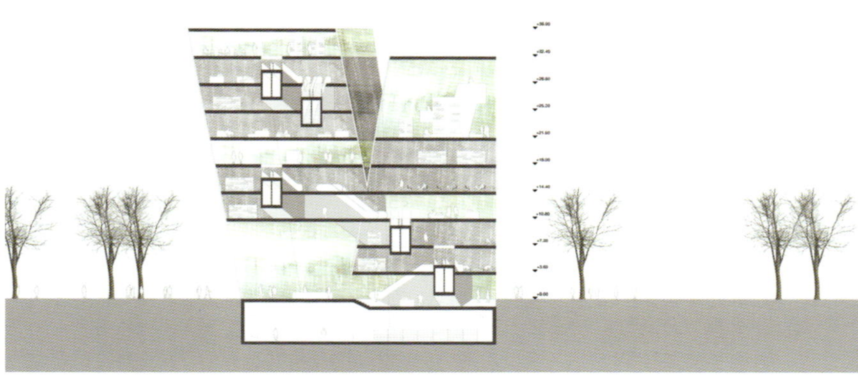

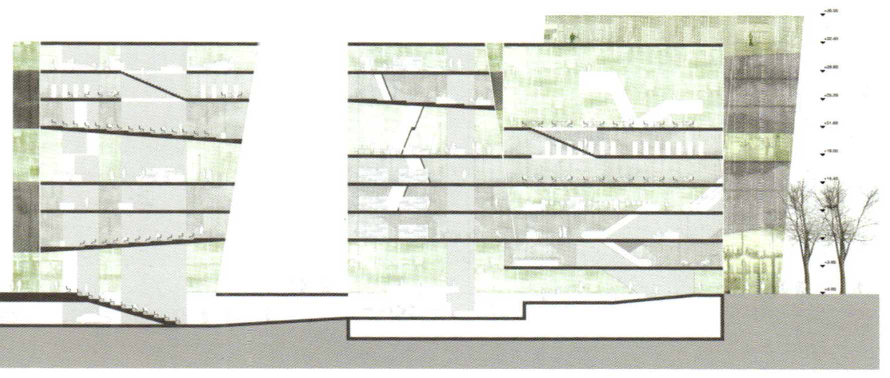
剖面图 SECTIONS

德赫蒙斯克大图书馆 · 奥斯陆

Deichmanske main library · Norway

入围 · Finalist

starfish (竞标代码)

Allmann Sattler Wappner Architekten (建筑师事务所)

Markus Allmann · Amandus Sattler · Ludwig Wappner (建筑师)

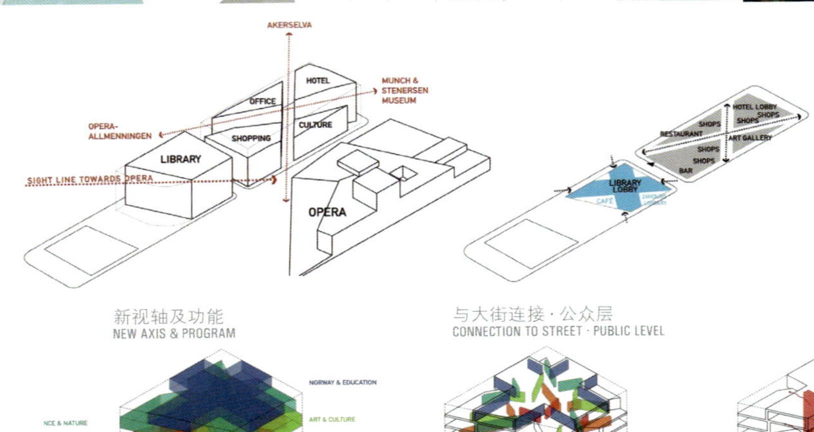

新视轴及功能
NEW AXIS & PROGRAM

与大街连接 · 公众层
CONNECTION TO STREET · PUBLIC LEVEL

向四个方向十字体量
4 DIRECTIONS CROSSING

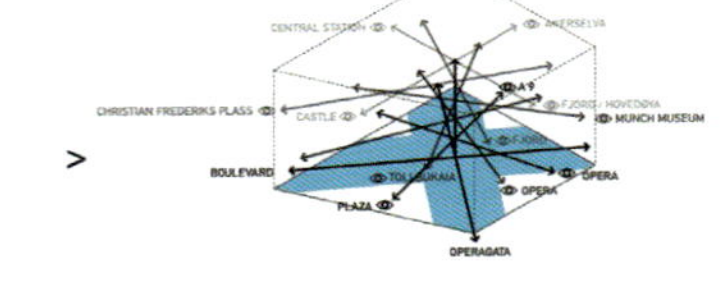

连接性 · 向指定视角垂直旋转
CONNECTIVITY · VERTICAL ROTATION TOWARDS SPECIFIC VIEWS

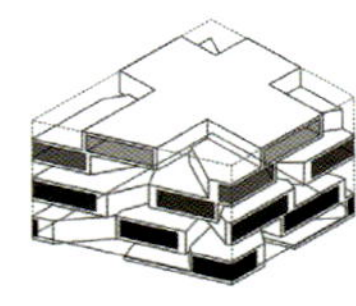

叠加十字体量 · 向视角旋转
STACKED CROSSES · RITATED ACCORDING TO VIEWS

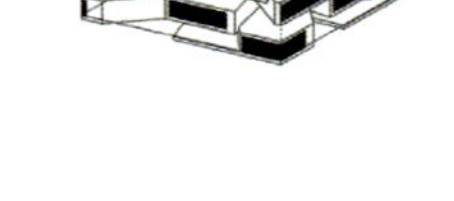

六个主题十字体量
SIX THEMATIC CROSSES

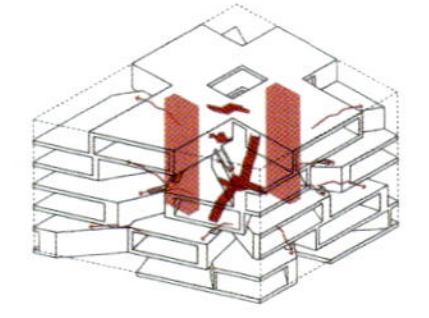

"图书墙" 3000平方米
"THE WALLS OF BOOKS" 3000SQM

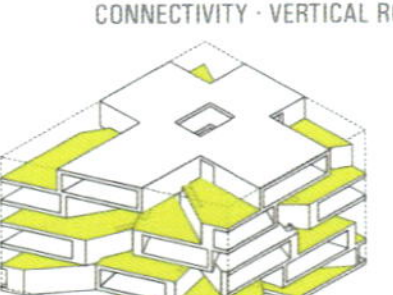

垂直流通
VERTICAL CIRCULATION

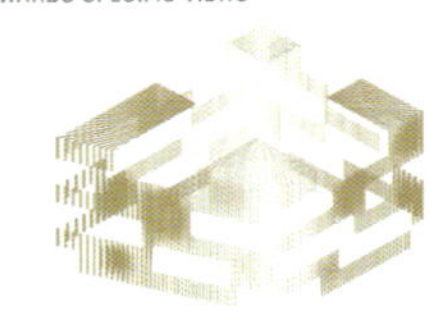

"高原"
"THE PLATEAUS"

"木质窗帘"
"THE WOODEN CURTAIN"

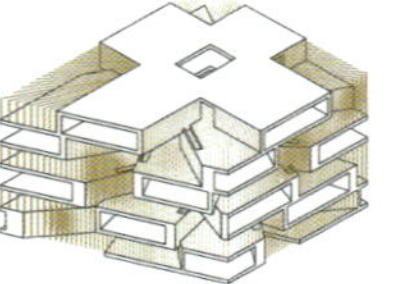

"新图书馆"
"THE NEW LIBRARY"

轴转结构

这里曾经矗立着一幢配备了一个极具表现力的造像的建筑，而今这个图书馆居然能选址于此？毗邻的剧院，其结构具有突出的意向性和联想性，在这一带尤为醒目。新图书馆面向剧院的风景线构成了其外立面。内部空间由多个围绕环中轴而建的十字形建筑结构组成，这些十字形结构相互叠加，与建筑轮廓分明的外立面完全重合。

ROTATION

How can the new library position itself at a location that seems formally saturated by a significant building with an expressive iconography? The neighbouring Opera dominates the site by its metaphoric and associative architecture. The important sight line towards the Opera defines the envelope for the new library. Volumes in the shape of crosses form interior spaces, which rotate around a central axis, and are stacked up and then trimmed by the boundary of the defined building envelope.

西南立面图 SOUTHWEST ELEVATION

东北立面图 NORTHEAST ELEVATION

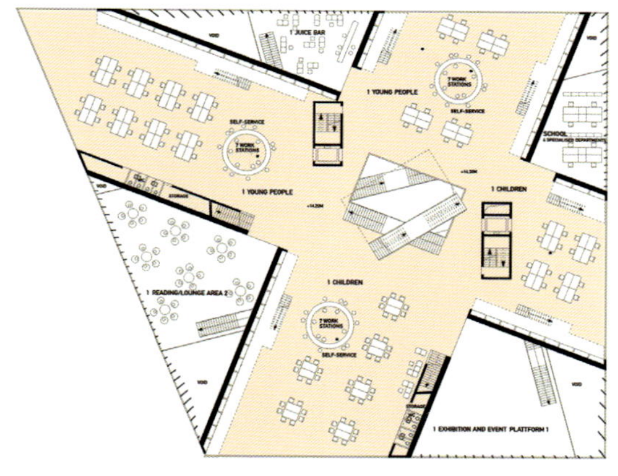
三层平面图·儿童及青少年 FLOOR PLAN 2 · CHILDREN AND YOUNG PEOPLE

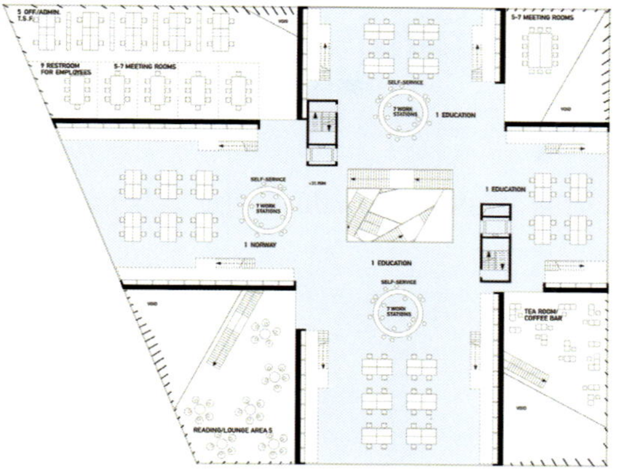
六层平面图·挪威及教育 FLOOR PLAN 5 · NORWAY AND EDUCATION

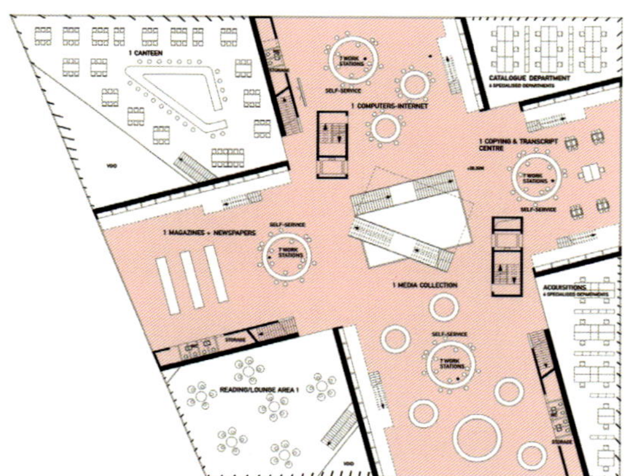
二层平面图·当代媒体 FLOOR PLAN 1 · CONTEMPORARY MEDIA

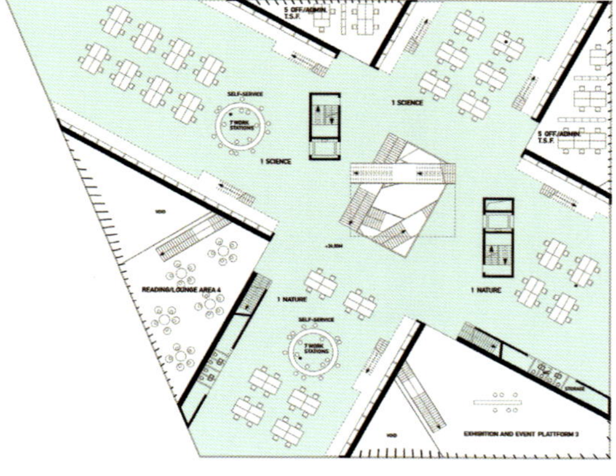
五层平面图·科学及大自然 FLOOR PLAN 4 · SCIENCE AND NATURE

剖面图 A-A' SECTION A-A'

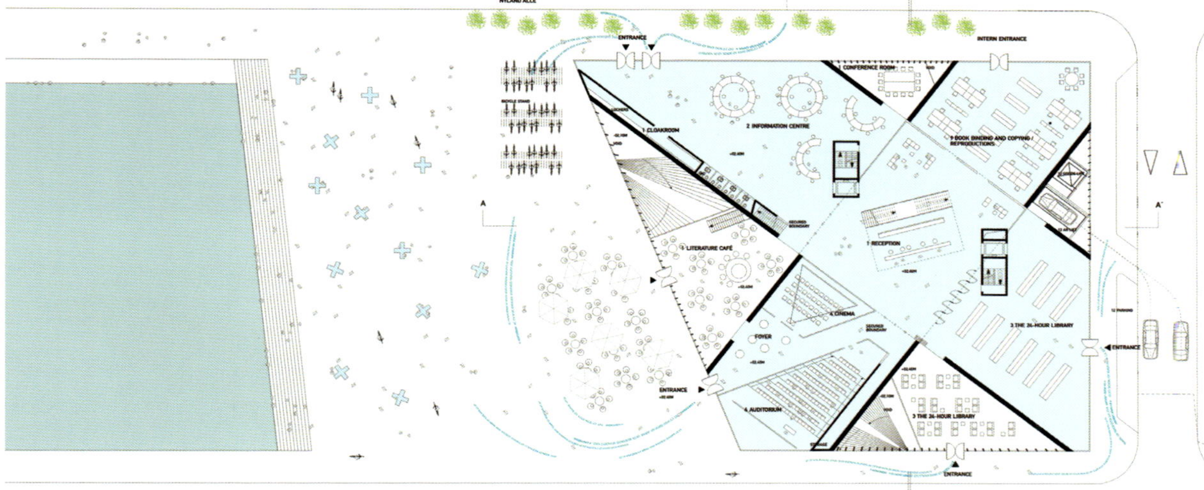
底层平面图 GROUND FLOOR PLAN

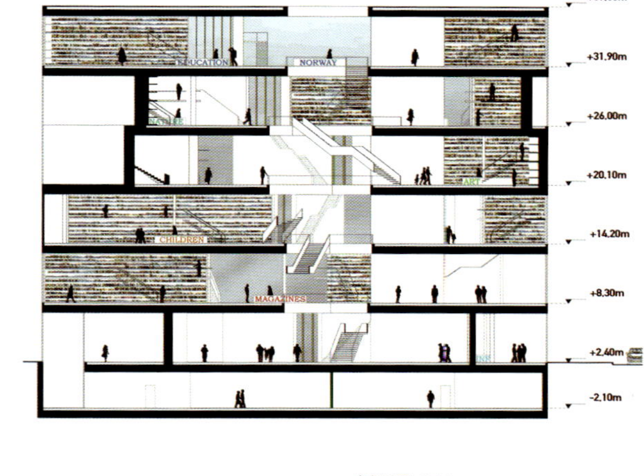

剖面图 B-E' SECTION B-B'

德赫蒙斯克大图书馆 · 奥斯陆

Deichmanske main library · Norway

入围 · Finalist

bookstand (竞标代码)

SeARCH i.c.w. CMA (建筑师事务所)

Bjarne Mastenbroek · Uda Visser · Christian Müller (建筑师)

日常生活空间

我们设计的图书馆两侧是如渗透膜一般的结构，中间是一个实体单元。首层是一个低于地面的空间，是都市街头生活的延续。入口大厅空间很大，在室内外各占一部分，适合举办大型会议和晚会。建筑的上层被设计为通风条件良好的平板结构，可作多项功能使用。在这里，人们可以俯瞰剧院的风景、峡湾的风光以及中央广场的景象。

ABSORBS DAILY LIFE

We have designed the library as a permeable membrane on both sides. In-between lays a more solid element. The ground floor is a lowered level of continuous urban street life. The entry hall/forum, partly inside and partly outside has sufficient size in order to host even large conferences and parties. The upper towers are conceived as well-ventilated urban slabs with functional flexibility. Here, one will be able to overlook the Opera, the fjord and the central square.

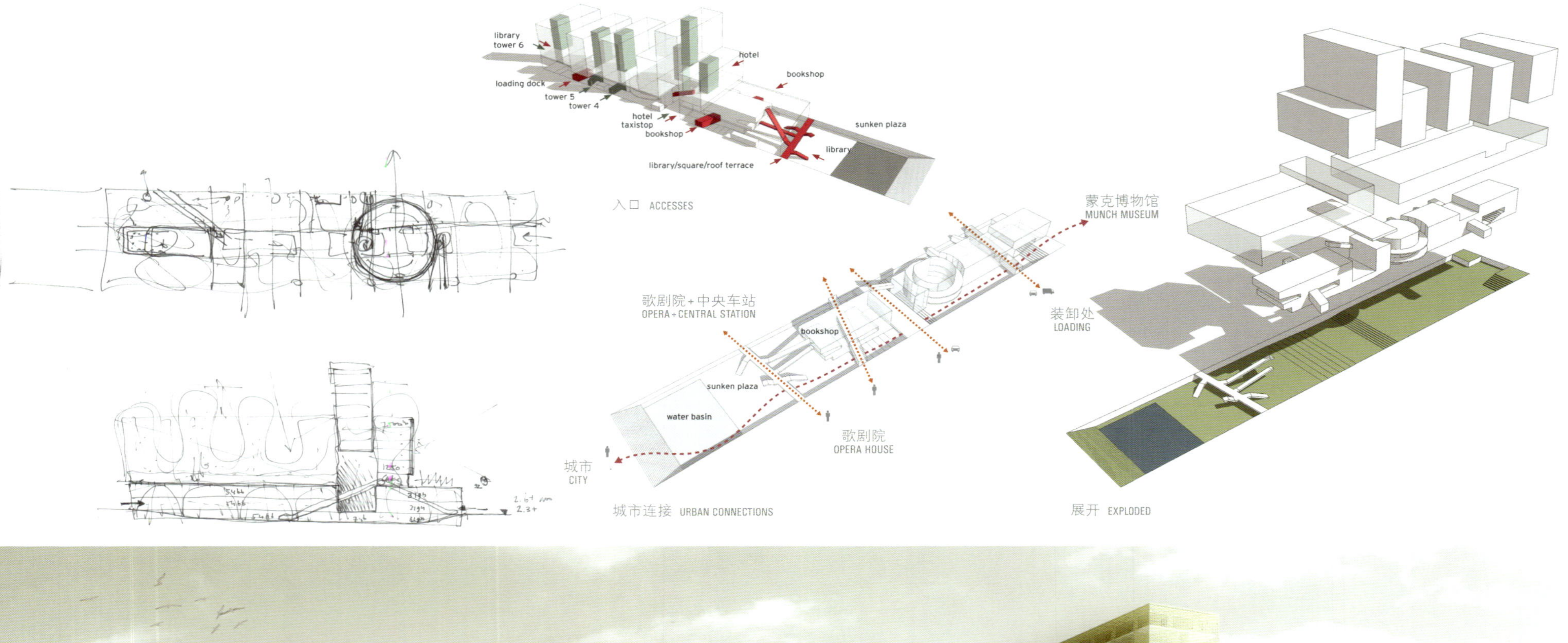

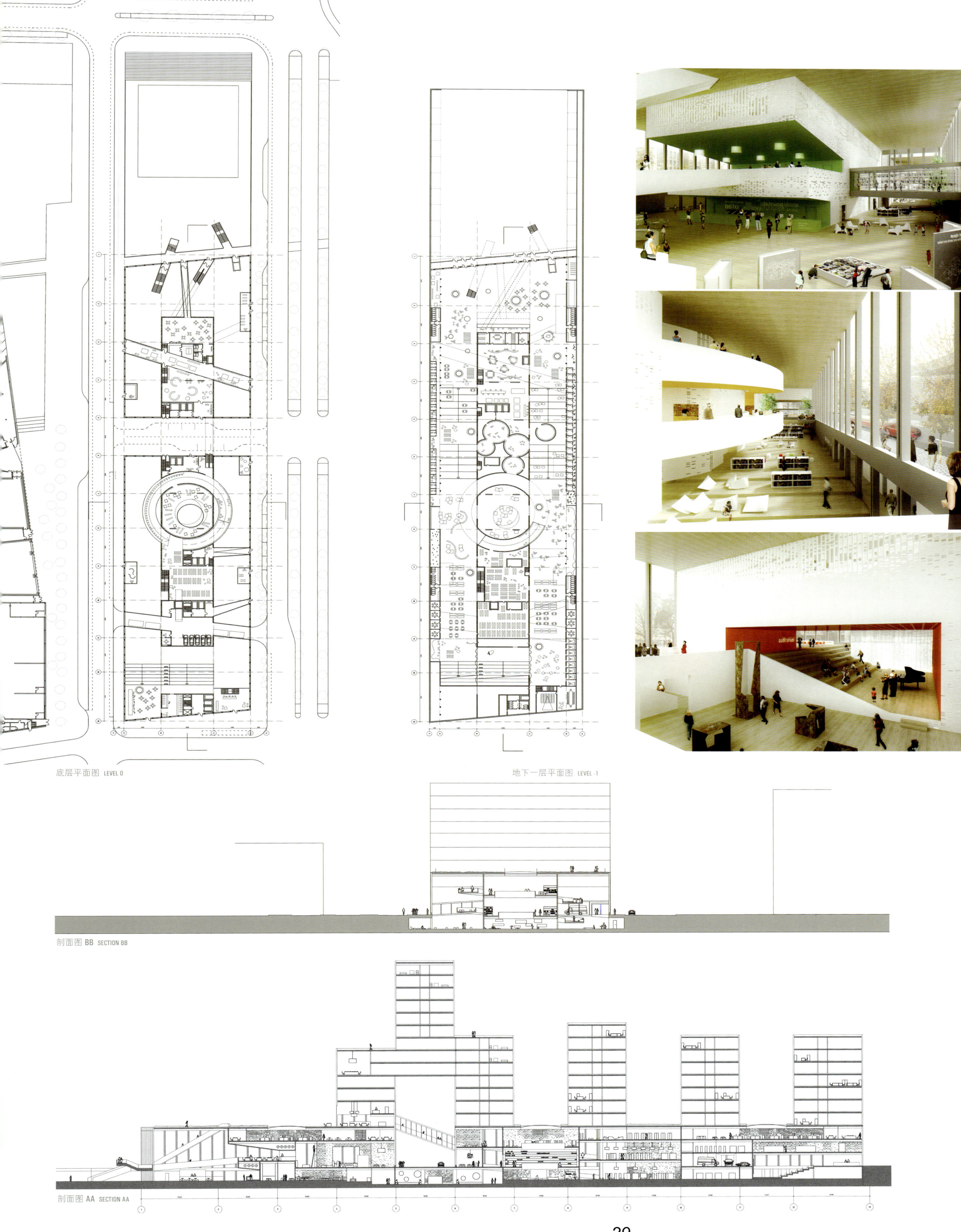

底层平面图 LEVEL 0

地下一层平面图 LEVEL -1

剖面图 BB SECTION BB

剖面图 AA SECTION AA

德赫蒙斯克大图书馆 · 奥斯陆

Deichmanske main library · Norway

入围 · Finalist

woven (竞标代码)

PLASMA Studio (建筑师事务所)

不同的采光条件

该设计与传统的图书馆建筑结构不同，它完美融合了建筑与风景、工程与环境。与其说建筑构成了风景的一部分，倒不如说建筑是一道延伸于风景之外的景观。我们设计了一种类似菱形的建筑结构，通过单点的水平位移将图书馆结构的透明度由0提升到90%，从而实现了从实本到通透的自然过渡。

DIFFERENT LIGHT CONDITIONS

The design challenges a library's traditional structure, fusing building and landscape, program and context. The building does not sit within a landscape, but is an extension of it. We have developed a rhomboidal component within the volumetric expression of the library that can change from 100% opaque to 90% transparent through the horizontal shift of a single point, enabling smooth transitions from solid to transparent.

地块位置 SITE PLAN

东立面图 EAST ELEVATION

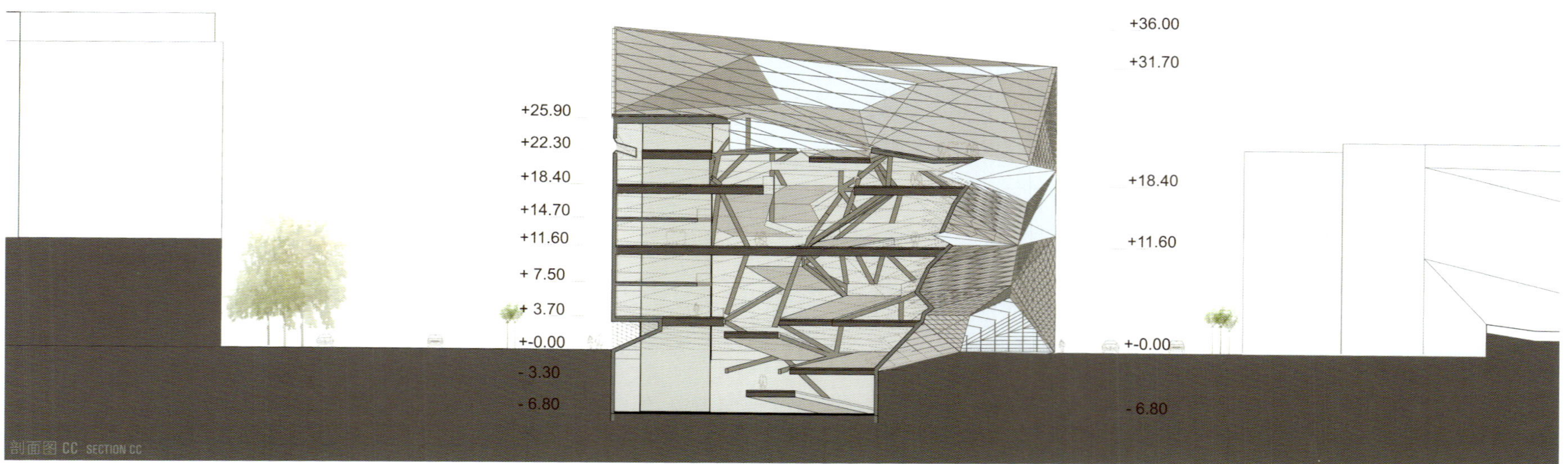

剖面图 CC SECTION CC

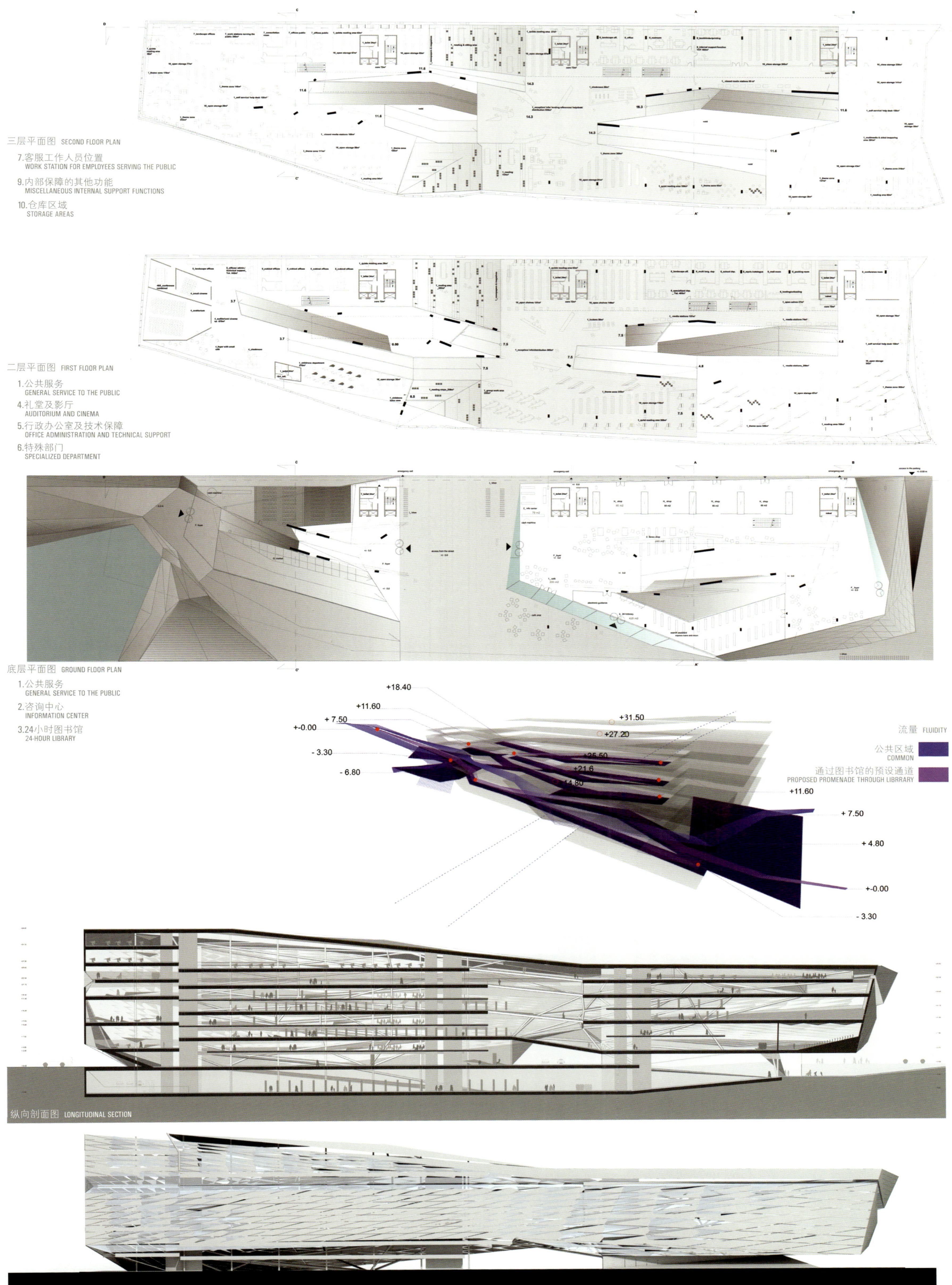

三层平面图 SECOND FLOOR PLAN

7.客服工作人员位置
WORK STATION FOR EMPLOYEES SERVING THE PUBLIC

9.内部保障的其他功能
MISCELLANEOUS INTERNAL SUPPORT FUNCTIONS

10.仓库区域
STORAGE AREAS

二层平面图 FIRST FLOOR PLAN

1.公共服务
GENERAL SERVICE TO THE PUBLIC

4.礼堂及影厅
AUDITORIUM AND CINEMA

5.行政办公室及技术保障
OFFICE ADMINISTRATION AND TECHNICAL SUPPORT

6.特殊部门
SPECIALIZED DEPARTMENT

底层平面图 GROUND FLOOR PLAN

1.公共服务
GENERAL SERVICE TO THE PUBLIC

2.咨询中心
INFORMATION CENTER

3.24小时图书馆
24-HOUR LIBRARY

纵向剖面图 LONGITUDINAL SECTION

北立面图 NORTH ELEVATION

德赫蒙斯克大图书馆 · 奥斯陆

Deichmanske main library · Norway

入围 · Finalist

space/scape (竞标代码)

70°N arkitektur + Dahl & Uhre arkitekter (建筑师事务所)

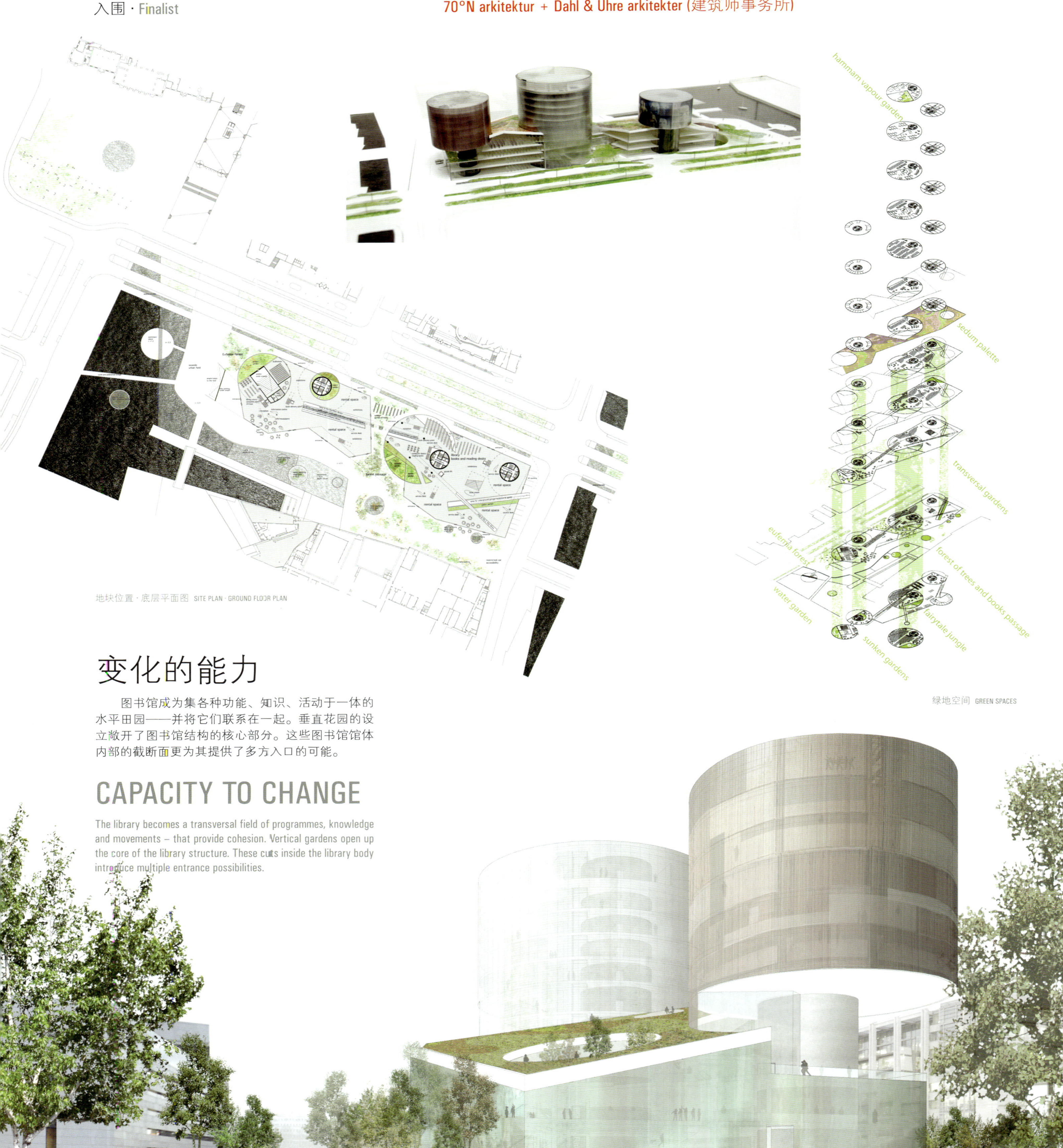

地块位置 · 底层平面图 SITE PLAN · GROUND FLOOR PLAN

绿地空间 GREEN SPACES

变化的能力

图书馆成为集各种功能、知识、活动于一体的水平田园——并将它们联系在一起。垂直花园的设立敞开了图书馆结构的核心部分。这些图书馆馆体内部的截断面更为其提供了多方入口的可能。

CAPACITY TO CHANGE

The library becomes a transversal field of programmes, knowledge and movements – that provide cohesion. Vertical gardens open up the core of the library structure. These cuts inside the library body introduce multiple entrance possibilities.

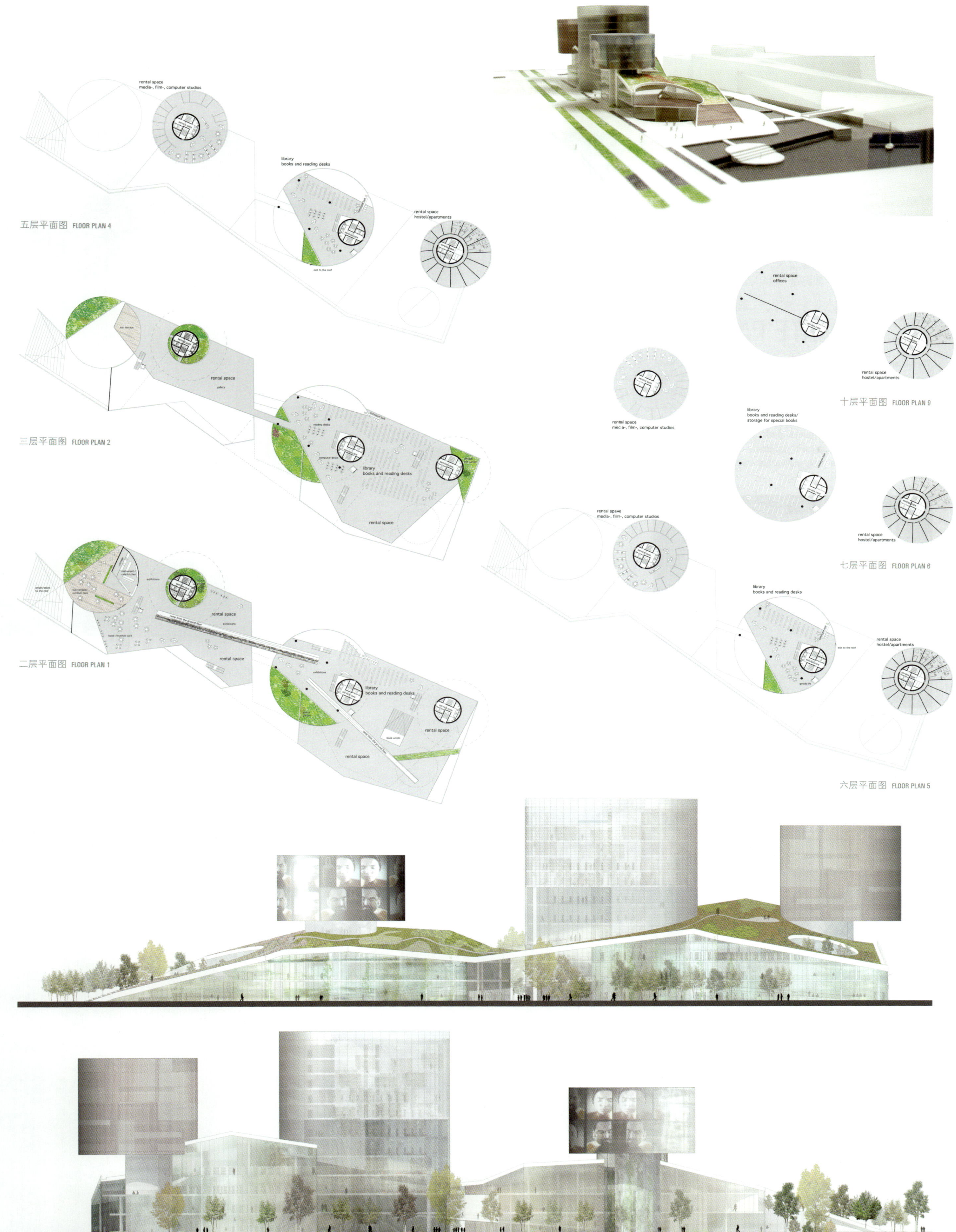

五层平面图 FLOOR PLAN 4
三层平面图 FLOOR PLAN 2
二层平面图 FLOOR PLAN 1
十层平面图 FLOOR PLAN 9
七层平面图 FLOOR PLAN 6
六层平面图 FLOOR PLAN 5
rental space
media-, film-, computer studios
library
books and reading desks
rental space
hostel/apartments
rental space
offices
library
books and reading desks/
storage for special books

入围 · Finalist

The City, the Green, the Library in between (竞标代码)

Lundgaard & Tranberg Architects (建筑师事务所)

私密公园

我们打算将Deichman Axis打造成一个林木茂密的私密公园，这将成为奥斯陆的又一个文化场所。新建的图书馆形似一个三面体，其一个开放的中心空间犹如一株枝叶层层生长的大树，勾勒出一个轮廓分明的建筑结构，打造出一个垂直结构的新都市空间。图书馆的中央空间呈辐状，正如公园里的大树连接着树根一样，仿佛要将都市生活融入馆内一般。

AN INTIMATE PARK

We envision the Deichman Axis transformed into an intimate, wooded park that will act as a new cultural forum for Oslo. The new library takes form as the primary figure in a composition of three faceted forms. The clear volume of the library is penetrated and set in relief by an open central room in the form of an inner tree branching up through the various levels, creating a new vertical urban space. The central space of the library turns and rotates, like a tree with roots in the park, pulling the urban life inside and up from the street.

南立面及剖面图 SOUTH SECTION AND ELEVATION
西南立面及剖面图 SOUTHWEST ELEVATION AND SECTION
北立面及剖面图 NORTH ELEVATION AND SECTION
地块位置 SITE PLAN

新火车中转站·蒙特德维苏威
New Interexchange Railway Station · Italy

竞标 · competition
新火车中转站
New Interexchange Railway Station

竞标类型 · competition type
限制性邀标
invited competition

项目地点 · site area
蒙特德维苏威 · 意大利 Monte Vesuvio · Italy

主办方 · promoter
意大利铁路交通集团
Rete Ferroviaria Italiana

日程安排 · schedule
招标 · Announcement 10.2008
评审结果 · Jury´s results 06.2009

评审团 · jury

Mario Botta
Carlo De Vito
Marina Habetswallner
Romeo La Pietra
David Nelson
Daniela Volpi
Stefano Reggio
Modestino Ferraro
Angelo Romeo

获奖者 · awards

中标 · winner
Philippe SAMYN and PARTNERS (建筑师事务所)
Philippe Samyn (建筑师)

主要合伙人 associate partner · Benedetto Calcagno
其他合伙人 associates · Carole Aspeslagh · Johan Berteloot · Mehdi Chtourou Asa Decorte · Zelda de Ruvo · Francesco Frontori · Marta Imperadori · Thomas Manche Manon Martin · Denis Mélotte · Marie Naudin · Chloe Stuerebaut · Bart Sueters · François Verrier
合伙建筑师 associate architect · Federico Bargone · S.b.arch. · Bargone Architetti Associati
结构工程 structural engineering · Philippe SAMYN and PARTNERS sprl with Mauro Eugenio Giuliani (Redesco Progetti srl)
建设服务 building services · Philippe SAMYN and PARTNERS sprl with Alessio Gatteschi (SETI Ingegneria srl)
景观设计 landscape architect · Philippe SAMYN and PARTNERS sprl with Enrico Auletta
交通分析 traffic study · Philippe SAMYN and PARTNERS sprl with Carlo Cecconi
文档管理 documentation management · A. Charon
模型 models · F. Van Hoye (AMA)
模型摄影 models photography · A. Fernandez
3D渲染 3D renderings · Polygon Graphics

入围 · finalist
5+1AA Femia Peluffo · Od'A Officina d'Architettura (建筑师事务所)
Alfonso Femia · Gianluca Peluffo · Giovanni Aurino · Bruno Discepolo Alessandra Fasanaro · Simonetta Cenci · Sara Gottardo (建筑师)

建筑及景观设计 architectural and landscape design · 5+1AA Alfonso Femia Gianluca Peluffo · Od'A Officina d'Architettura
整体工程 integrated engineering · AI Studio, AI Engineering
设计团队 design team · Sara Gottardo · Giovanni Aurino · Annalisa Beghelli · Emanuela Bartolini · Toti Di Dio · Vincenzo Fiorillo · Francesca Mazzoni · Emanuele Mondin · Michelangelo Pavia
渲染 rendering · Antonella Rossi (5+1AA & Troppart)
模型 model · Edoardo Miola

入围 · finalist
Foreign Office Architects (建筑师事务所)
Farshid Moussavi and Alejandro Zaera-Polo (建筑师)

合作 (c) Elliott Hodges · Robert Johnson · James Khamsi · Yukinobu Toyama
结构工程 structural engineering · Halcrow Yolles
交通工程 traffic engineering · Halcrow Yolles
大楼服务及环境策划 building services and environmental strategy · Halcrow Yolles
工程预算及项目经理 quantity surveyors and project management · Cyril Sweet

新车站位于Circumvesuviana铁路和Monte del Vesuvio高速铁路专线的相交处。**其设计规划了以下三个功能阶段：**

—第一个阶段，到2015年，建成一个建筑面积约为3,000平方米的车站（站台不包括在内），并投入使用，日客流量为3,000人，其中每小时的最大客流量为450人。
—第二个阶段，到2020年，在原有基础上将车站总面积扩建为6,500平方米。
—第三个阶段，到2030年，完善车站设施，并将车站总面积扩建为8,000平方米。

该工程计划将地方铁路网、城市铁路和其他公共交通体系连为一个整体，以此增强各级交通系统的机动性。

The area provided for the construction of the new station is located in the crossing between the Circumvesuviana railway and the High-Speed railway line Monte del Vesuvio. **The design shall envisage the development of the station in 3 functional stages:**

- The first stage, coinciding with the activation of the station - time horizon 2015 - shall envisage an area of about 3,000 sqm gross floor space, excluding platform level, considering a flow of 3,000 users per day, with a maximum peak of 450 users per hour.

- The second stage - time horizon 2020 - shall envisage a possible extension of the surfaces up to about 6,500 sqm.

- The third stage - time horizon 2030 - shall provide a further development of the station and an increase in the total surface up to a maximum of 8,000 sqm.

The goal is to make a system of regional railways networks, urban railways and other systems of public transport as a whole, in order to improve regional and sub-regional transport mobility.

新火车中转站·蒙特德维苏威

New Interexchange Railway Station · Italy

Philippe SAMYN and PARTNERS (建筑师事务所)

一等奖 · First Prize

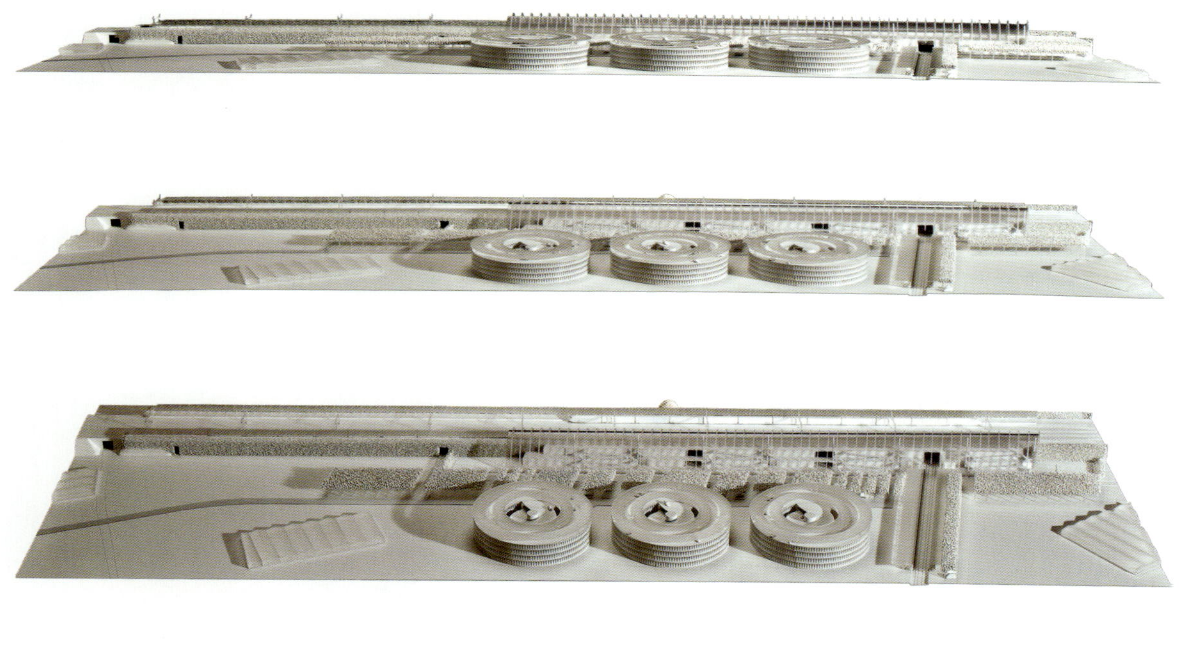

低调的建筑

根据规划，火车站的主要区域位于一个斜坡上，这样，其周边区域形成了鲜明的空间层级结构——前区的主要功能是疏通客流，后区的主要功能是提供各项服务及食物饮料供应，还开设了专门通道。正厅的屋顶呈弧形结构，线条流畅，使得火车站在不同的工程阶段均能自由扩建。如此完美的结构设计不仅能覆盖大面积的空间，而且不会浪费材料。用一系列4.5米的金属模块状弧形拱顶支撑次生镀铝结构，而这些镀铝结构被固定在多孔聚碳酸酯板上。这些圆形结构特别适合用来建造停车区，而三个完全相同的圆柱体又符合不同工程阶段的要求。

A REDUCED-FOOT-PRINT ON THE SITE

The decision to concentrate the principal area of the station on one slope creates a clear spatial hierarchy of the surrounding areas with a front zone destined for the flow of passengers and a back zone destined for services, provisions and technical access. The roof of the main hallway is a continuous parabolic section that enables freedom in the extension of the station in different phases; an ideal structural concept which can cover a large space without excessive material consumption. A series of metallic 5,4m module parabolic arches support a secondary structure in aluminum profiles fixed to multi-cellular polycarbonate panels. The circular volumes are particularly adapted to function as parking areas and each of the three identical cylinders corresponds to the different phases of construction.

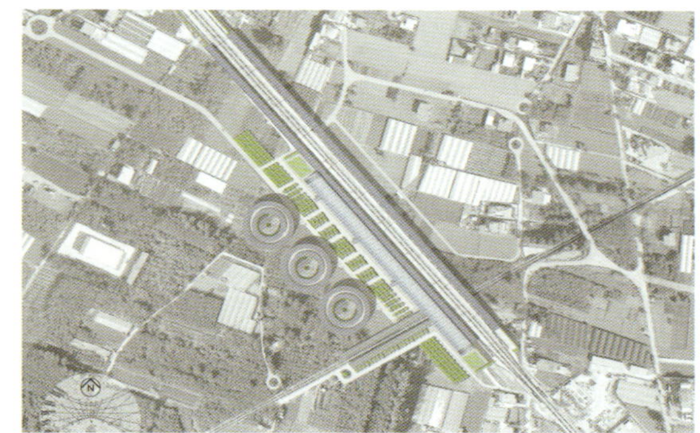

地块位置 SITE PLAN

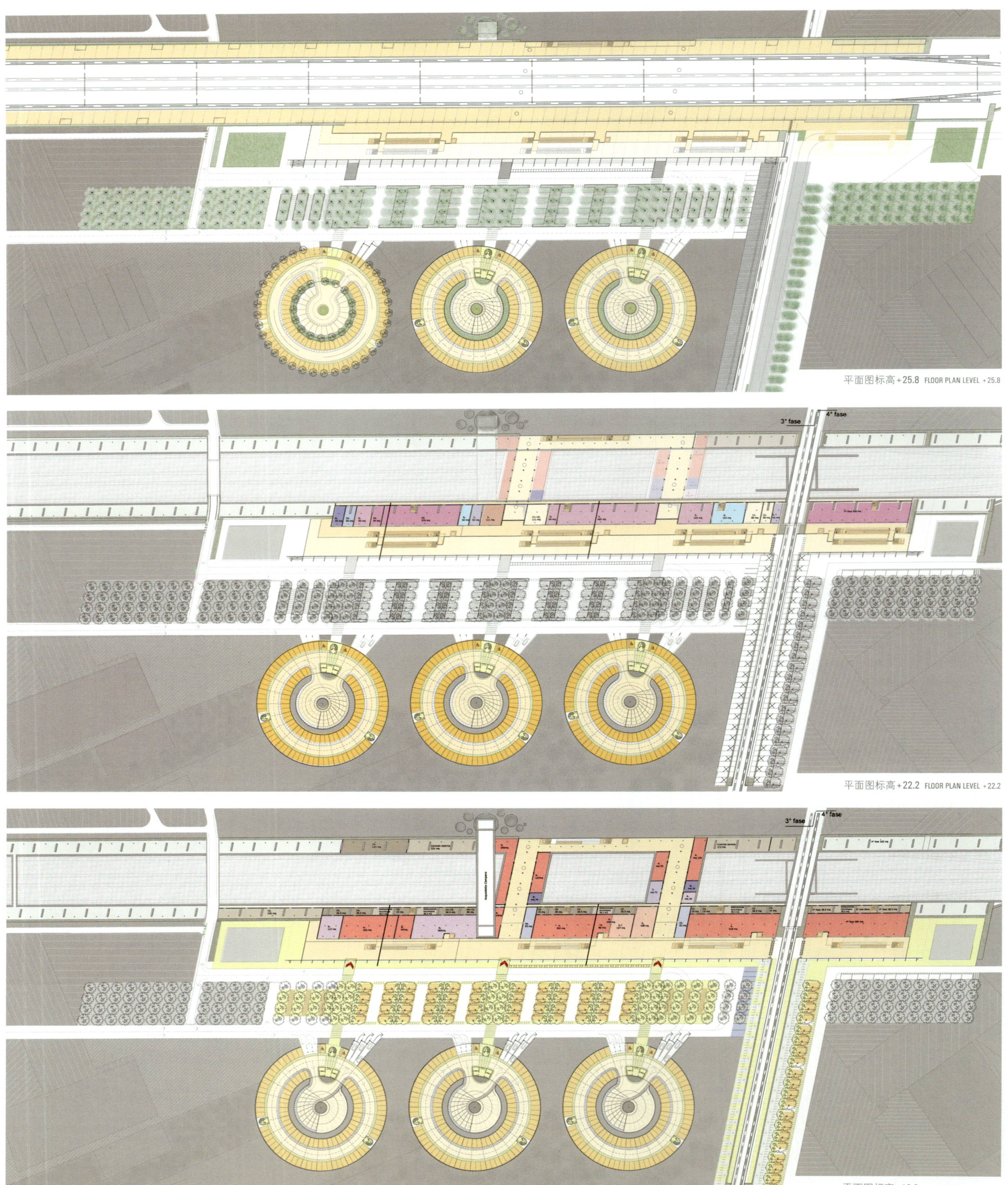

平面图标高＋25.8 FLOOR PLAN LEVEL ＋25.8

平面图标高＋22.2 FLOOR PLAN LEVEL ＋22.2

平面图标高＋18.6 FLOOR PLAN LEVEL ＋18.6

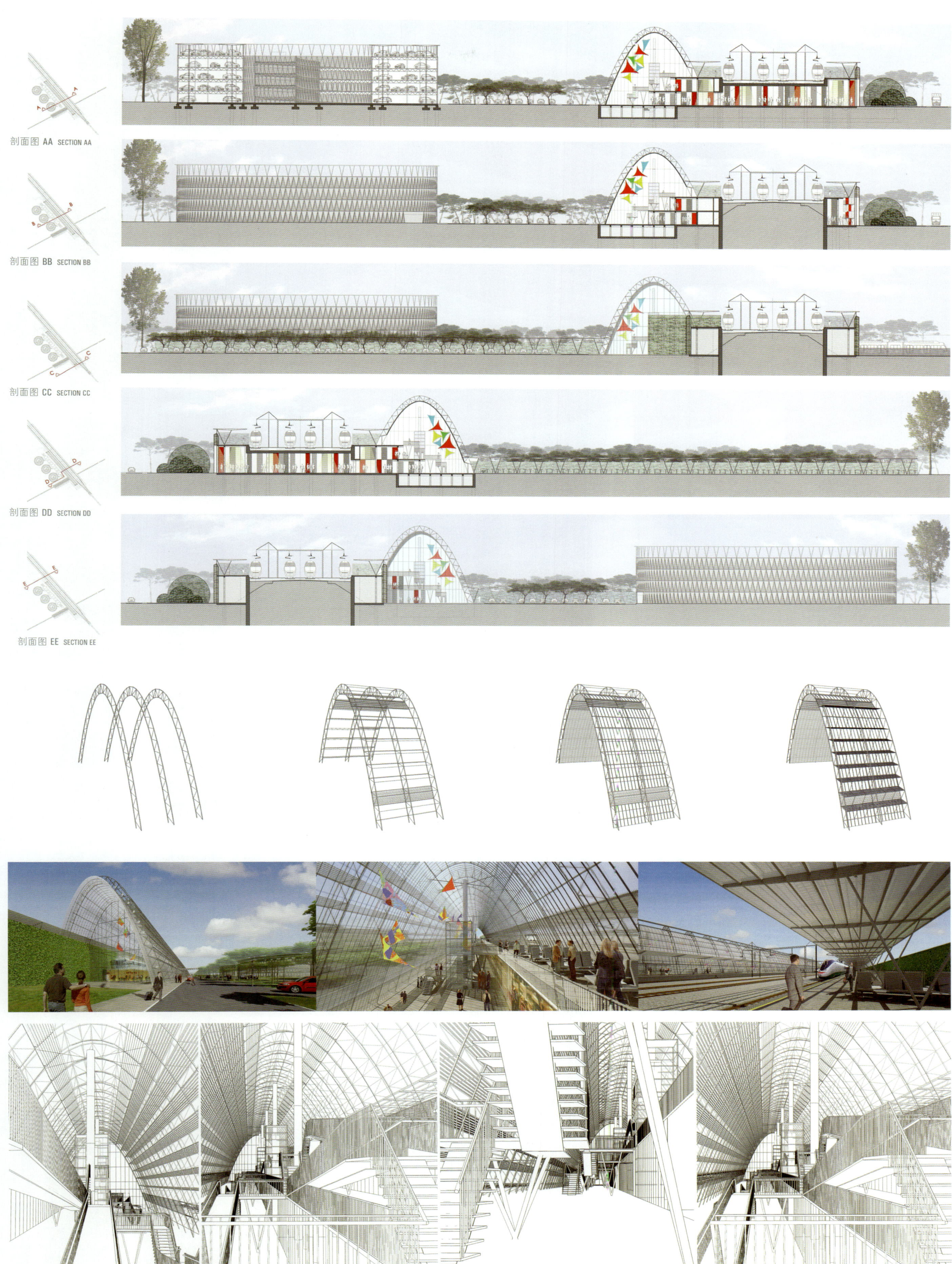
剖面图 AA SECTION AA
剖面图 BB SECTION BB
剖面图 CC SECTION CC
剖面图 DD SECTION DD
剖面图 EE SECTION EE

新火车中转站 · 蒙特德维苏威

New Interexchange Railway Station · Italy

入围　Finalist

5+1AA Femia Peluffo (建筑师事务所)

Alfonso Femia · Gianluca Peluffo (建筑师)

平面图标高 +6.98 FLOOR PLAN LEVEL +6.98

平面图标高 +3.50 FLOOR PLAN LEVEL +3.50

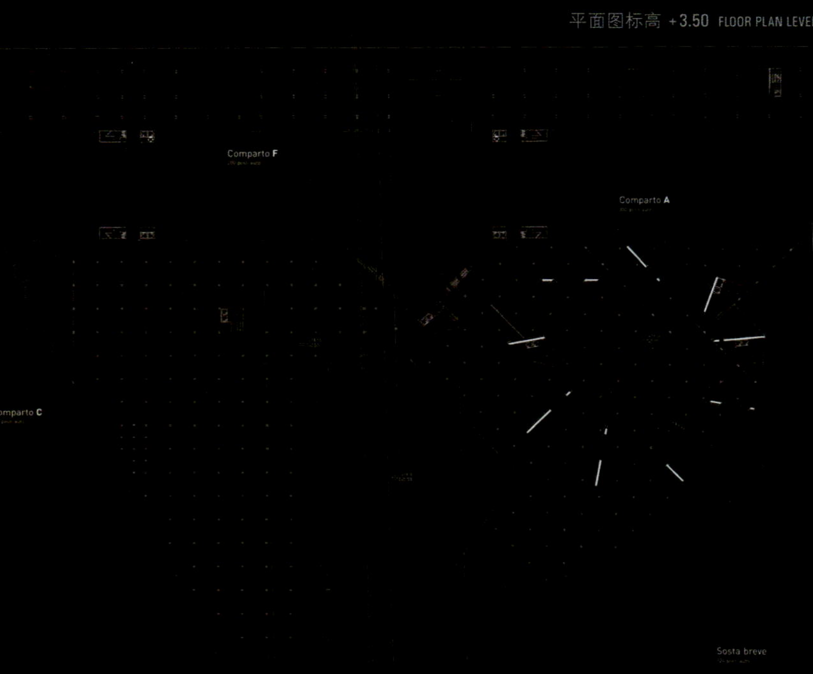

平面图标高 +0.00 FLOOR PLAN LEVEL +0.00

结构剖面图 CONSTRUCTION SECTION

平面图标高 +17.58 FLOOR PLAN

平面图标高 +13.98 FLOOR PLAN L

平面图标高 +10.48 FLOOR PLAN L

新火车中转站·蒙特德维苏威

New Interexchange Railway Station · Italy

入围 · Finalist

Foreign Office Architects (建筑师事务所)

Farshid Moussavi · Alejandro Zaera-Polo (建筑师)

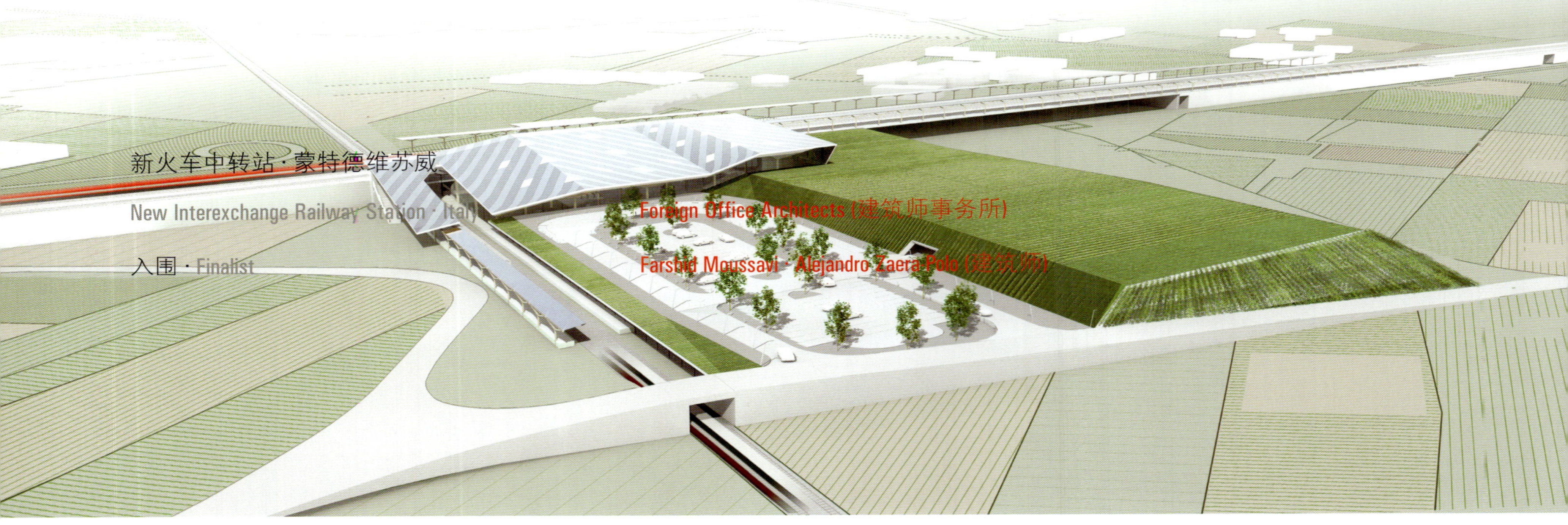

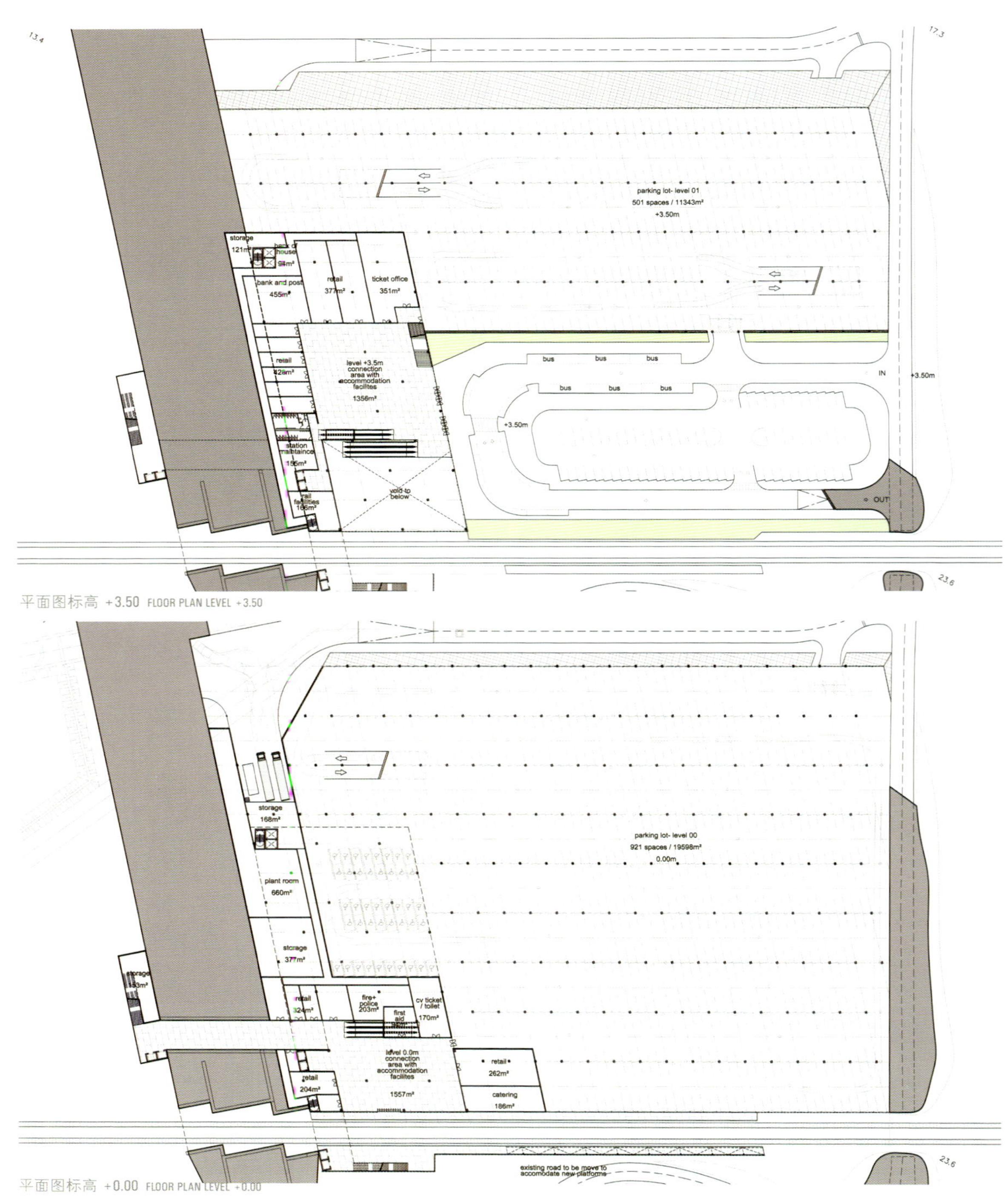

平面图标高 +3.50 FLOOR PLAN LEVEL +3.50

平面图标高 +0.00 FLOOR PLAN LEVEL +0.00

基础设施和风景相得益彰

火车站的选址具有两个显著特征：第一，该区域有美丽的田园风光，能看到维苏威火山。为了使火车站基础设施和周围风光浑然一体，而不显得突兀，我们将不建造一座高大的庞然大物，而是将火车站设计为一个在水平方向上延伸的建筑物。第二，这个火车站的两条列车线分别位于两个不同的水平面上，相互垂直排列，从景观带中穿行而过。这就给我们提供了机会，可以将火车站修建于这两条列车线所在的平面之间。

若停车区是一个独立的多层建筑，那么其结构将比火车站大四倍。因此，必须通过致力打造其高度，掩饰其外形，让其结构与火车站融为一体，尽量使其不独占鳌头。该结构没有采用单跨大棚，因为单跨大棚在样式和大小上均具有局限性。我们的结构体系采用的是模块化系统。这些模块呈树形——树枝部分从圆柱顶部伸出，与附近顶棚上的树枝状结构相互交搭。这种模块结构确保了火车站在各个工程阶段的结构完整性，不影响将来的扩建规划。

INTEGRATION OF INFRASTRUCTURE AND LANDSCAPE

The site of the station presents two unique features. Firstly, the area is a beautiful agricultural landscape with views of Mount Vesuvius. In order to integrate the station infrastructure with the landscape and avoid a monumental approach, we have developed the station as a horizontal plane with a minimum footprint. The second unique feature of the site is that it presents the new station with two train lines which pass through the landscape at two different levels, and perpendicularly to one another. We have approached this sitting of the train lines as an opportunity to locate the station at a level in-between.

If the parking is built as a stand-alone multi-storey building, the parking structure as specified in the brief would be four times larger than the train station. Therefore, a significant effort must be made to minimize its impact on the site by shaping its height, disguising its form and incorporating it into the form of the station itself. As opposed to a large single span shed, which has a finite form and size, our structural system proposal is to use a modular system. These modular elements take on a tree-like form - branches of structure spring from the tops of columns and interlock with the branches from the adjacent canopy. This modular approach ensures that the station image is complete in all its phases and does not depend on future extensions.

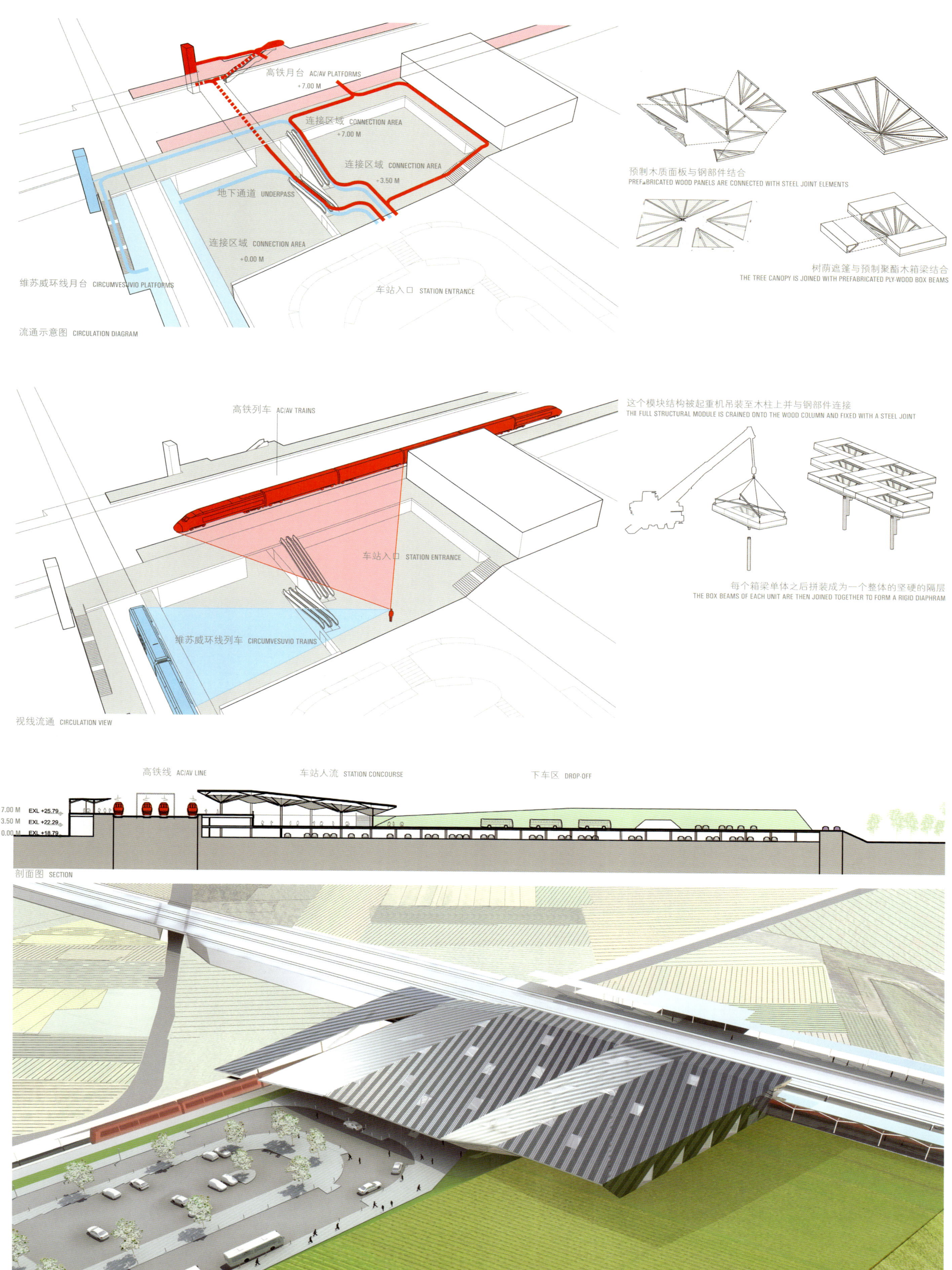
高铁月台 AC/AV PLATFORMS
+7.00 M
连接区域 CONNECTION AREA
+7.00 M
连接区域 CONNECTION AREA
+3.50 M
地下通道 UNDERPASS
连接区域 CONNECTION AREA
+0.00 M
维苏威环线月台 CIRCUMVESUVIO PLATFORMS
车站入口 STATION ENTRANCE
流通示意图 CIRCULATION DIAGRAM
预制木质面板与钢部件结合
PREFABRICATED WOOD PANELS ARE CONNECTED WITH STEEL JOINT ELEMENTS
树荫遮篷与预制聚酯木箱梁结合
THE TREE CANOPY IS JOINED WITH PREFABRICATED PLY-WOOD BOX BEAMS
高铁列车 AC/AV TRAINS
车站入口 STATION ENTRANCE
维苏威环线列车 CIRCUMVESUVIO TRAINS
视线流通 CIRCULATION VIEW
这个模块结构被起重机吊装至木柱上并与钢部件连接
THE FULL STRUCTURAL MODULE IS CRAINED ONTO THE WOOD COLUMN AND FIXED WITH A STEEL JOINT
每个箱梁单体之后拼装成为一个整体的坚硬的隔层
THE BOX BEAMS OF EACH UNIT ARE THEN JOINED TOGETHER TO FORM A RIGID DIAPHRAM
高铁线 AC/AV LINE
车站人流 STATION CONCOURSE
下车区 DROP-OFF
站 STATION +7.00 M EXL +25.79
站 STATION +3.50 M EXL +22.29
站 STATION +0.00 M EXL +18.79
剖面图 SECTION

艺术文化中心(黎巴嫩-阿曼人中心)·贝鲁特

House of Arts and Culture (the Lebanese-Omani Centre) · Lebanon

竞标 · competition
艺术文化中心(黎巴嫩-阿曼人中心)
House of Arts and Culture (the Lebanese-Omani Centre)

竞标类型 · competition type
国际公开招标
international open competition

项目地点 · site area
贝鲁特 · 黎巴嫩 Beirut · Lebanon

主办方 · promoter
黎巴嫩共和国文化部及文化馆国际建筑竞标委员会 Ministry of Culture of the Republic of Lebanon, Att. International Architectural Competition for the House of Arts and Culture

日程安排 · schedule
招标 · Announcement 09.2008
评审结果 · Jury´s results 03.2009

评审团 · jury

Frederic Husseini, 黎巴嫩共和国文化部长 The Minister of Culture of the Republic of Lebanon
Said Harith al Barashdi, 阿曼苏丹代表 Representative of the Sultanate of Oman
Assem Salam, 黎巴嫩建筑工程署代表 Representative of the Orders of Engineers and Architects of Lebanon
Angus Gavin, 黎巴嫩城市开发及贝鲁特市中心重建公司 Representative of the Lebanese Company for the Development and Reconstruction of Beirut Central District
Magda Hosam Eldin Mostapha, 埃及建筑师 Architect (Egypt), 国际建筑师协会代表 representative of the International Union of Architects (UIA)
Suha Özkan, 建筑师、历史学家、理论家 Architect, Historian and Theorist (土耳其 Turkey)
Momoyo Kaijima, Atelier Bow Wow (日本 Japan)
Izaskun Chinchilla, 建筑师 Architect (西班牙 Spain)
Okwui Enwezor (尼日利亚 Nigeria), 旧金山艺术学院高级副院长及学术主管 Dean of Academic Affairs and Senior Vice President at San Francisco Art Institute

获奖者 · awards

一等奖 · first prize (美元 USD 75,000)
TECKNOARCH (建筑师事务所)
ALBERTO CATALANO (建筑师)

合作 (c) Giulia Iurcotta · Barbarangelo Licheri · Mariangela Murgia · Daniele Piludu
Celestino Sanna · Souraya Frem · Emanuela Forcolini

二等奖 · second prize (美元 USD 40,000)
STAR STRATEGIES + ARCHITECTURE (建筑师事务所)
Beatriz Ramo with Jean-Vianney Deleersnyder · Simone de Iacobis
Iñigo Paniego de la Cuesta (建筑师)

三等奖 · third prize (美元 USD 25,000)
"PROJECT MEGANOM" (建筑师事务所)
Yuri Grigorian · Natalia Tatunashvili · Tatiana Kornienko
Yuri Kuznezov · Elena Uglovskaya · Irina Livieva
Artem Staborovskiy · Ruben Grigoryan (建筑师)

荣誉提名奖 · honourable mention
DORELL.GHOTMEH.TANE (建筑师事务所)
Dan Dorell · Lina Ghotmeh · Tsuyoshi Tane (建筑师)
合作 (c) Abdulrahman Ajeenah · Daisuke Sekine · Emma Bush · Hélène Brisard
Marie de Laportaliere · Ross Perkins · Sarah Castle
联合设计 in collaboration with:
经济预算 business strategist: Joumana Al-Jabri
结构工程 structural engineering: Klaas De Rycke / BOLLINGIER+GROHMANN SARL

荣誉提名奖 · honourable mention
POLY.M.UR (建筑师事务所)
Chris S. Yoo · Homin Kim · Heekyung Moon · Ji-yeon Kim
(建筑师)

荣誉提名奖 · honourable mention
BOARD (建筑师事务所)
设计团队 design team: Bernd Upmeyer · Fotini Gouveli · David Sebastian Martin

参标 · proposal
LAR/FERNANDO ROMERO (建筑师事务所)
设计团队 design team: Fernando Romero · Prabhu Sugumar · Ana Medina
Johanna Huang · Tiago Pinto de Carvalho · Joshua Petrie · Cornelis van Almsick · Pedro Lechuga

艺术文化馆(黎巴嫩Omani中心)将是一个孕育文化和艺术产品的空间，是黎巴嫩创意产业发展的温床。该项目位于繁华地带，是黎巴嫩首都首屈一指的工程。**该中心被定位为艺术文化的先驱，其建立意义深远。**

The House of Arts and Culture (the Lebanese-Omani Centre) will be a space for cultural and artistic production, an incubator for the development of creativity in the country. The site of the project is located in a full expansion district. The first of its kind in the capital city of Lebanon, **the centre holds an important responsibility: that of being the leading place in arts and culture.**

艺术文化中心（黎巴嫩–阿曼人中心）· 贝鲁特

House of Arts and Culture · Lebanon

一等奖 · First Prize

Tecknoarch（建筑师事务所）

Alberto Catalano（建筑师）

融合

该项目最显著的特点在于其和周边都市环境的融合，将建筑打造成一个公共广场。该项目实现了一个重要目标——与公共领域的衔接。该项目巧妙地运用了几近“非建筑”的建筑手法，使其与附近大厦的行政管理区域和工作区域交相辉映。

INTEGRATION

The most prominent aspect of this project was its sensitive approach to the integration with the surrounding urban fabric, as represented in the concept of the building as a public plaza. The project achieved one of the important objectives - namely the engagement of the public realm. The subtle almost “non - building” approach of this project, paralleled with the highly functional concentration of administrational and workshop spaces in the adjacent tower.

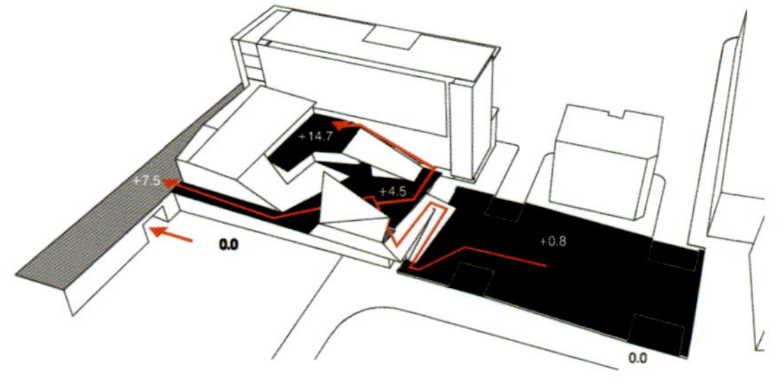

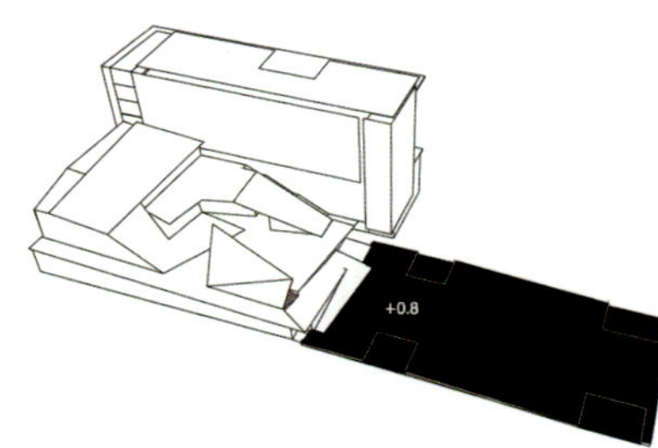

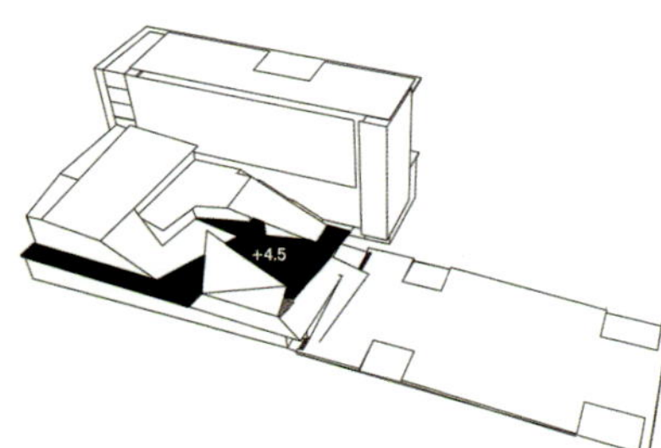

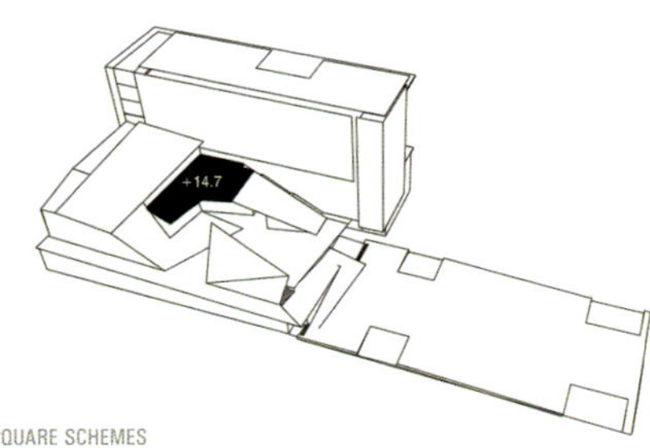

广场构造 SQUARE SCHEMES

地块位置 SITE PLAN

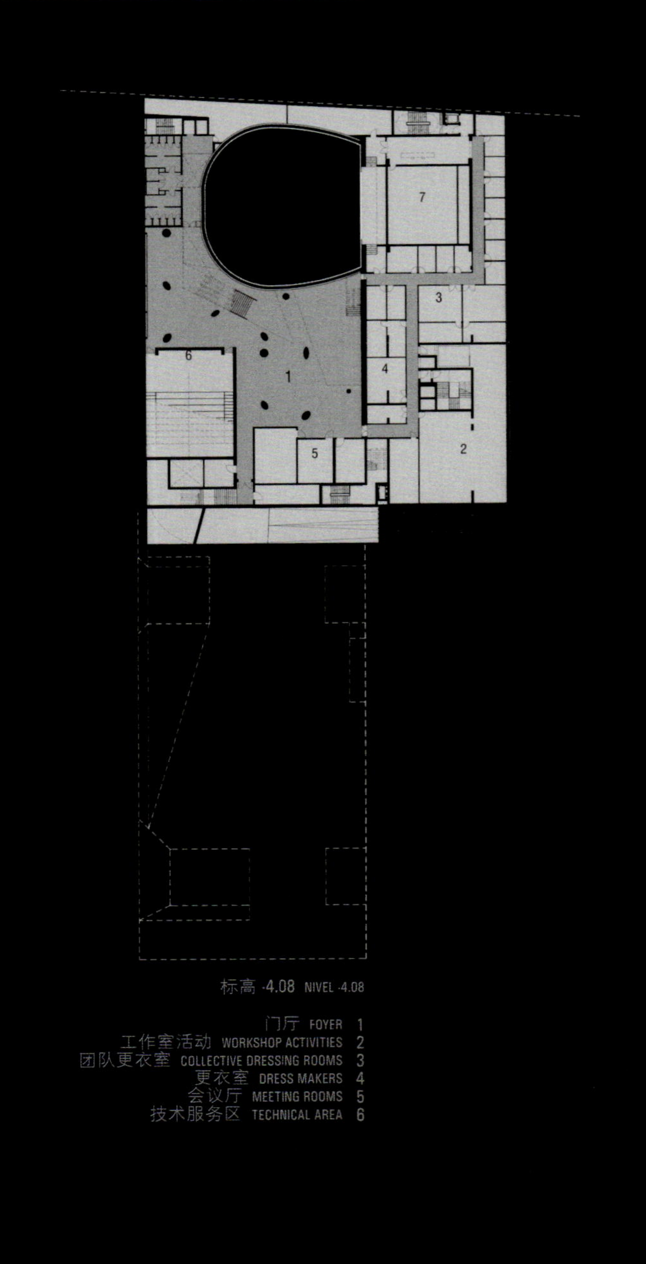
标高 -4.08 NIVEL -4.08
门厅 FOYER 1
工作室活动 WORKSHOP ACTIVITIES 2
团队更衣室 COLLECTIVE DRESSING ROOMS 3
更衣室 DRESS MAKERS 4
会议厅 MEETING ROOMS 5
技术服务区 TECHNICAL AREA 6

标高 +0.00 NIVEL +0.00
售票处 TICKET OFFICE 1
邮寄 DELIVERY 2
演艺厅 CAJA ESCÉNICA 3
电影戏剧 MOVIE THEATRE 4
舞台 STAGE 5
后台 BACKSTAGE 6
标高 +14.62 NIVEL +14.62
办公室 OFFICES 1
公共空间 OPEN SPACE 2
歌剧 OPERA
舞台 ARENA
歌舞表演 CABARET
平地 FLAT
报告厅 CONGRESS

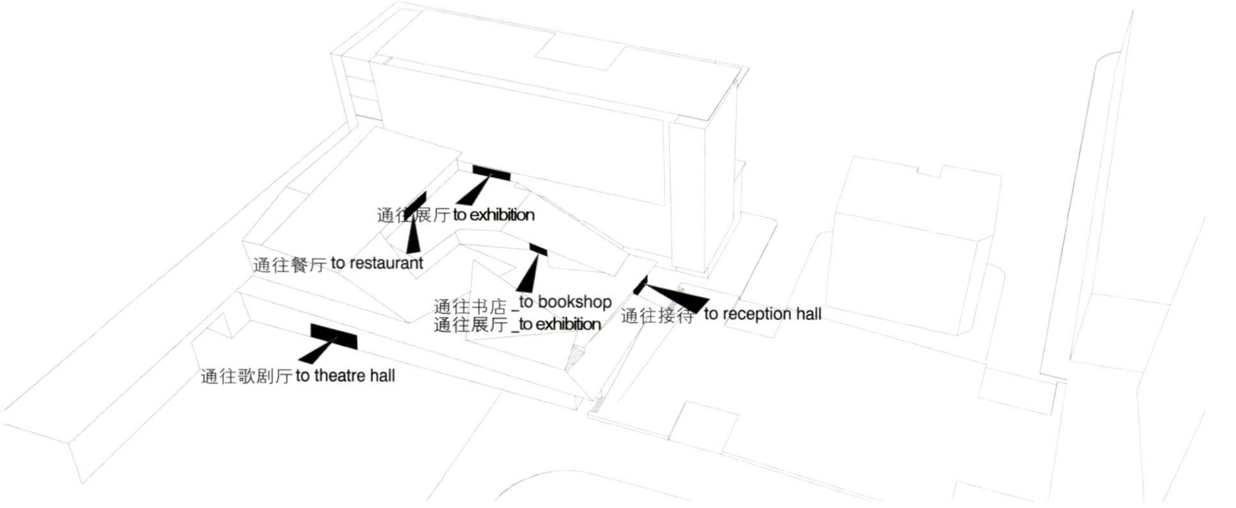

通往展厅 to exhibition

通往餐厅 to restaurant

通往书店 _to bookshop
通往展厅 _to exhibition

通往接待 to reception hall

通往歌剧厅 to theatre hall

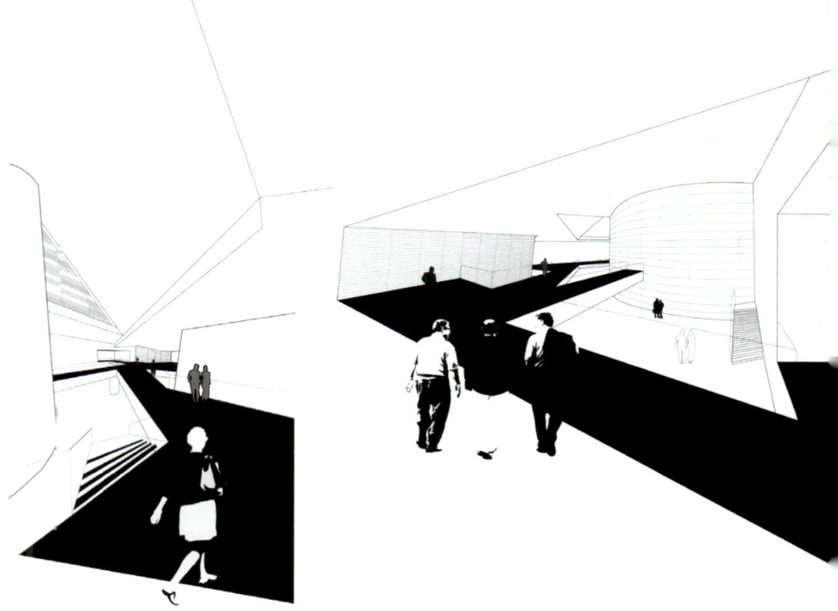

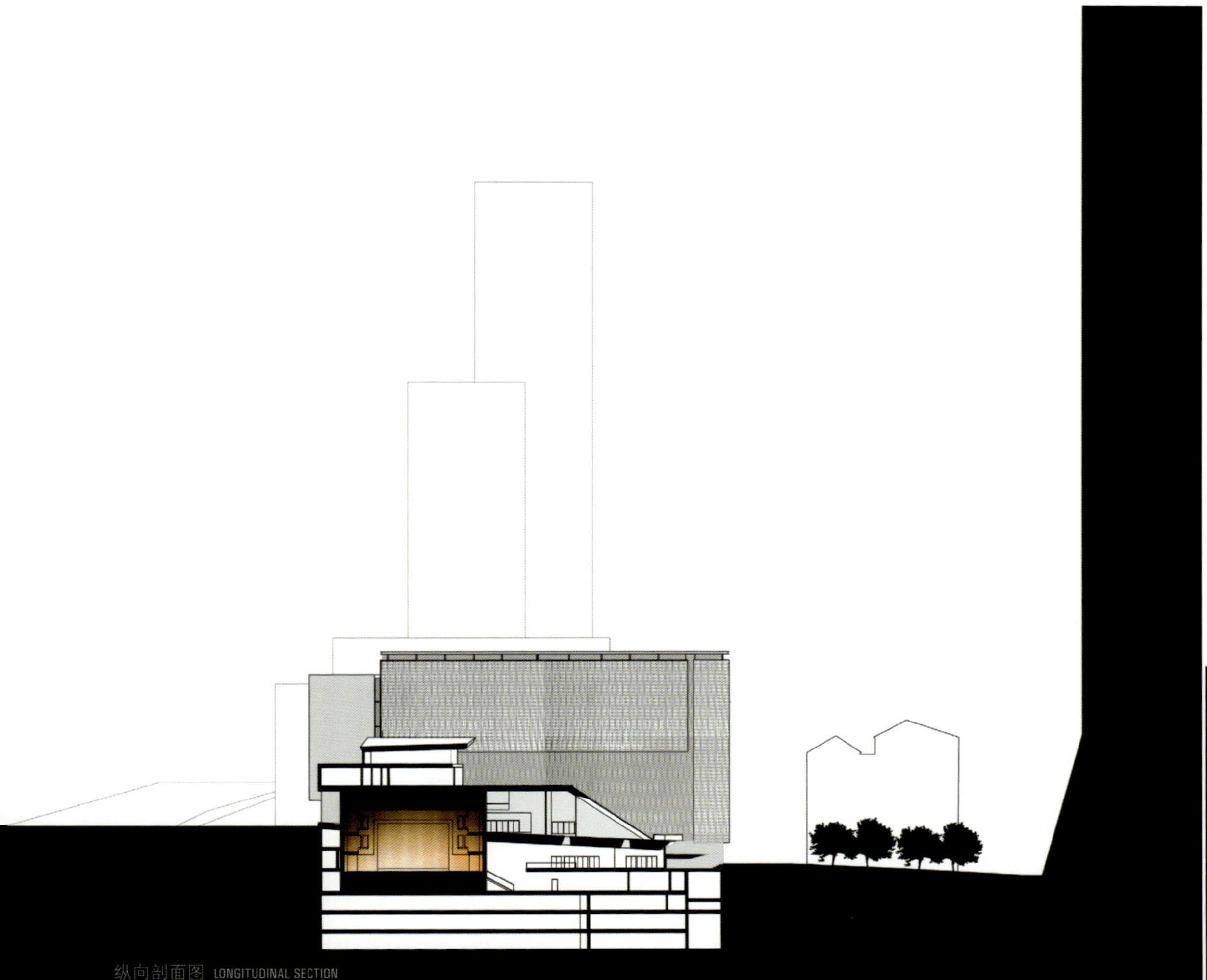

纵向剖面图 LONGITUDINAL SECTION

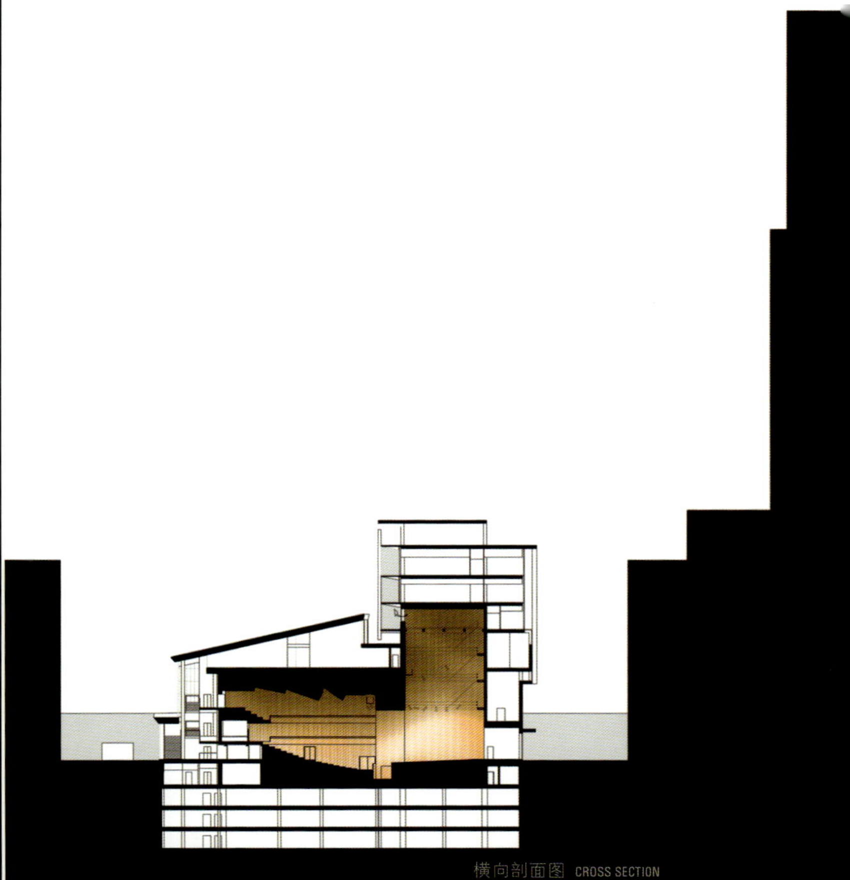

横向剖面图 CROSS SECTION

艺术文化中心(黎巴嫩–阿曼人中心)·贝鲁特

House of Arts and Culture · Lebanon

二等奖 · Second Prize

STAR strategies + architecture (建筑师事务所)

谁说过图标？

看了竞赛的概要后，不禁为没有见到“图标”这个词而惊喜。图标死板而生硬，常常很快就成为了平常之物。该项目不是死板地反映时代中的某一瞬间。相反，它必须活力无限、变化层出不穷；正如一台机器时刻在运转，不断地带来变化。根据其关于高度和互动的要求，我们将建筑功能分成了两类，依此打造了两个建筑结构体：一个高大的盒状结构和一个低矮的基座结构。我们将高大而垂直的建筑安排在盒状结构中，而将需要与街道直接连通的建筑安排在基座结构中。

WHO SAID ICON?

After studying the brief of the competition, it was a pleasant surprise not to encounter the word icon once. Icons are rigid and static, and somehow quickly become part of the routine. The project cannot be a static depiction of a single moment in time. It must be dynamic, changeable; like a machine, it should reflect transformation. By dividing the functions into two categories according to their requirements on height and interaction, we obtain two volumes: a tall box and a low plinth. In the box, we place the programme that requires big heights and vertical circulation, while in the plinth we accommodate the programme that could enjoy a direct connection to the street.

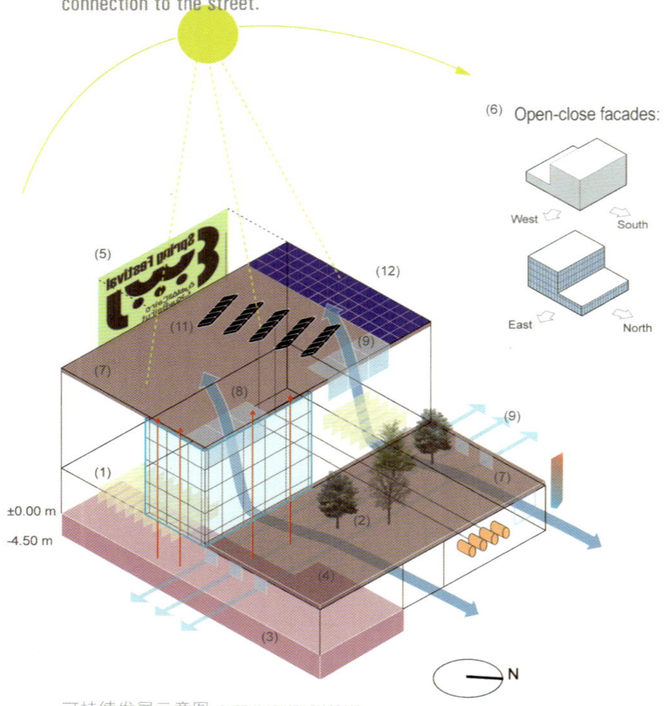

可持续发展示意图 SUSTAINABILITY DIAGRAM

1 自然合法生态材料 ECOLOGICALLY CERTIFIED AND NATURAL MATERIALS
2 原生植物 VERNACULAR VEGETATION
3 -1层室外空间超过800平方米 MORE THAN 800 M2 OF EXTRA SPACE AT -1 LEVEL
4 1500平方米外部公共空间 1500 M2 OF OUTDOOR PUBLIC SPACE
5 1200平方米的艺术、广告、展示面积 1200 M2 OF SURFACE FOR ART, ADVERTISEMENTS OR PROJECTIONS
6 南面闭合+西立面 · 东面开放+北立面 CLOSED SOUTH+WEST FAÇADE · OPEN EAST+NORTH SIDES
7 地面开挖部分填补与屋顶 PART OF THE SOIL FROM THE EXCAVATION WILL BE LAYERED ON THE ROOFS
8 门厅的玻璃幕墙面起到烟囱的作用 THE GLASS FAÇADE OF THE FOYER WILL ACT AS A CHIMNEY
9 空气水平流通 THE AIR WILL FLOW HORIZONTALLY
11 太阳能板保障热水需求 SOLAR PANELS TO COVER THE DEMAND FOR WARM WATER
12 光伏板提供用电保障 PHOTOVOLTAIC PANELS TO COVER PART OF ELECTRICITY NEEDS

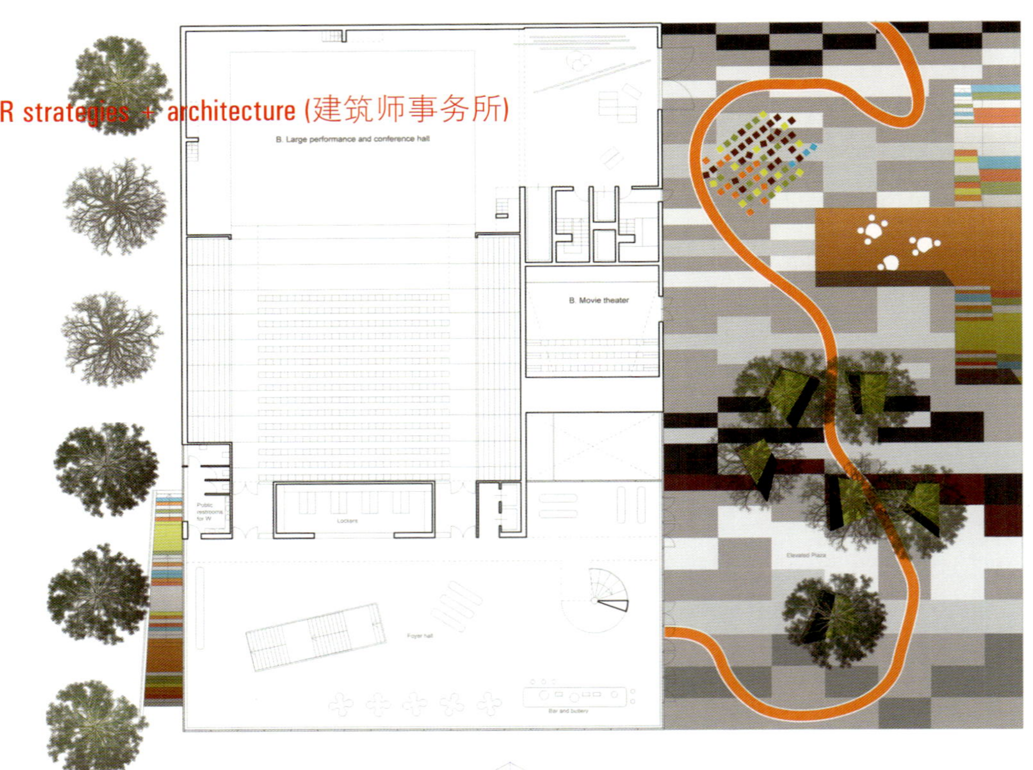

平面图 2 标高 +10.00 LEVEL 2 +10.00

平面图 1 标高 +6.00 LEVEL 1 +6.00

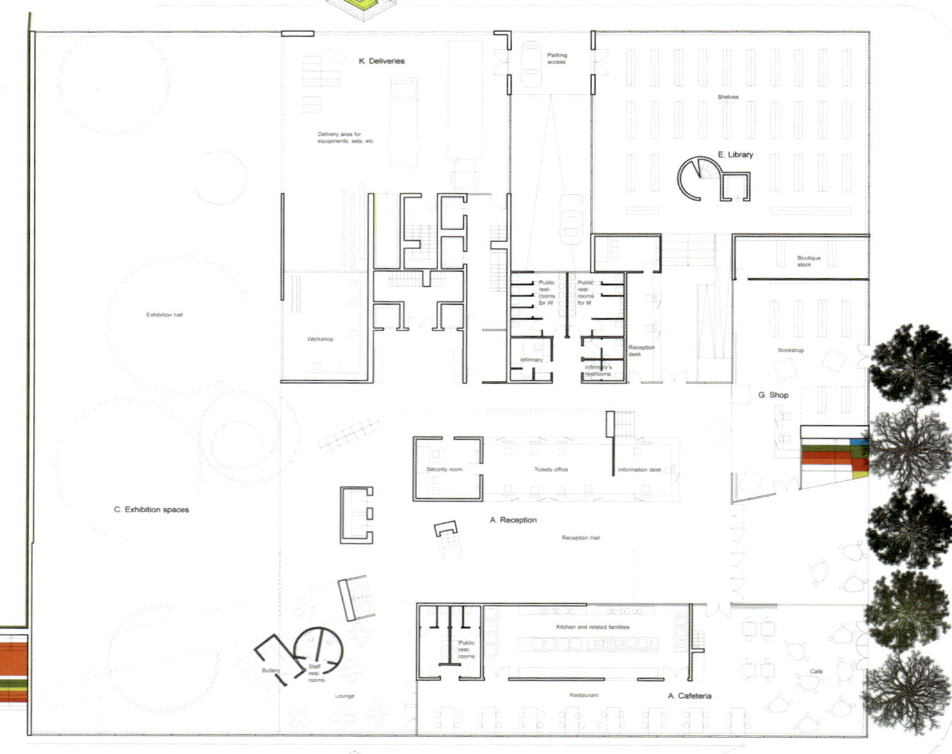

底层平面图 GROUND FLOOR PLAN

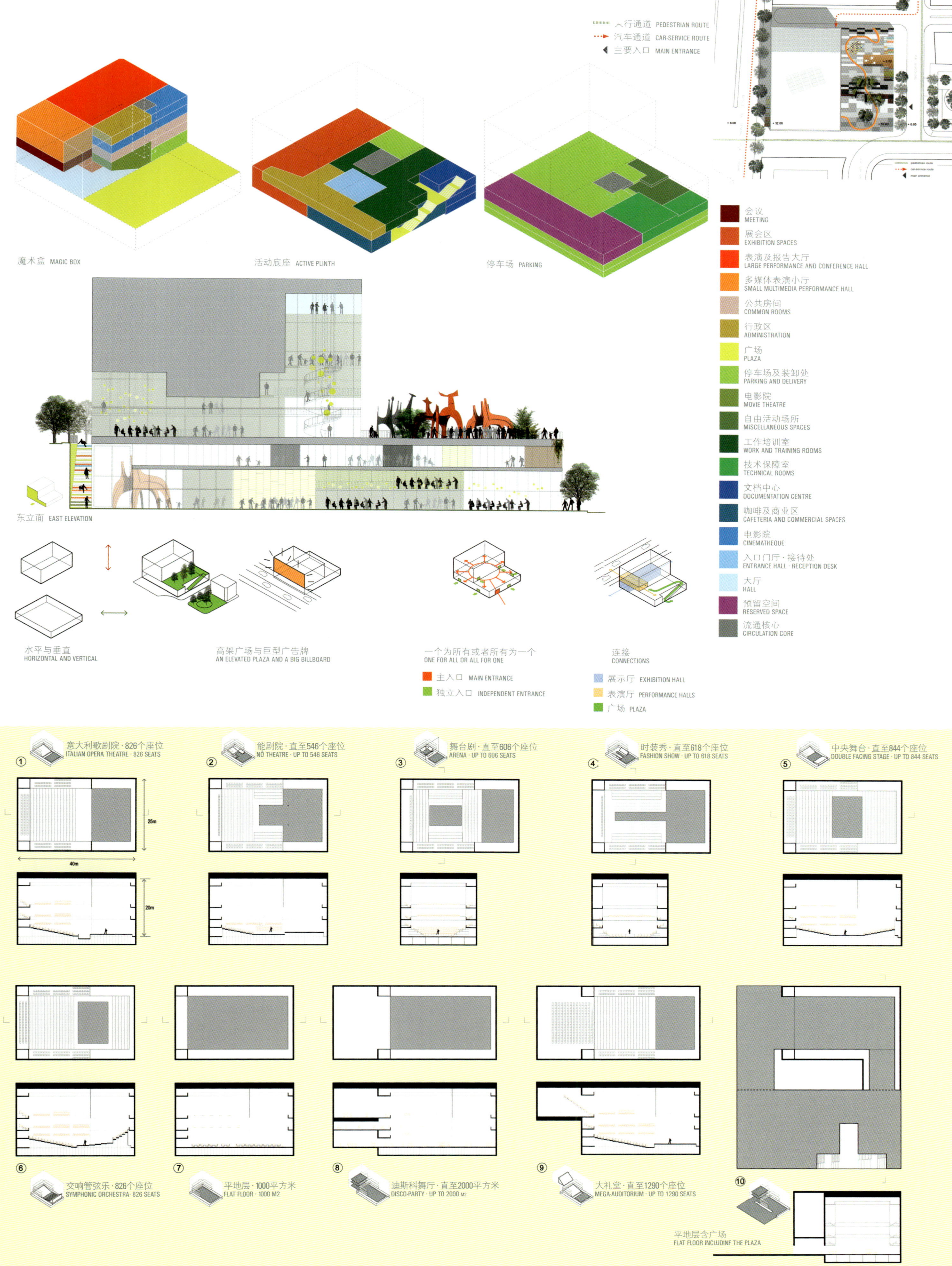
人行通道 PEDESTRIAN ROUTE
汽车通道 CAR-SERVICE ROUTE
主要入口 MAIN ENTRANCE
魔术盒 MAGIC BOX
活动底座 ACTIVE PLINTH
停车场 PARKING
会议 MEETING
展会区 EXHIBITION SPACES
表演及报告大厅 LARGE PERFORMANCE AND CONFERENCE HALL
多媒体表演小厅 SMALL MULTIMEDIA PERFORMANCE HALL
公共房间 COMMON ROOMS
行政区 ADMINISTRATION
广场 PLAZA
停车场及装卸处 PARKING AND DELIVERY
电影院 MOVIE THEATRE
自由活动场所 MISCELLANEOUS SPACES
工作培训室 WORK AND TRAINING ROOMS
技术保障室 TECHNICAL ROOMS
文档中心 DOCUMENTATION CENTRE
咖啡及商业区 CAFETERIA AND COMMERCIAL SPACES
电影院 CINEMATHEQUE
入口门厅 · 接待处 ENTRANCE HALL · RECEPTION DESK
大厅 HALL
预留空间 RESERVED SPACE
流通核心 CIRCULATION CORE
东立面 EAST ELEVATION
水平与垂直 HORIZONTAL AND VERTICAL
高架广场与巨型广告牌 AN ELEVATED PLAZA AND A BIG BILLBOARD
一个为所有或者所有为一个 ONE FOR ALL OR ALL FOR ONE
主入口 MAIN ENTRANCE
独立入口 INDEPENDENT ENTRANCE
连接 CONNECTIONS
展示厅 EXHIBITION HALL
表演厅 PERFORMANCE HALLS
广场 PLAZA
① 意大利歌剧院 · 826个座位 ITALIAN OPERA THEATRE · 826 SEATS
25m
40m
20m
② 能剧院 · 直至546个座位 NO THEATRE · UP TO 546 SEATS
③ 舞台剧 · 直至606个座位 ARENA · UP TO 606 SEATS
④ 时装秀 · 直至618个座位 FASHION SHOW · UP TO 618 SEATS
⑤ 中央舞台 · 直至844个座位 DOUBLE FACING STAGE · UP TO 844 SEATS
⑥ 交响管弦乐 · 826个座位 SYMPHONIC ORCHESTRA · 826 SEATS
⑦ 平地层 · 1000平方米 FLAT FLOOR · 1000 M2
⑧ 迪斯科舞厅 · 直至2000平方米 DISCO-PARTY · UP TO 2000 M2
⑨ 大礼堂 · 直至1290个座位 MEGA-AUDITORIUM · UP TO 1290 SEATS
⑩ 平地层含广场 FLAT FLOOR INCLUDINF THE PLAZA

艺术文化中心(黎巴嫩–阿曼人中心)· 贝鲁特

House of Arts and Culture · Lebanon

荣誉提名奖 · Honourable Mention

DORELL.GHOTMEH.TANE (建筑师事务所)

Dan Dorell · Lina Ghotmeh · Tsuyoshi Tane (建筑师)

等待发现的空间

该项目呈现的是一个开放的建筑，反映出贝鲁特的持续在建状态。这是一个开放式的结构，一个公共性的场所，并在持续不断地变化着。此建筑通过空中花园的四季变化展现时光的流逝，并随着在贝鲁特不同盛事的举行呈现不同的立面外观。

A SPACE TO DISCOVER

The project is an open house that reflects the continuous under-construction state of Beirut. It is an open structure, a public space, under continuous transformation. It is a place which expresses the passing of time through the seasonal changes of its hanging gardens, and a house where its face evolves with the changing events in Beirut.

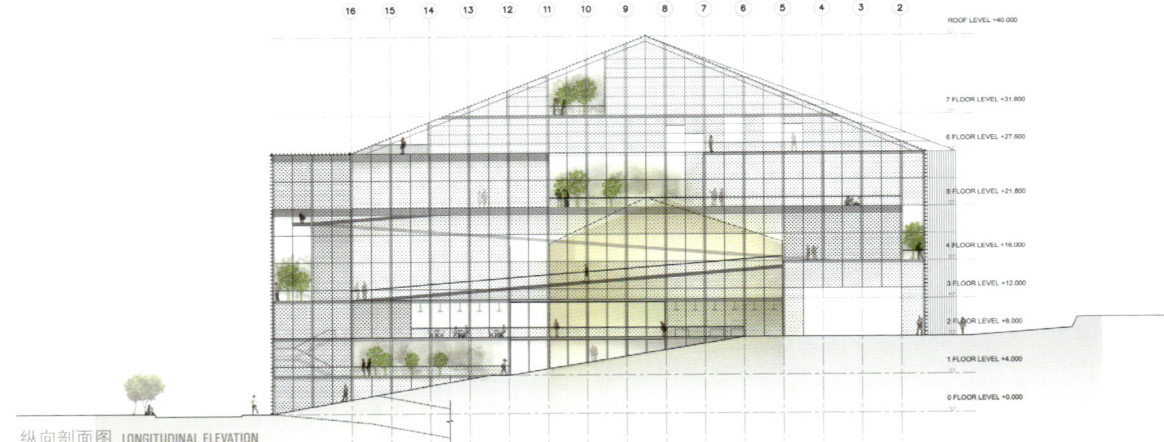

纵向剖面图 LONGITUDINAL ELEVATION

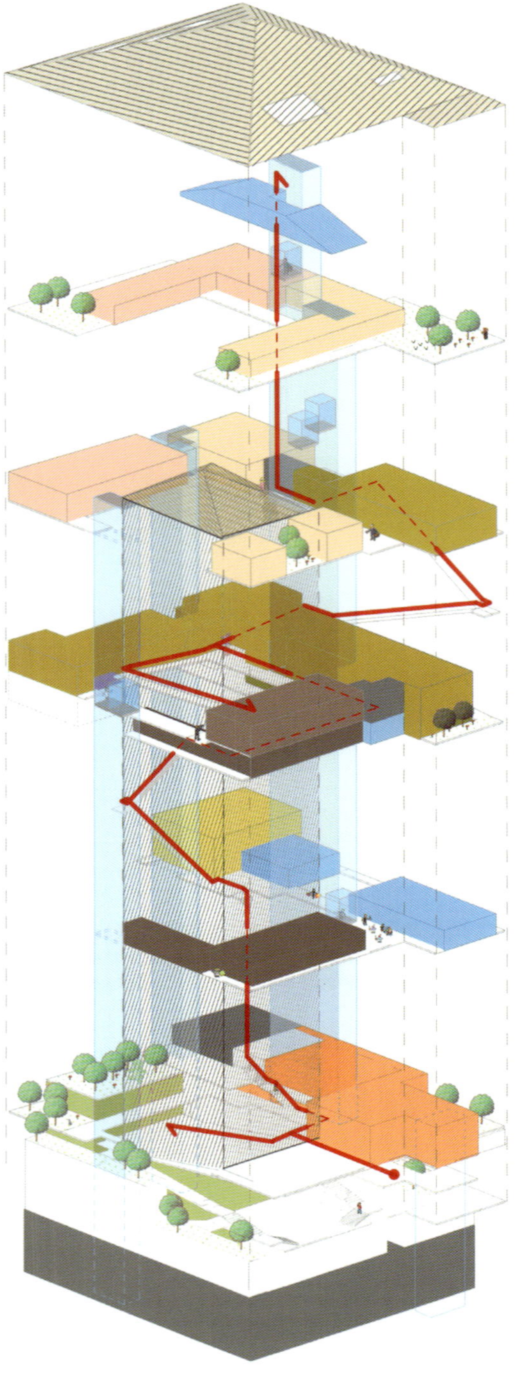

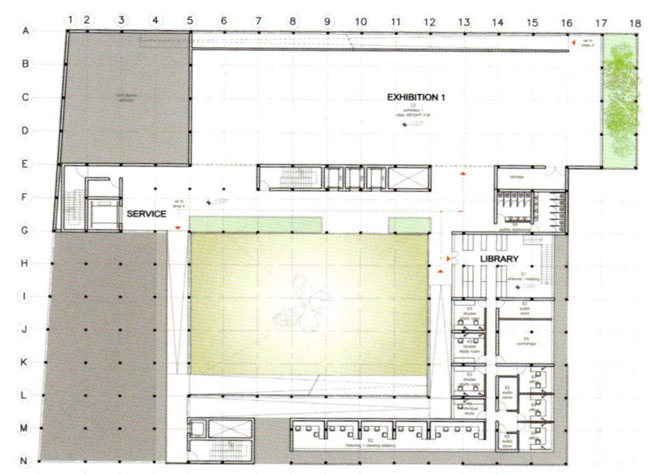

四层平面图 FLOOR PLAN 3

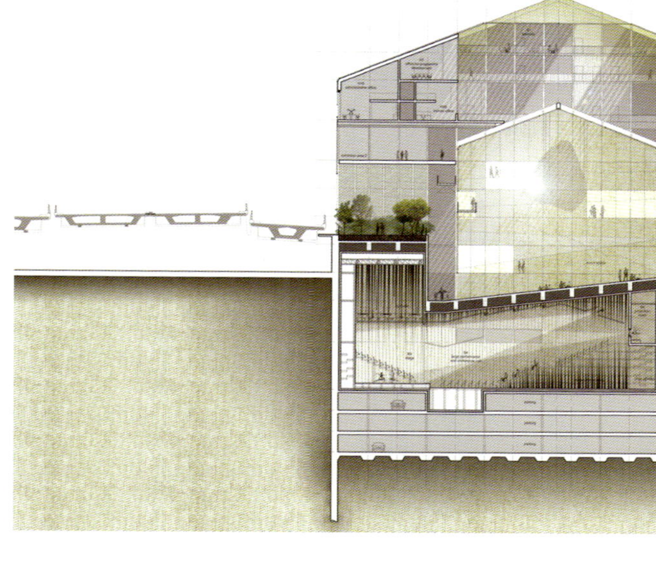

INDEPENDANT ACCESS FOR SHOP + CAFE
DELIVERY HALL
CAFE
CAFE TERRACE
LIBRARY
CHALCHOUL

三层平面图 FLOOR PLAN 2

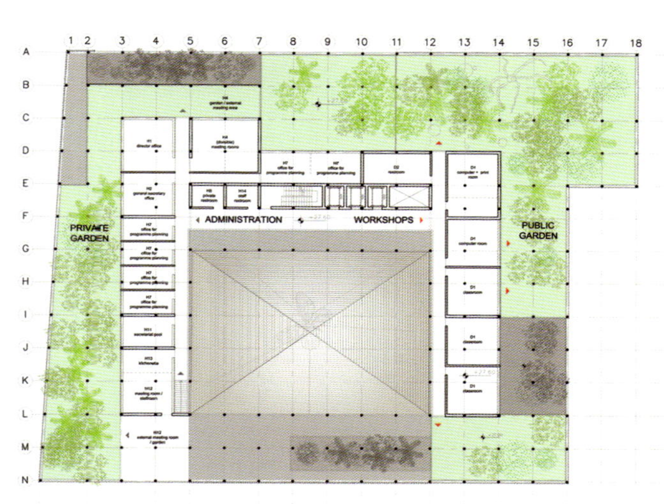

七层平面图 FLOOR PLAN 6

二层平面图 FLOOR PLAN 1

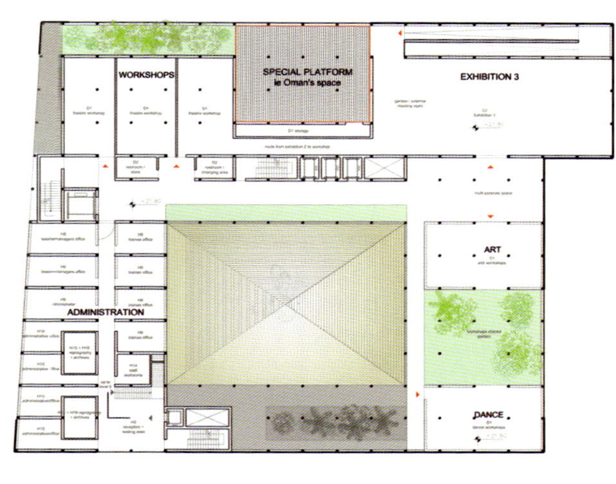

六层平面图 FLOOR PLAN 5

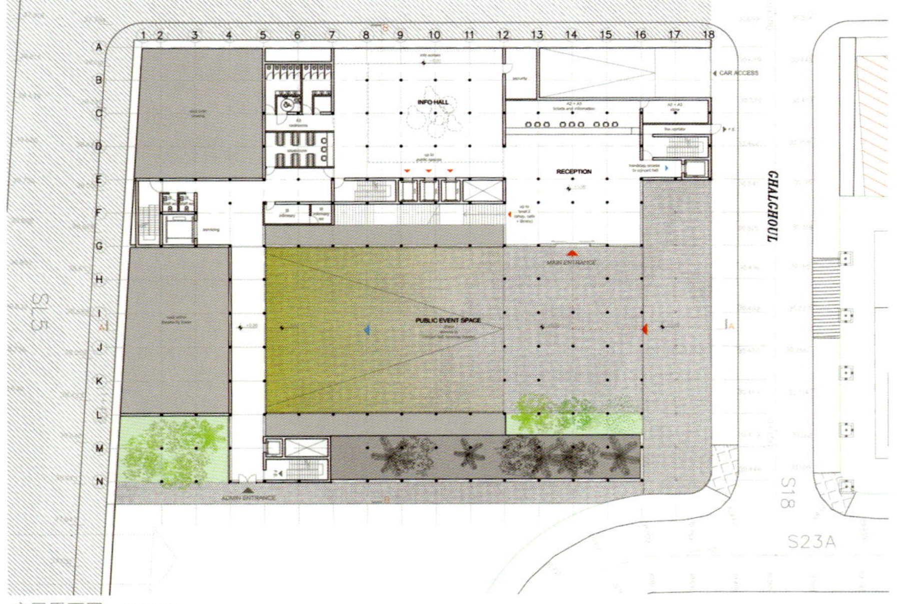

底层平面图 FLOOR PLAN 0

EXHIBITION 2
CINEMATHQ

五层平面图 FLOOR PLAN 4

艺术文化中心 (黎巴嫩–阿曼人中心) · 贝鲁特

House of Arts and Culture · Lebanon

荣誉提名奖 · Honourable Mention

poly.m.ur (建筑师事务所)

Chris S. Yoo · Homin Kim · Heekyung Moon · Ji-yeon Kim (建筑师)

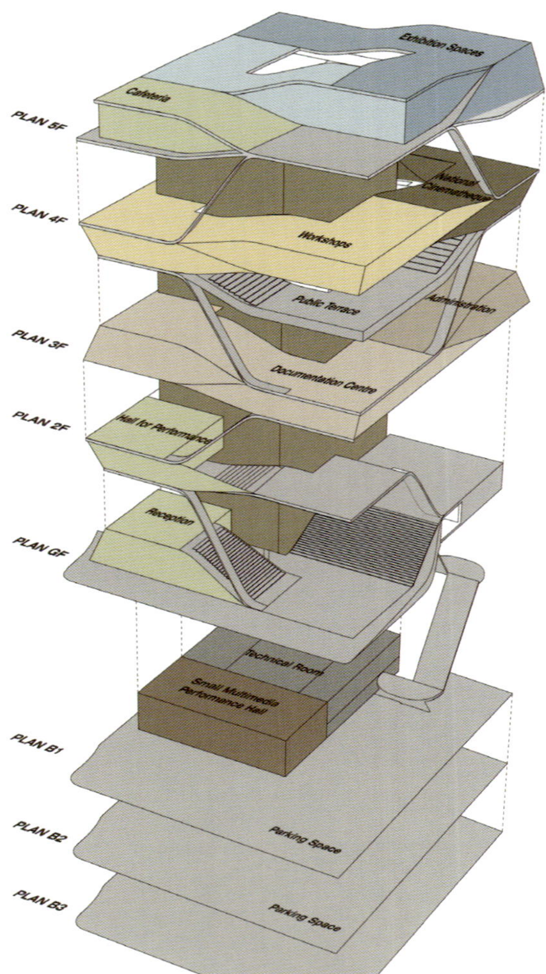

视线立体展开 EXPLODED AXONOMETRIC VIEW

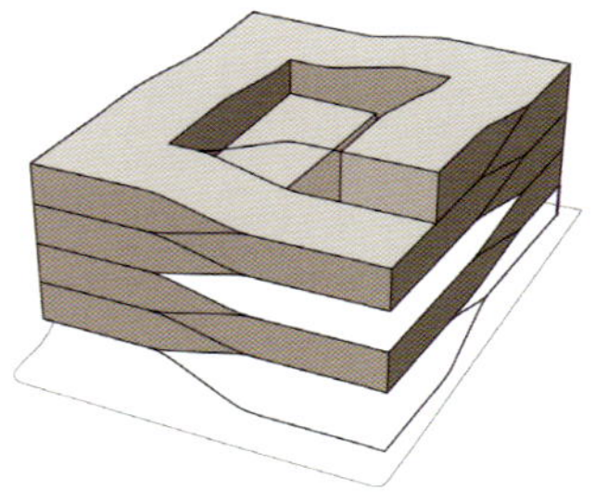

大块实体 SOLID MASSING

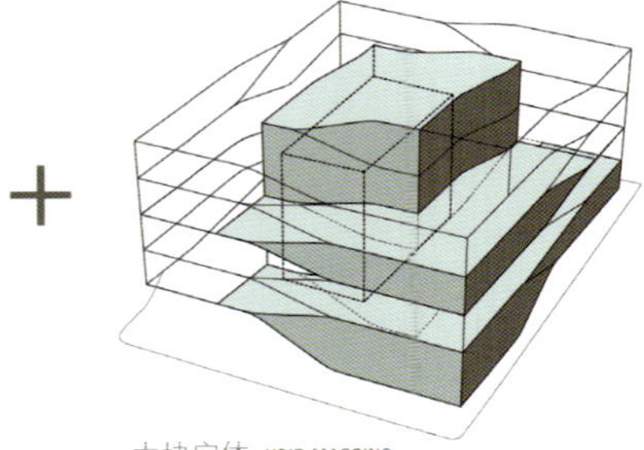

大块空体 VOID MASSING

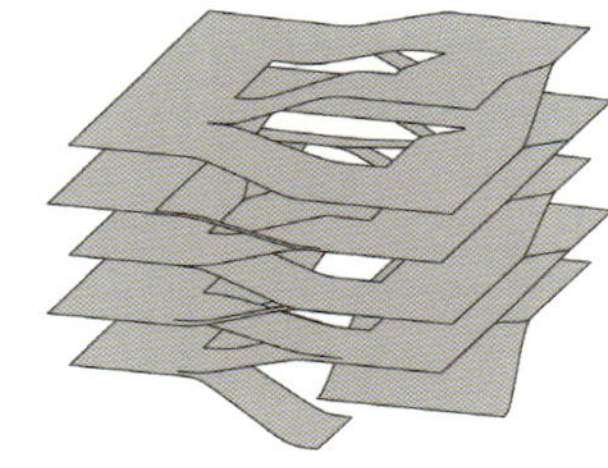

楼板 FLOOR PLATES

流体

我们旨在建造一幢坚固而渗透性好、注重形象塑造而兼具吸收力的建筑，让其建筑空间在公共空间的贯穿下相互交错。将公共空间与建筑打造为一体的灵感源自中东地区庭院的特色——吸收开放的空间使其成为半公共空间。传统的庭院扩张为一个三维结构。地板纵向蜿蜒开去，取得更流畅的连接效果。中东建筑常采用外部概念和明暗手法在建筑的外立面上打造和谐的图案屏幕。

A FLUID

Our aim is to create a building that is solid yet porous, iconic yet absorptive where spaces are intertwined with public realm throughout. Our strategy to incorporate the public spaces into the building is based on the Middle East courtyard houses, where the open space is interiorized as semi public space. The traditional courtyard typology is expanded into a three dimensional version. The floors are zigzagging vertically in order to produce much more fluent connection. The exterior concept and shading devices are commonly used in Middle-Eastern architecture in the form of integrated patterned screens in the facade.

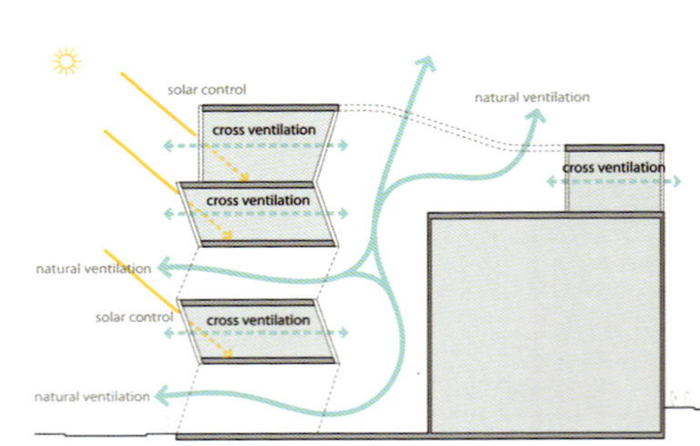

南立面 SOUTH ELEVATION

西立面 WEST ELEVATION

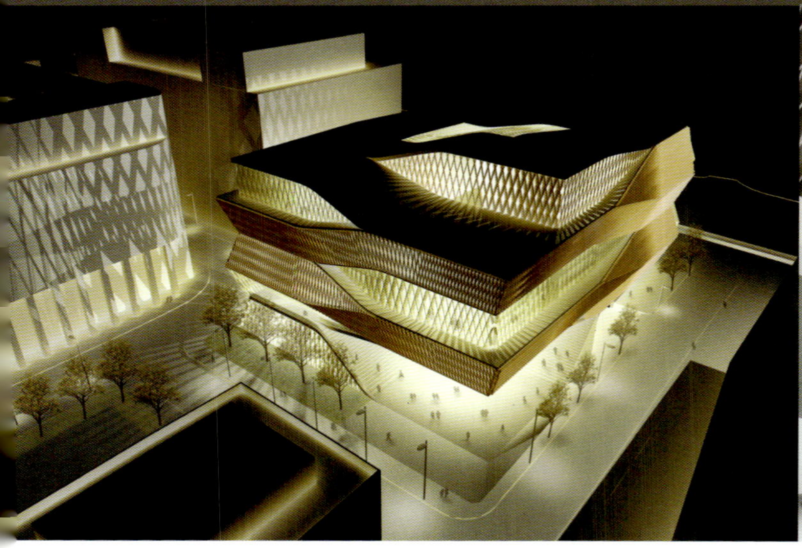

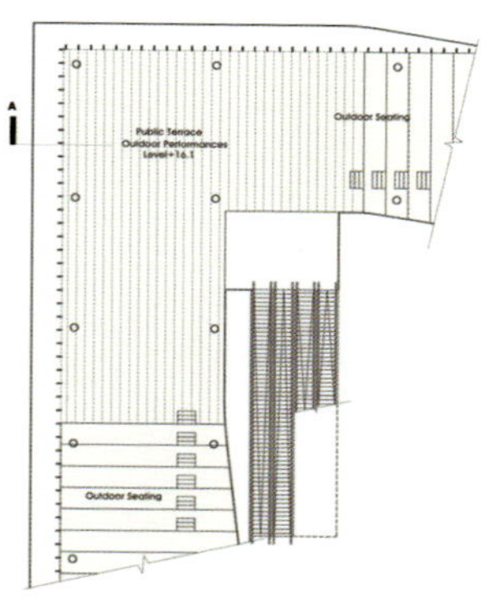

四层平面图 FLOOR PLAN 3

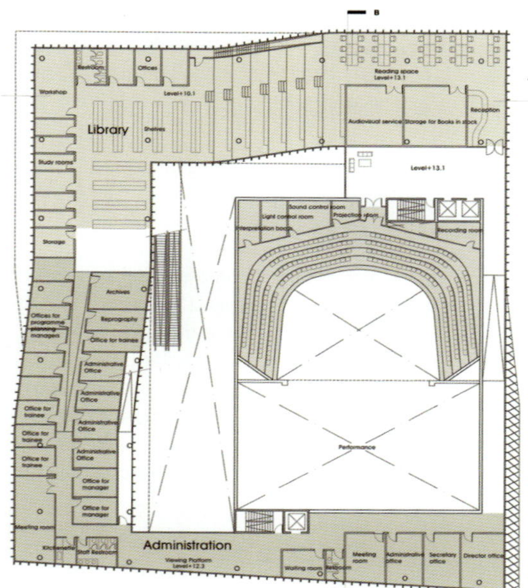

三层平面图 FLOOR PLAN 2

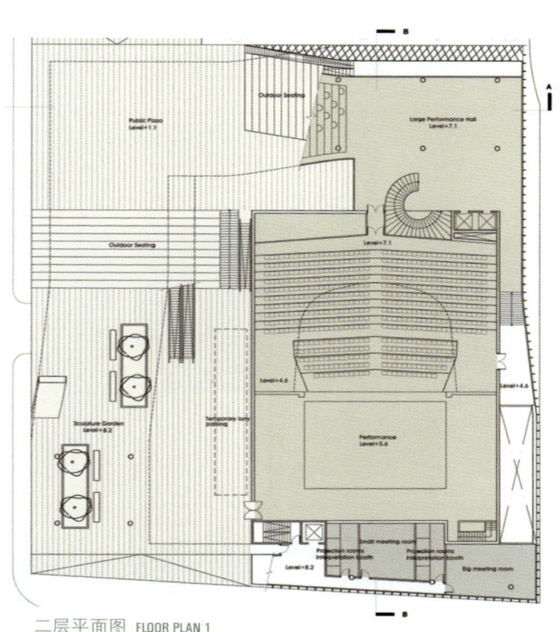

二层平面图 FLOOR PLAN 1

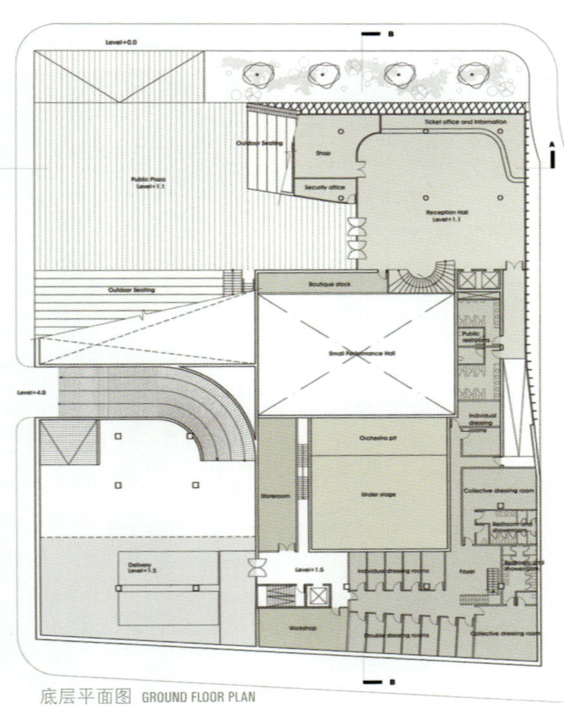

底层平面图 GROUND FLOOR PLAN

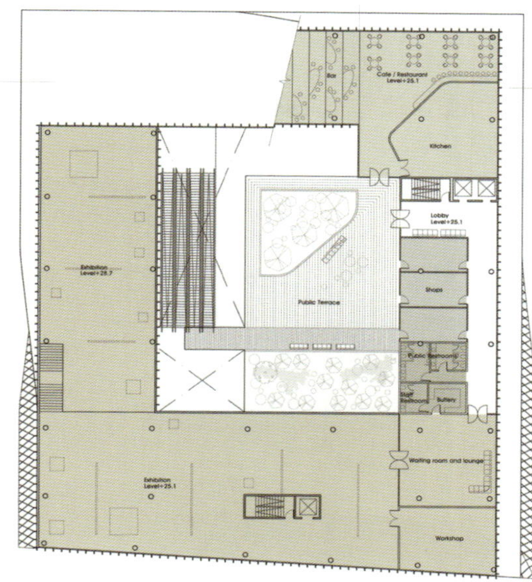

六层平面图 FLOOR PLAN 5

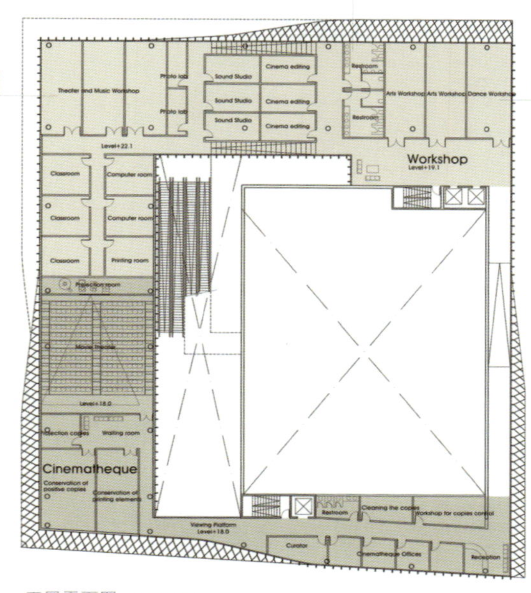

五层平面图 FLOOR PLAN 4

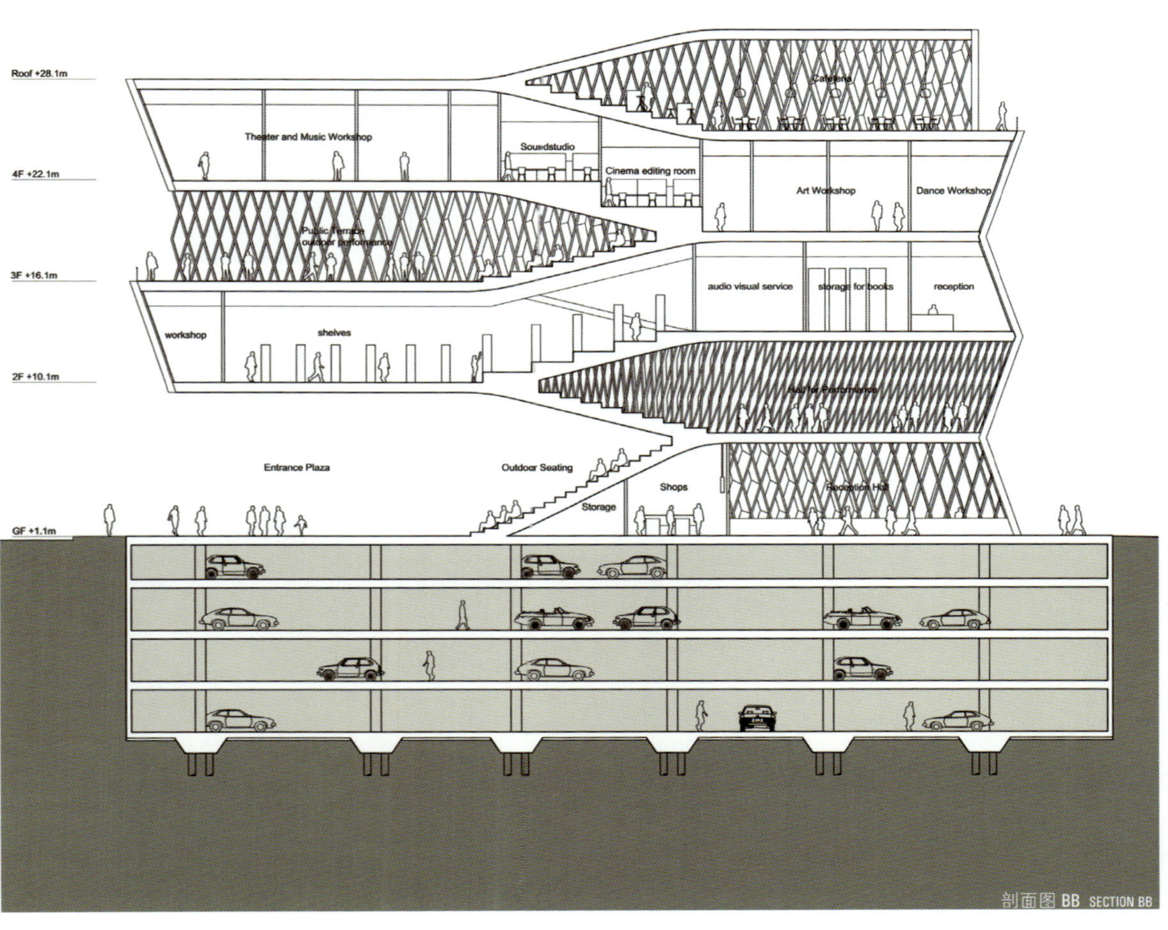

剖面图 BB SECTION BB

艺术文化中心（黎巴嫩–阿曼人中心）· 贝鲁特

House of Arts and Culture · Lebanon

BOARD（建筑师事务所）

荣誉提名奖 · Honourable Mention

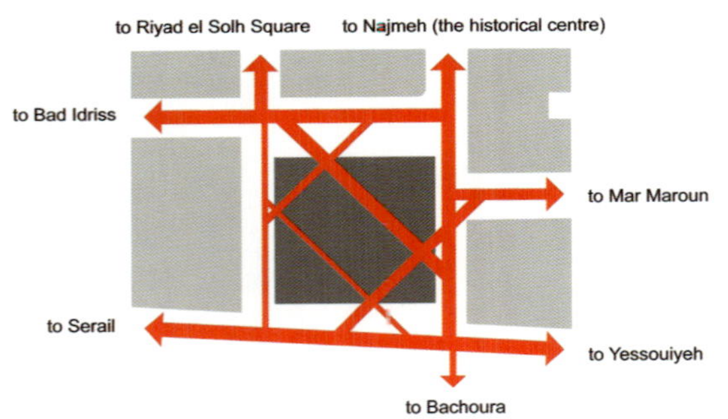

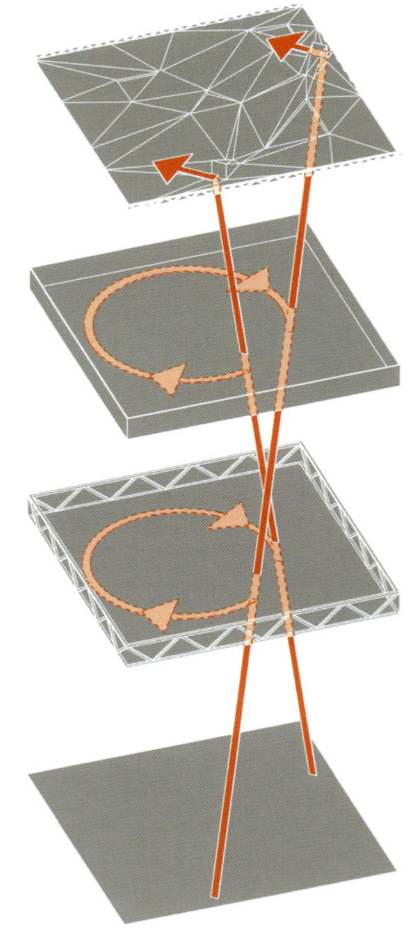

自由空间

该设计方案旨在表现许多不同形式的自由：从四面八方穿越该地点和建筑物，并将后者与都市背景融为一体的自由；通过两架从建筑底层到屋顶花园的电动扶梯，与建筑中功能各异的各层交互的自由；通过无柱结构呈现建筑平面设计的自由；利用可调节空间实现表演和展览区域使用的自由。

FREE SPACE

The proposal tries to establish freedom in a lot of different ways: freedom to cross the site and the building in all directions and to integrate the building into the urban context; freedom to interact with the different functional layers of the building through two escalators that penetrate the entire building from the ground floor to the roof garden; freedom of the floor plans through a construction that does not require any column; and freedom to use the performance and exhibition areas through adaptable spaces.

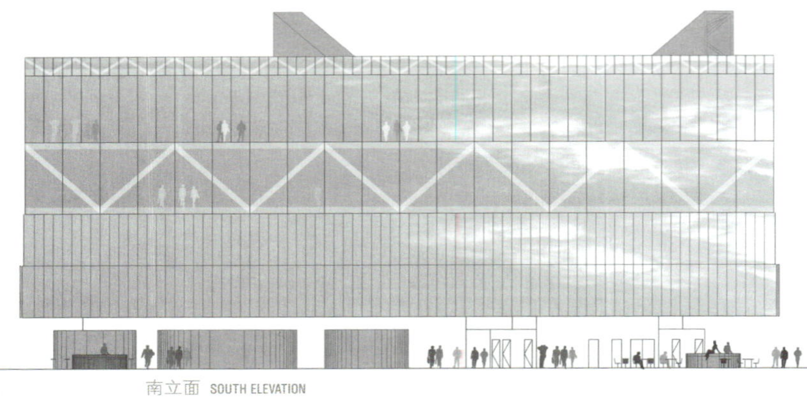

南立面 SOUTH ELEVATION

剖面图 B SECTION B

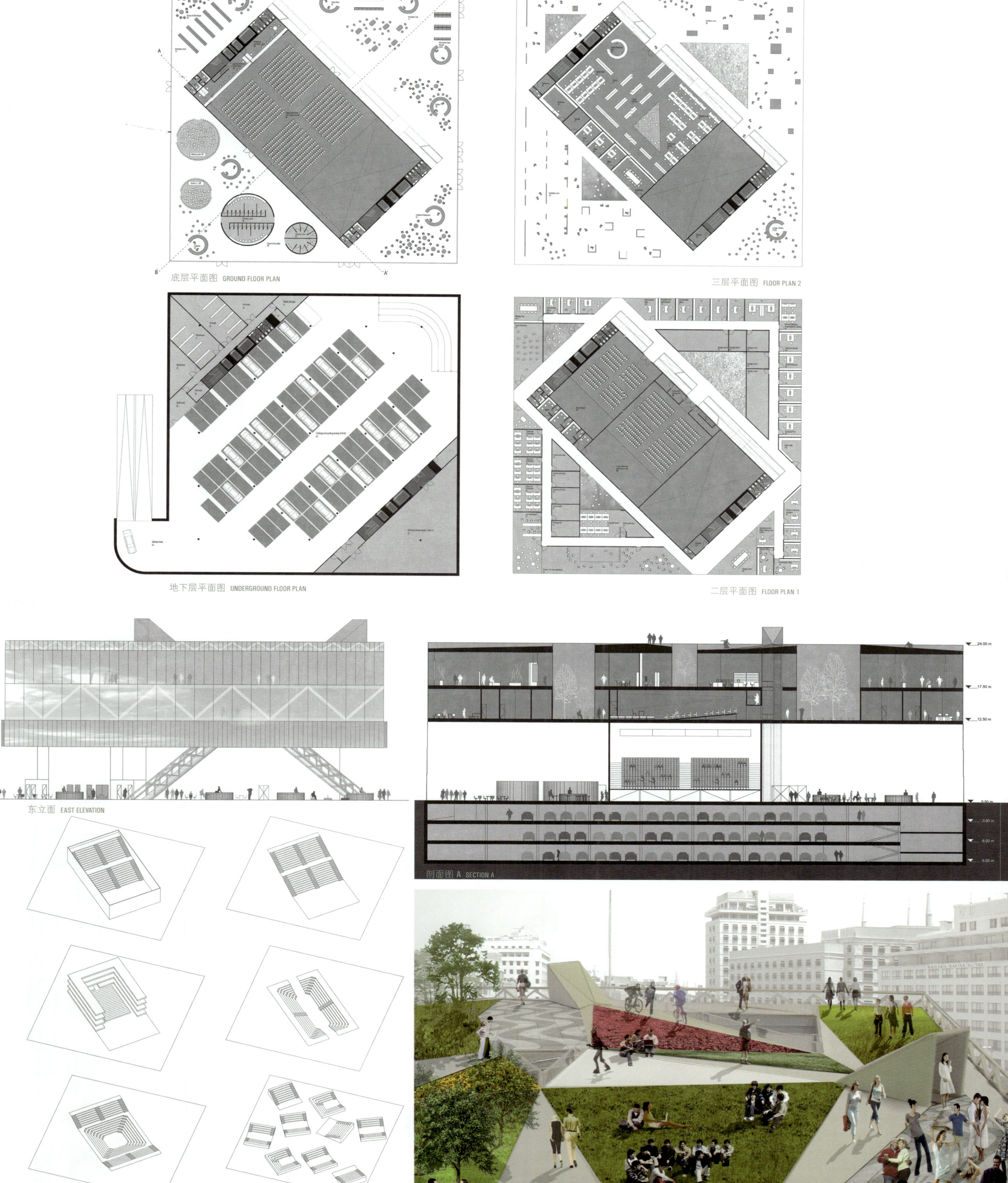

底层平面图 GROUND FLOOR PLAN

三层平面图 FLOOR PLAN 2

地下层平面图 UNDERGROUND FLOOR PLAN

二层平面图 FLOOR PLAN 1

东立面 EAST ELEVATION

剖面图 A SECTION A

自由空间的可能性 FREE SPACE POSSIBILITIES

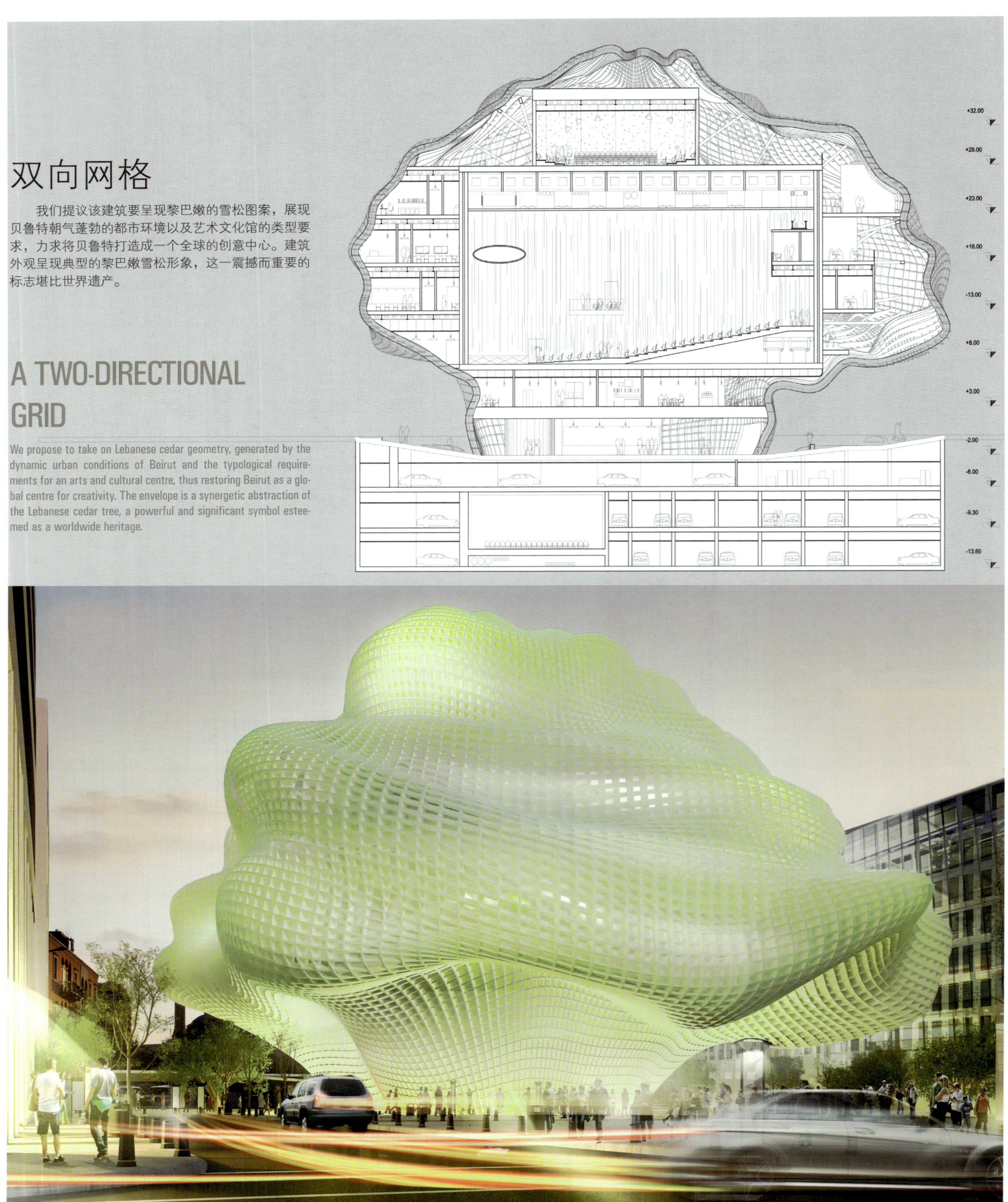

双向网格

我们提议该建筑要呈现黎巴嫩的雪松图案，展现贝鲁特朝气蓬勃的都市环境以及艺术文化馆的类型要求，力求将贝鲁特打造成一个全球的创意中心。建筑外观呈现典型的黎巴嫩雪松形象，这一震撼而重要的标志堪比世界遗产。

A TWO-DIRECTIONAL GRID

We propose to take on Lebanese cedar geometry, generated by the dynamic urban conditions of Beirut and the typological requirements for an arts and cultural centre, thus restoring Beirut as a global centre for creativity. The envelope is a synergetic abstraction of the Lebanese cedar tree, a powerful and significant symbol esteemed as a worldwide heritage.

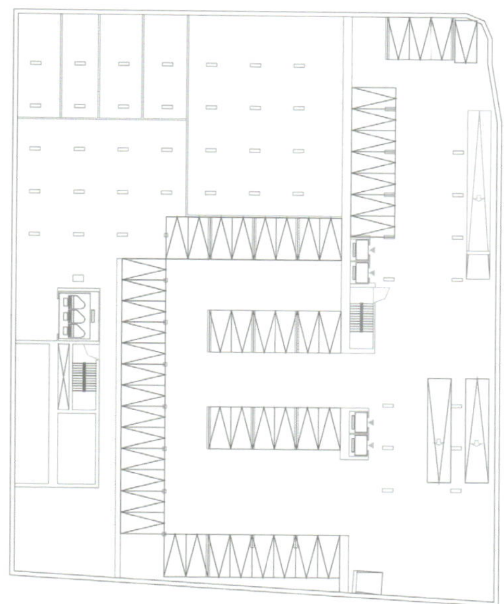
平面图 标高 -6.00 FLOOR PLAN LEVEL -6.00

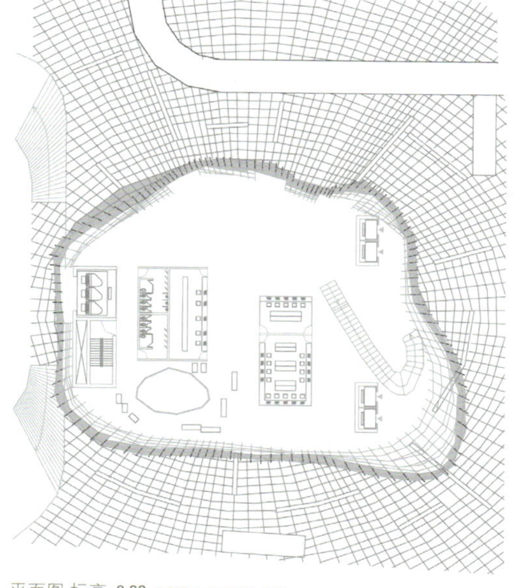
平面图 标高 -2.00 FLOOR PLAN LEVEL -2.00

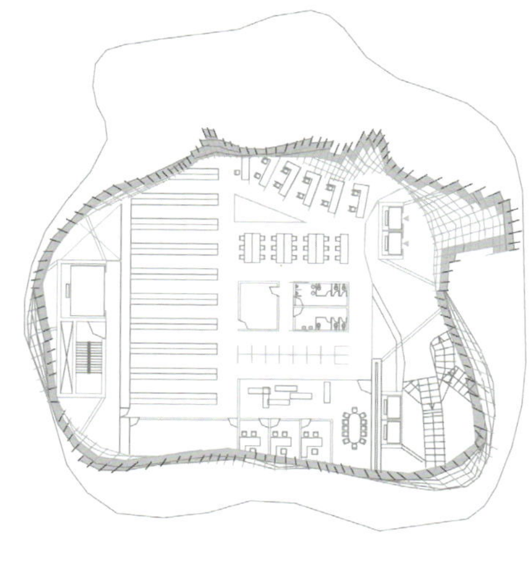
平面图 标高 +3.00 FLOOR PLAN LEVEL +3.00

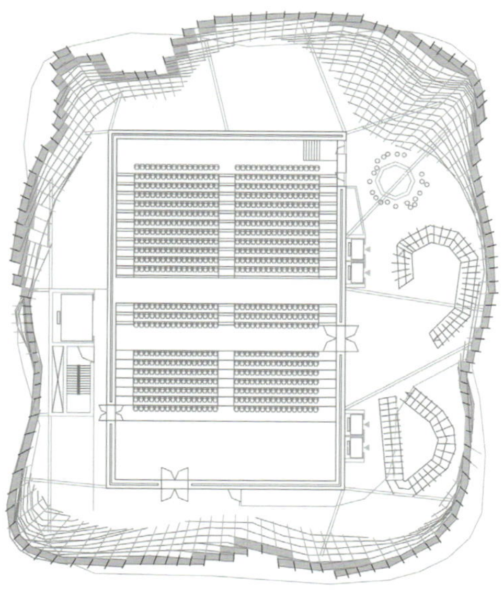
平面图 标高 +8.00 FLOOR PLAN LEVEL +8.00

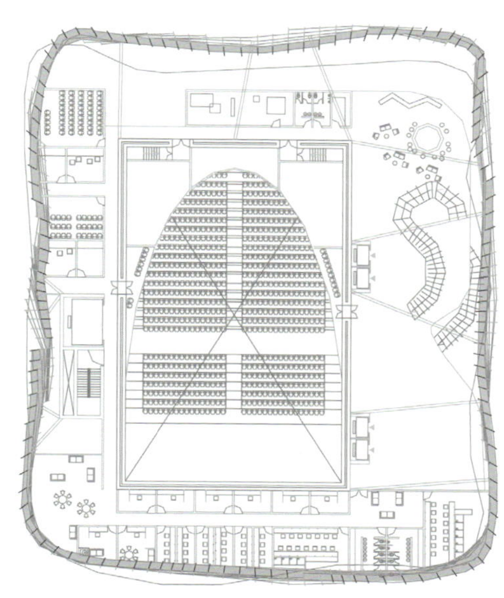
平面图 标高 +13.00 FLOOR PLAN LEVEL +13.00

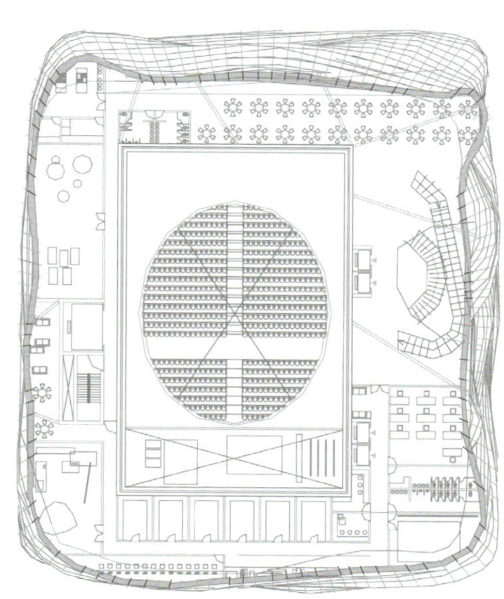
平面图 标高 +18.00 FLOOR PLAN LEVEL +18.00

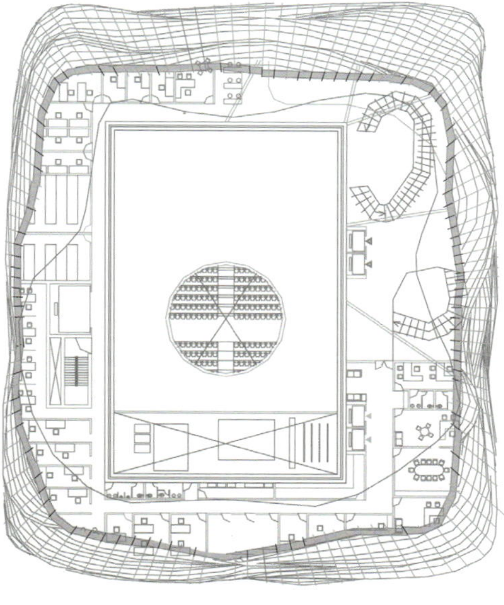
平面图 标高 +23.00 FLOOR PLAN LEVEL +23.00

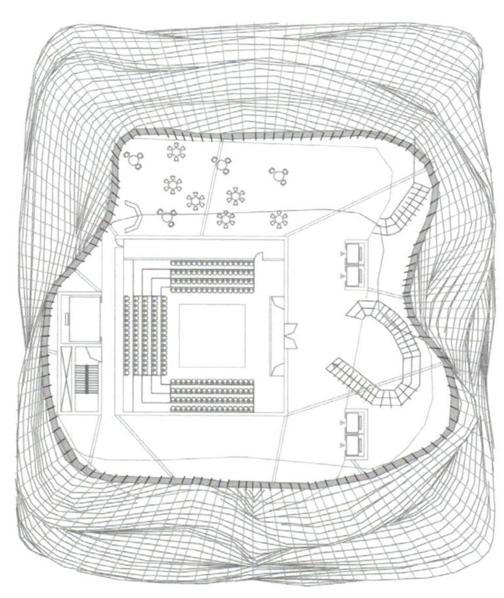
平面图 标高 +28.00 FLOOR PLAN LEVEL +28.00

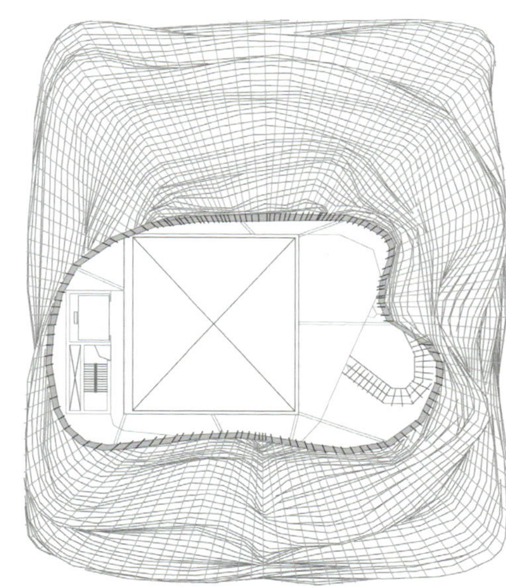
平面图 标高 +32.00 FLOOR PLAN LEVEL +32.00

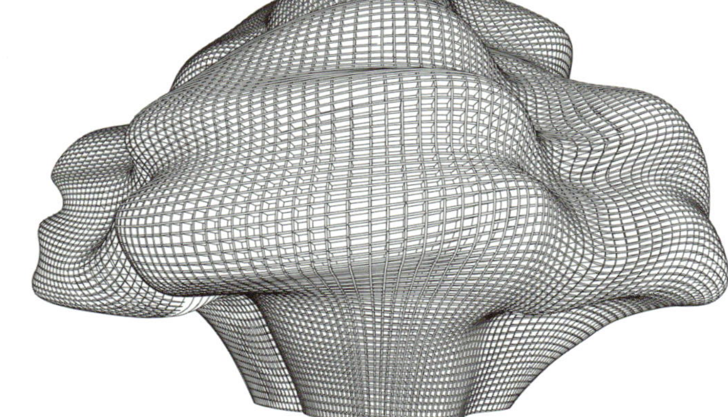
立体图 AXONOMETRY

音乐会展中心·维多利亚加斯特斯

Music, Congress and Exhibitions Palace · Vitoria Gasteiz, Spain

竞标 · competition
音乐会展中心
Music, Congress and Exhibitions Palace

竞标类型 · competition type
国际公开竞标
international open competition

项目地点 · site area
维多利亚加斯特斯 · 西班牙 Vitoria Gasteiz · Spain

主办方 · promoter
维多利亚加斯特斯市政府城市发展公司
Municipal Urban Society of Vitoria Gasteiz

日程安排 · schedule
招标 · Announcement 11.2008
评审结果 · Jury´s results 04.2009

获奖者 · awards

中标 · first prize
BAYON ARQUITECTURA Y URBANISMO (建筑师事务所)
Mariano Bayón (建筑师)

合作 (c) Pablo Bayón Villamor · José Manuel Blázquez Serrano · Jorge Escobar Chavero Guadalupe Sierra Arévalo · Marco García Moratalla · Julio Rodríguez Pareja · Daniel Martínez Díaz · Carmen Ortega Izquierdo · Nacho Román Santiago · Laura Pérez Lupi · Ángel Lallana Diez-Canseco · Ignacio Rubio Prieto
工程预算顾问 economic and project consulting · Apartec Colegiados S.L.
安装顾问 installation consulting · Prointec
舞台设备顾问 stage equipment consulting · Stolle
结构顾问 structural consulting · E.T.E.S.A.
工业工程学生 student of industrial engineering · Daniel Bayón
计算机及经济预算 computing and economic affairs · Teresa Villamor
行政 administration · María José Corcho
行政财务 administration-accountant · Mercedes Ordóñez

第一提名奖 · first accesit
Andrés Perea Ortega · Euroestudios (建筑及结构师事务所)
合作 (c) Alfredo García Horstmann · Ángel Muñoz Pérez · Alicia Bedmar Perlado · Salvador Benimeli López · María Carmona Díaz · Jae-Hyoung Park · Elisa Herranz Contreras · Javier Gutiérrez Rodríguez · Teresa Joven González · Viviana Peña Suárez · Diego Ceresuela Wiesmann · Jorge López Hidalgo · Lucio Luque Restrepo · José María Oruña San Emeterio Félix Manzano Alonso
工程 engineering · Euroestudios s.l. Ingenieros de consulta
声学 acoustics · Santiago Valero
场景、音响、灯光 scenic facilities, sound and lighting · Francisco Revilla Saavedra
预算 budget · ODV. Arquitectos Técnicos

第二提名奖 · second accesit
S&Aa (建筑师事务所)
Federico Soriano, S&Aa · Dolores Palacios, S&Aa
Esteban Rodríguez, IDOM (建筑师)

第三提名奖 · third accesit
cmArquitectos (建筑师事务所)
Javier Camacho · Mª Eugenia Maciá (建筑师)

合作 (c) María Navascués · Cristina Müller-Hillebrand · Juan Santana · Antho Kosmos Beatriz Noves
监理 quantity surveyor · Álvaro Rivera
安装 installation · Grupo JG Ingenieros
特殊安装 special installation · DITEC
结构 structure · NB35, S.L.
安全 security · Pedro Beguería
舞台设施 stage equipment · García Dieguez
园艺 gardening · Teresa Gali

入围 · finalist
JAAM · ARRANZ (建筑师事务所)

短名单 · shortlisted
SELGASCANO · FHECOR (建筑师事务所)
合作 (c) Carlos Chacón · José Jaraiz · Jae Hoon Yook · Lorena del Rio · Andrea Carbajo

维多利亚拟建造一个能容纳450人的中型会议厅。但是，该会议厅预计将能容纳近一千人。会议厅选址于阿拉瓦省首府的新开发地，位于未来火车站和巴斯克政府总部大楼之间。

Vitoria bets on placing on the track of the congresses of medium size, with an influx of about 450 participants. Still, the conference hall will be equipped to assume a capacity close to a thousand people. It will be located in a plot in one of the new nodes of the capital of Alava, between the future rail station and the headquarters of the Basque Government.

音乐会展中心·维多利亚加斯特斯

Music, Congress and Exhibitions Palace · Spain

一等奖 · First Prize

BAYON ARQUITECTURA Y URBANISMO (建筑师事务所)

Mariano Bayón (建筑师)

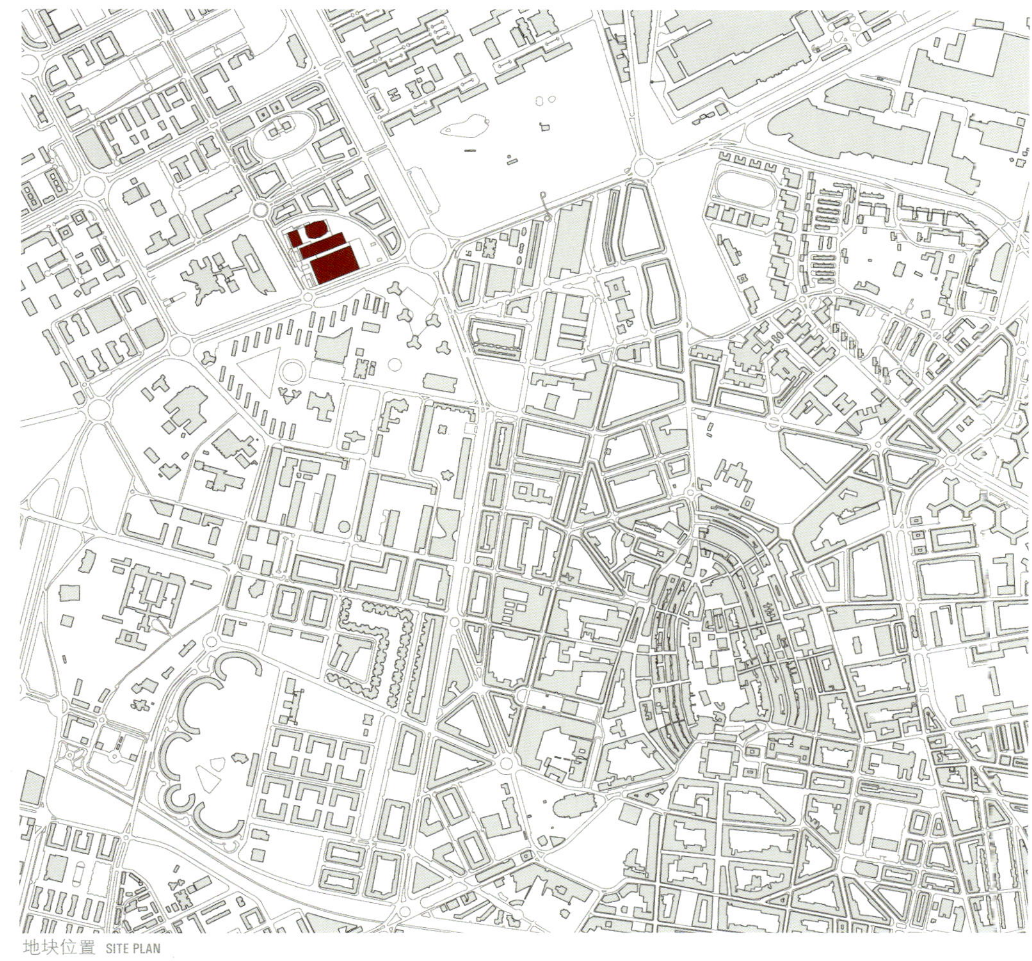

地块位置 SITE PLAN

一个崭新的“高架广场”

该设计并非针对单幢建筑物，而是针对一个多元化的小镇。这是一个特色分明的建筑广场，同处在两个平台顶部，一个室内，一个室外。在整个项目设计中，我们优先考虑了节能环保问题，以利于规划的自由分合，且最大化地减少维护工作量。此外，尽可能让开阔的露天空地面向圣塞瓦斯蒂安和巴斯克政府。该项目由各个不同的部分组成，这些组成部分将第一层和第二层转化为面向街道的基座结构，从而打造出了“阶梯”效果。

A NEW "ELEVATED PLAZA"

The proposal raises a diverse little town rather than a single building: an agora of buildings with their own character, juxtaposed, which cohabit on top of two platforms: one which is covered and another outdoor. There is an environmental priority of energy-saving throughout the project which can easily separate or join programs and with a minimum maintenance. The proposal also recognizes the importance of the large open urban esplanade facing calle Donostia and the Basque Government. It creates a "staircase" with various sectors which turn the ground and first floor plans into a plinth facing this street.

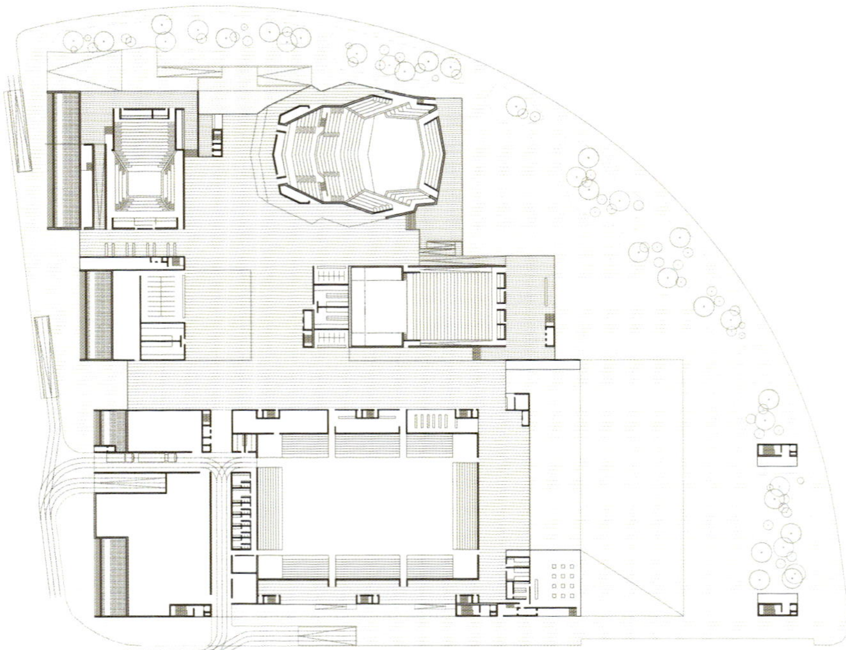

平面图标高 +1.50 FLOOR PLAN LEVEL +1.50

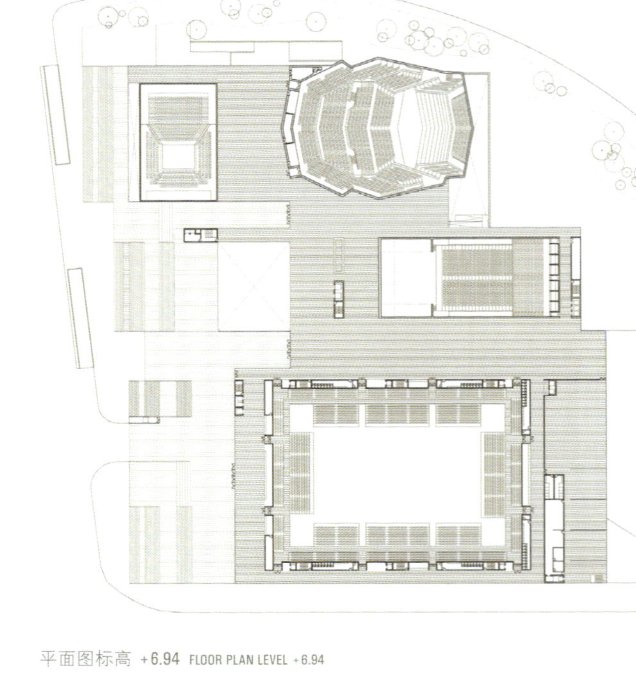

平面图标高 +6.94 FLOOR PLAN LEVEL +6.94

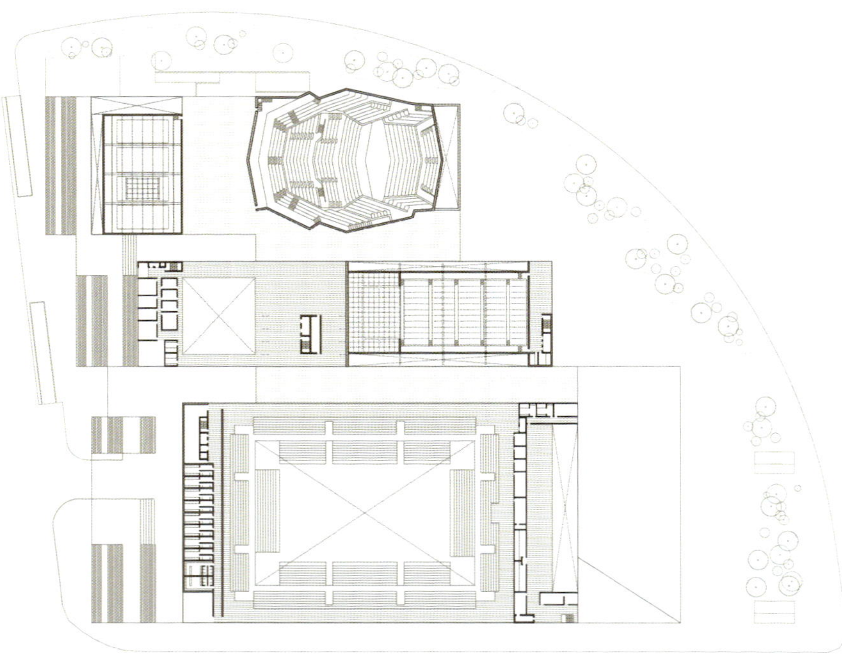

平面图标高 +11.34 FLOOR PLAN LEVEL +11.34

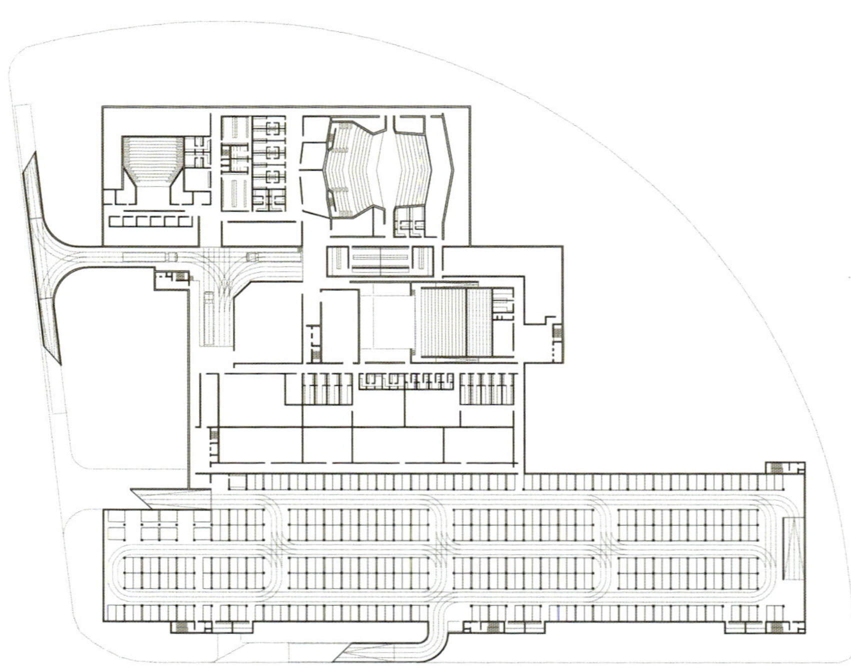

平面图标高 -3.00/-4.00 FLOOR PLAN LEVEL -3.00/-4.00

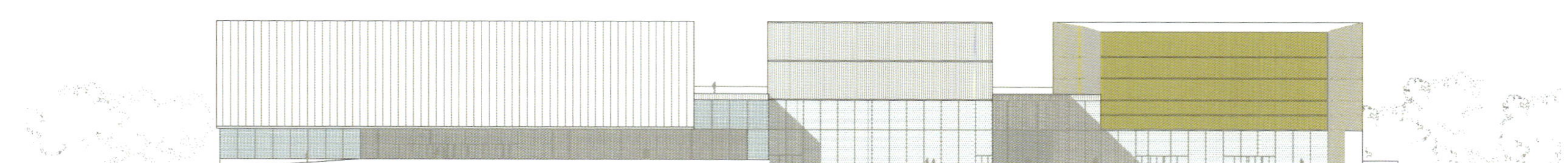

东北立面图 NORTHEAST ELEVATION

交响乐大厅内部景观 INTERIOR VIEW OF SYMPHONY HALL

前厅景观 VIEW OF LOBBY

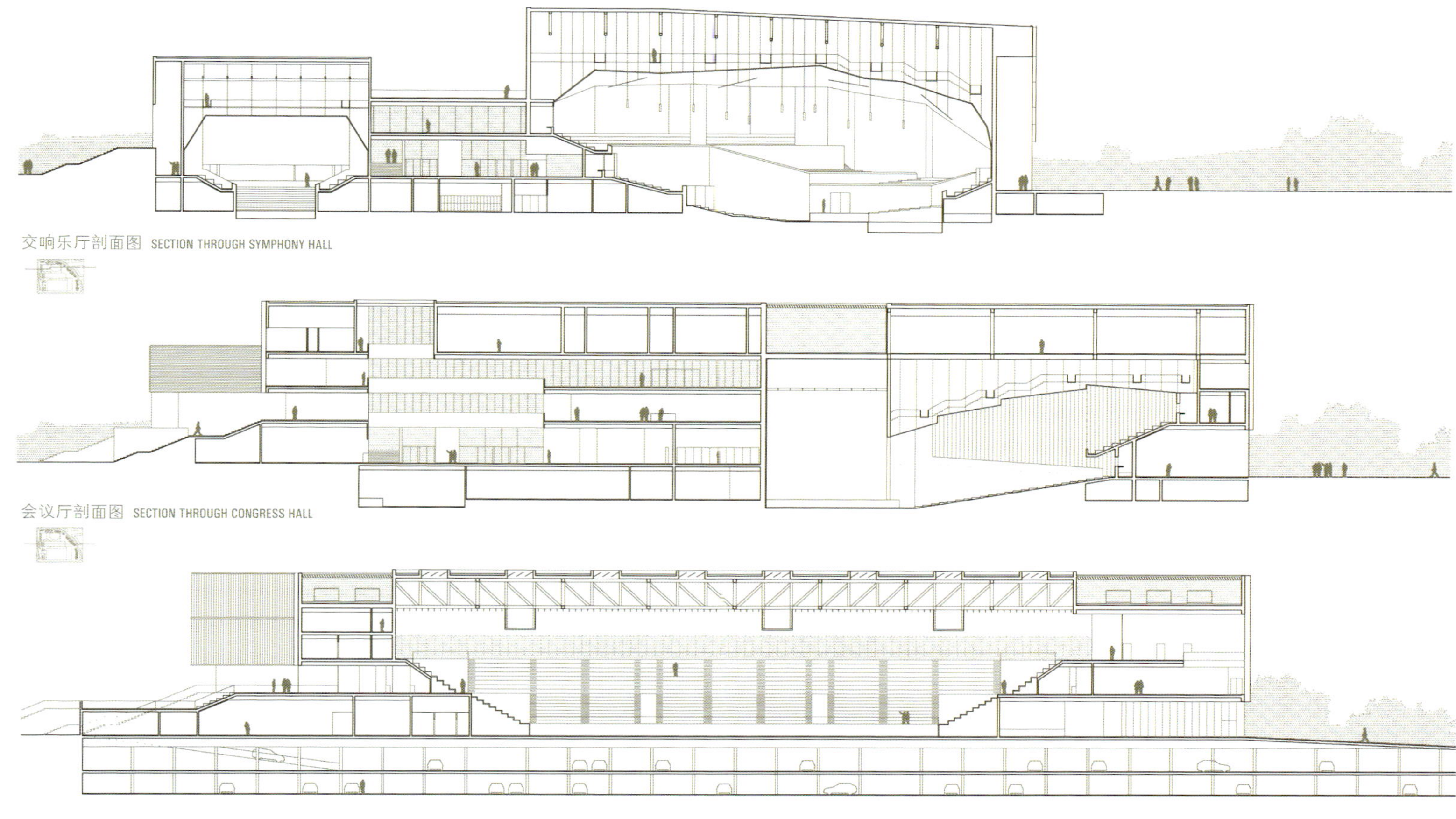

文响乐厅剖面图 SECTION THROUGH SYMPHONY HALL

会议厅剖面图 SECTION THROUGH CONGRESS HALL

展厅剖面图 SECTION THROUGH EXHIBITION HALL

1. 太阳能收集真空管道群
   EVACUATED TUBES SOLAR THERMAL COLLECTORS
2. 欧标240 IPE型钢
   IPE 240
3. 欧标120 IPE型钢(切断)
   IPE 120 PROFILE (CUT)
4. TRAMEX 50.50.4混凝土技术过道
   TECHNICAL FOOTBRIDGE OF TRAMEX 50.50.4
5. 前厅固定STADIP隔音玻璃
   FIXED STADIP GLASS 12.12.12 IN LOBBIES
6. PLADUR装饰天花板，安装过道
   FALSE CEILING OF PLADUR. FLOW OF INSTALLATIONS
7. 持续照明
   CONTINUOUS LIGHTING
8. 地板
   FLOORING
9. 混凝土楼板
   CONCRETE SLAB
10. L 50.50.4
    L 50.50.4
11. 欧标120 HEB型钢
    HEB 120
12. 铝薄板
    ALUMINIUM SHEET
13. 金色玻璃陶瓷圆柱外包的不锈钢管道
    CERAMIC GLASS CILYNDER COVERING TUBE OF STAINLESS STEEL
14. 乳浊区的预制混凝土墙
    PREFABRICATED CONCRETE WALL IN OPAQUE AREAS
15. 隔热
    THERMAL INSULATION
16. PLADUR镶板
    PLADUR PANELLING

1. 100.100.5钢管
   STEEL TUBE 100.100.5
2. 欧标160 IPE型钢
   IPE 160
3. 6.6.6 STADIP全息隔音玻璃
   STADIP GLASS 6.6.6 SECRITEX+HOLOGRAPHIC
4. 60CM TRAMEX混凝土技术过道
   TECHNICAL FOOTBRIDGE OF TRAMEX 60CM
5. 25MM圆形扶手钢杆
   HANDRAIL OF 25MM STEEL RODS
6. PLADUR
   PLADUR
7. 用于房间和预制板之间固定的金属构造
   METALLIC STRUCTURE TO FIX PREFABRICATED PANELS AND CHAMBER
8. 预制混凝土板
   PREFABRICATED CONCRETE PANEL
9. 混凝土楼板
   CONCRETE SLAB
10. 地板
    FLOORING
11. 持续照明
    CONTINUOUS LIGHTING

1. SECRITEX 10.10曲面超亮STADIP隔音玻璃
   EXTRA-BRIGHT AND CURVED STADIP GLASS 10.10 SECRITEX
2. 带螺丝45.50.5 U型钢
   U 45.50.5 WITH SCREWS
3. 40.4薄板分隔管道
   40.4 SHEET TO SEPARATE TUBE
4. 6.6.12.6.6可美透玻璃
   CLIMALIT GLASS 6.6.12.6.6
5. 100.100.7钢管
   STEEL TUBE 100.100.7
6. 80.80.3钢管
   STEEL TUBE 80.80.3
7. 60.60.3管道
   TUBE 60.60.3
8. 50.50.4 TRAMEX混凝土技术过道
   TECHNICAL FOOTBRIDGE OF TRAMEX 50.50.4
9. 60 60.4 L型钢
   L 60.60.4
10. 欧标180 IPE型钢
    IPE 180
11. 欧标240 IPE型钢
    IPE 240
12. 背光
    REVERSE LIGHTING

1. 内侧33/60/7 双层U型半透明隔离玻璃
   U-GLASS: DOUBLE LAYER 33/60/7 TRANSLUCID INSULATION IN THE INTERIOR
2. 100.50.5钢管
   STEEL TUBE 100.50.5
3. 欧标120 IPE型钢(切断)
   IPE PROFILE 120 (CUT)
4. 50.50.4 TRAMEX混凝土技术过道
   TECHNICAL FOOTBRIDGE OF TRAMEX 50.50.4
5. 可美透超亮玻璃及6.6.12.6.6 STADIP曲面玻璃
   EXTRA-BRIGHT AND CURVED STADIP GLASS 6.6.12.6.6
6. PLADUR装饰天花板，安装过道
   FALSE CEILING OF PLADUR. FLOW OF INSTALLATIONS
7. 电动遮阳布 控制光线
   MOTORIZED FABRIC - SOLAR CONTROL
8. 持续照明灯座
   LAMP HOLDER WITH CONTINUOUS LIGHT
9. 混凝土楼板
   CONCRETE SLAB
10. 50.50.4 L型钢
    L 50.50.4
11. 欧标120 HEB型钢
    HEB 120
12. 铝薄板
    ALUMINIUM SHEET

外墙的通风处理 VENTILATED FAÇADE SOLUTION IN:
交响乐大厅 SYMPHONY HALL

横向剖面图 HORIZONTAL SECTION

纵向剖面图 VERTICAL SECTION

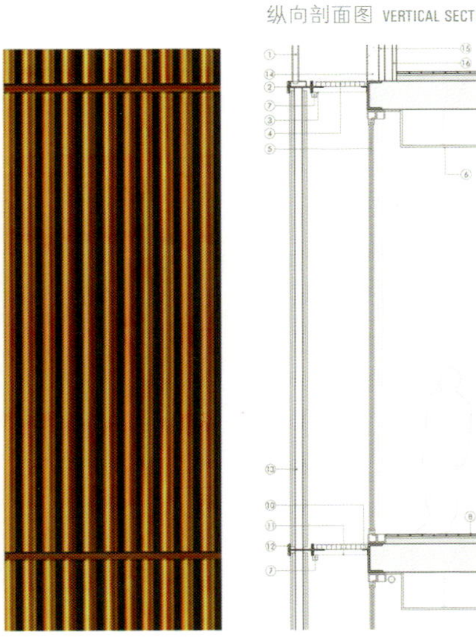

外墙的通风处理 VENTILATED FAÇADE SOLUTION IN:
表演展示厅 SPECTACLES-EXHIBITION HALL

横向剖面图 HORIZONTAL SECTION

纵向剖面图 VERTICAL SECTION

外墙的通风处理 VENTILATED FAÇADE SOLUTION IN:
餐厅 RESTAURANT

横向剖面图 HORIZONTAL SECTION

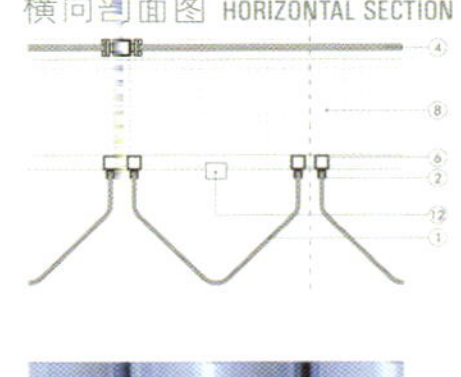

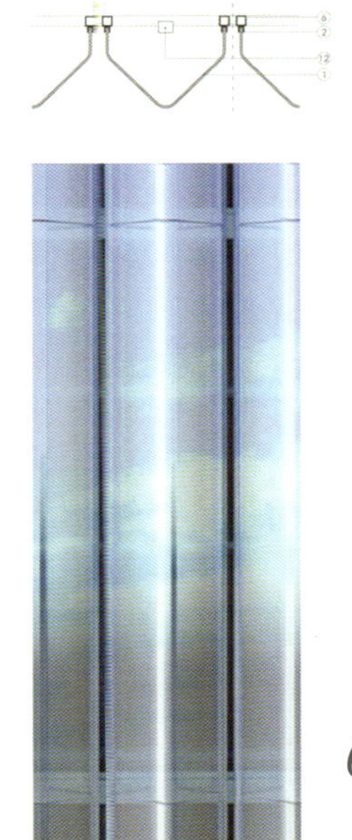

纵向剖面图 VERTICAL SECTION

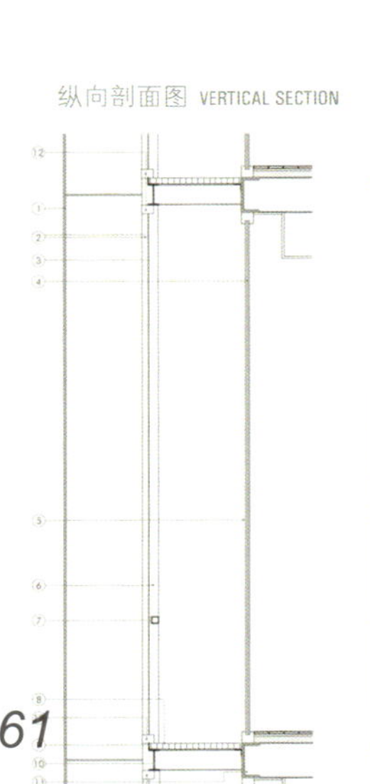

外墙的通风处理 VENTILATED FAÇADE SOLUTION IN:
会议厅 CONGRESS

横向剖面图 HORIZONTAL SECTION

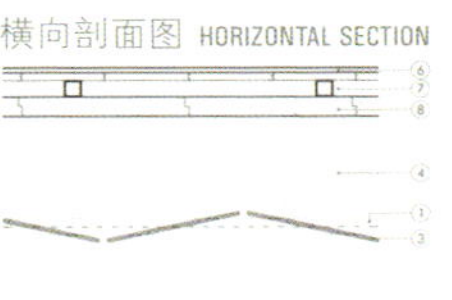

纵向剖面图 VERTICAL SECTION

音乐会展中心·维多利亚加斯特斯

Music, Congress and Exhibitions Palace · Spain

第一提名奖 · First Accesit

Andrés Perea Ortega · Euroestudios (建筑师)

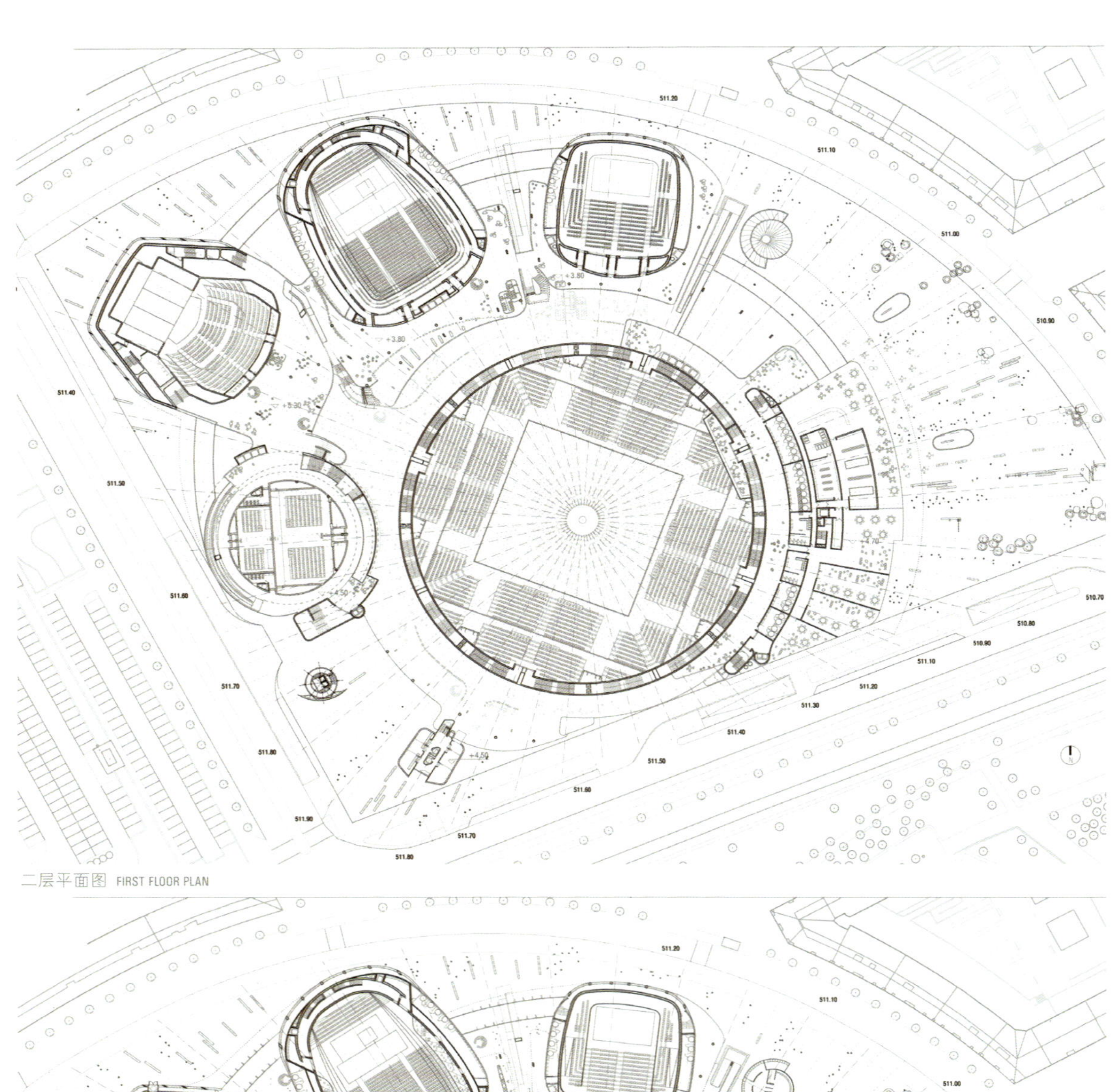
二层平面图 FIRST FLOOR PLAN

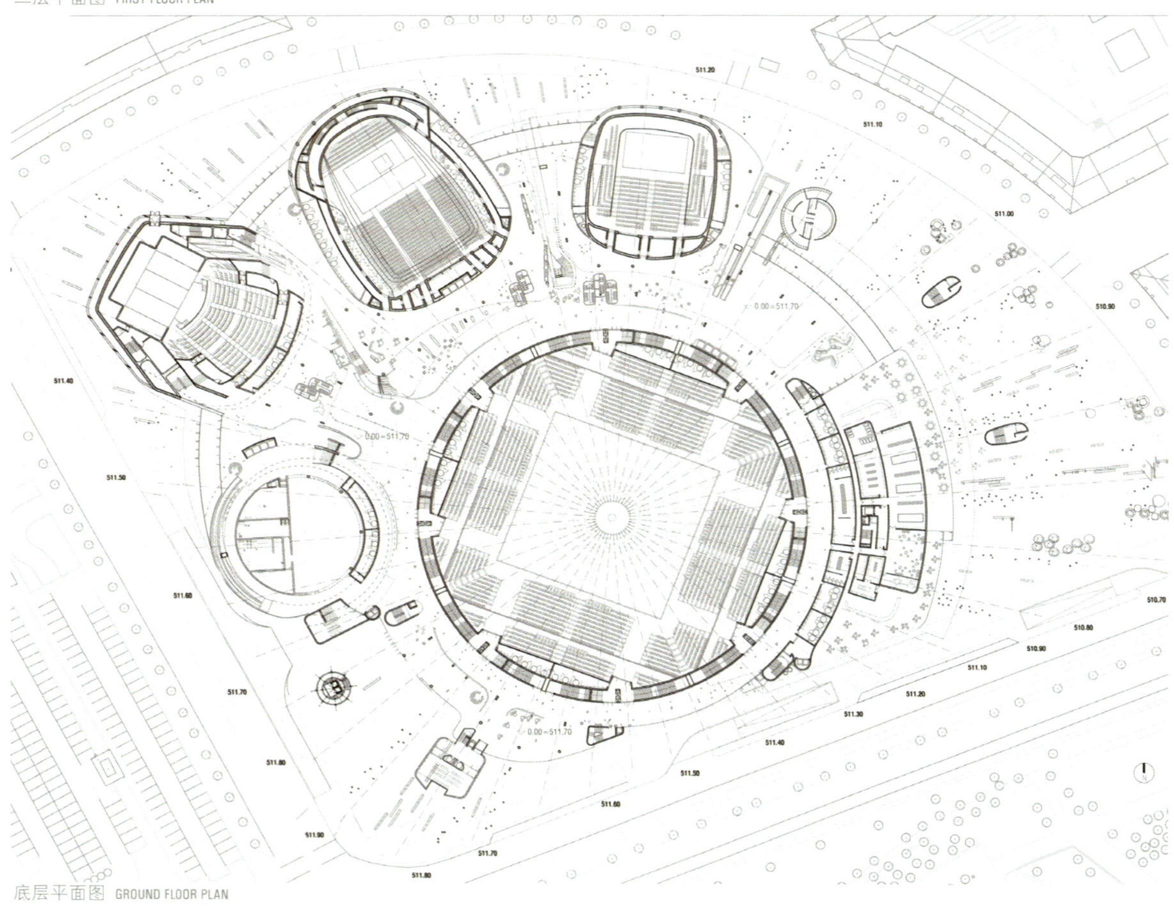
底层平面图 GROUND FLOOR PLAN

城市建设

我们将位于城市边缘的这个流畅空间几乎融入了建筑物内部，彻底颠覆了"外立面"的概念，也摒弃了其他项目所采用的层次结构。而"外立面"的概念则是通过连续而变化的外立面和内外部的层次叠加突显的。

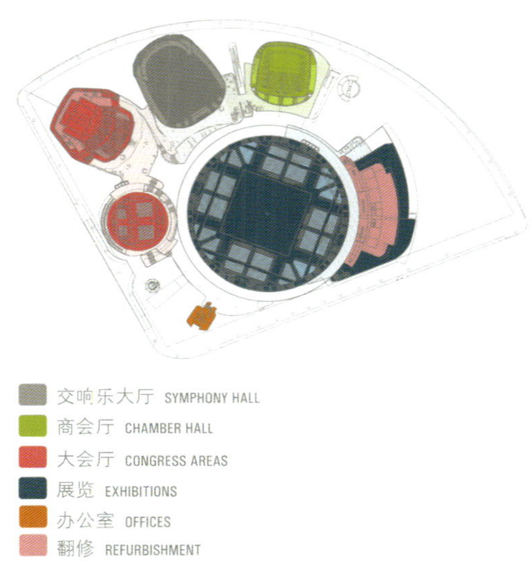

交响乐大厅 SYMPHONY HALL
商会厅 CHAMBER HALL
大会厅 CONGRESS AREAS
展览 EXHIBITIONS
办公室 OFFICES
翻修 REFURBISHMENT

BUILDING-CITY

The fluid status of the space at the urban edge , but mostly inside the building's envelope, is a radical commitment to the denial of the concept of "façade" and the denial of the hierarchy in some programs over others. The concept of "façade" is outweighed by a continuous and changing façade and by an internal and external superimposition of layers.

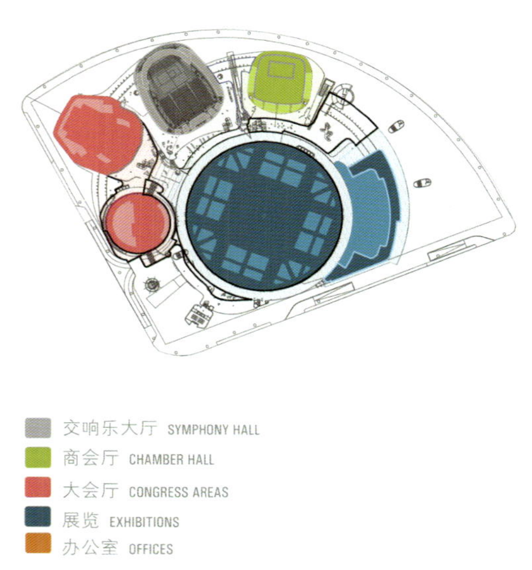

交响乐大厅 SYMPHONY HALL
商会厅 CHAMBER HALL
大会厅 CONGRESS AREAS
展览 EXHIBITIONS
办公室 OFFICES

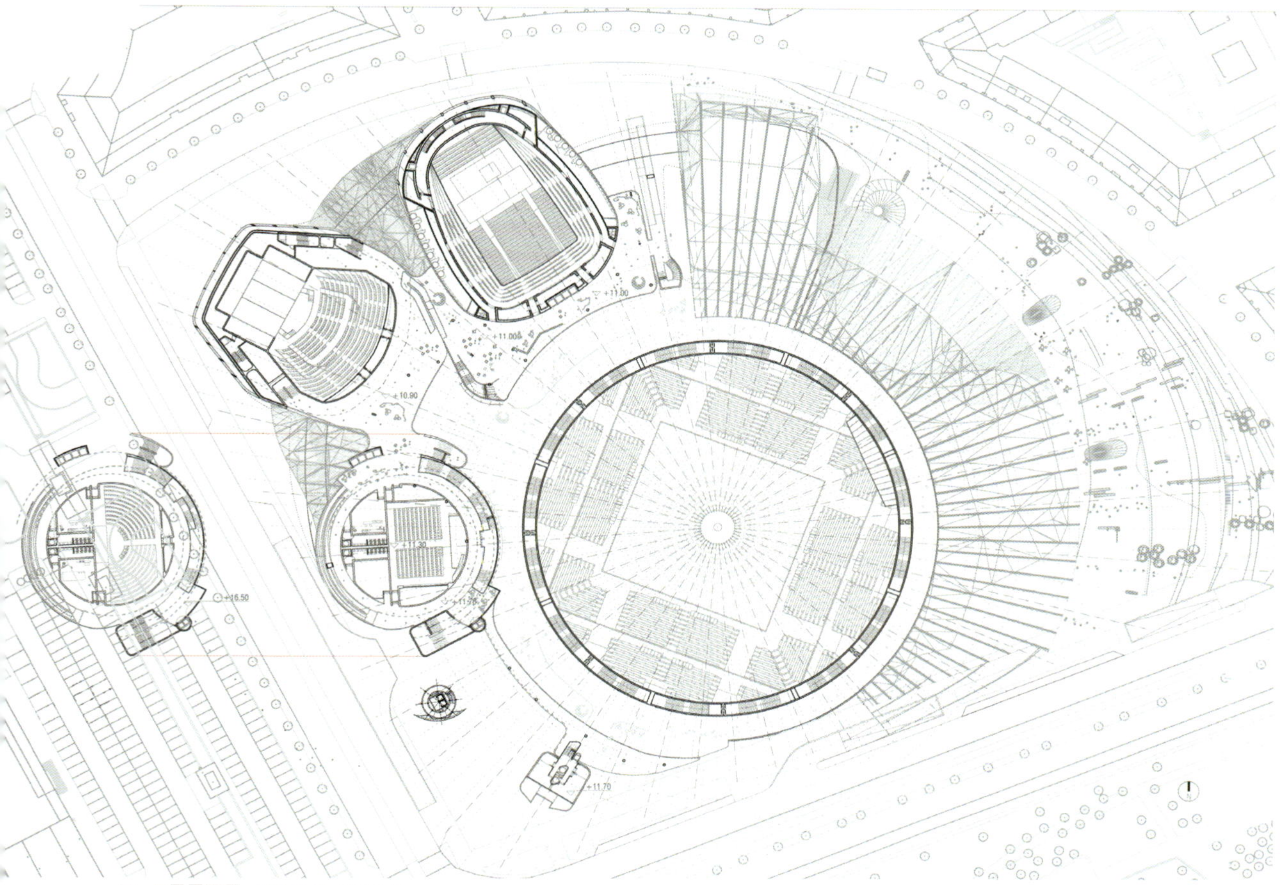

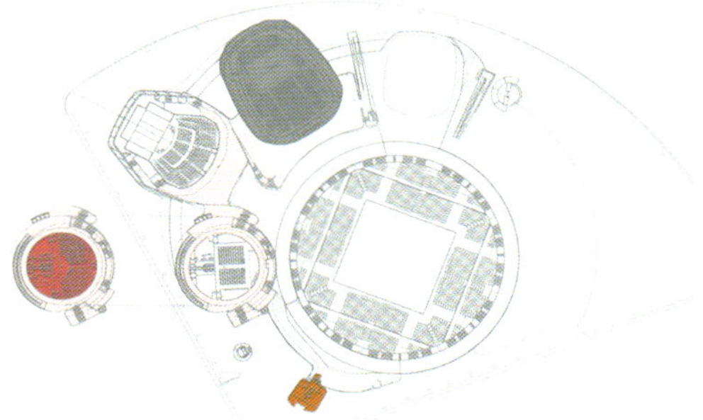

交响乐大厅 SYMPHONY HALL
办公室 OFFICES
大会厅 CONGRESS PALACE

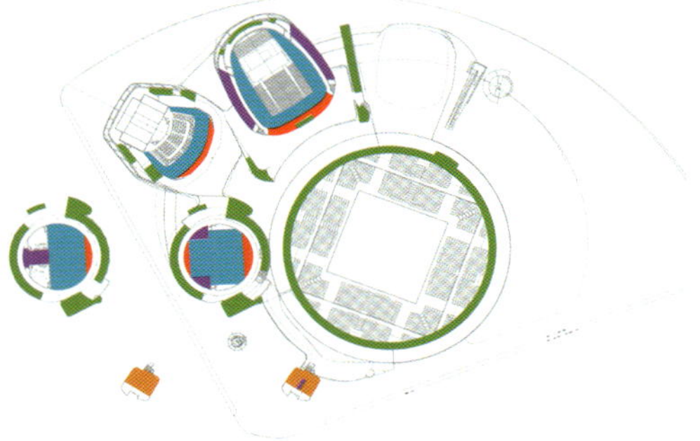

控制中心 CONTROL
商会 CHAMBER
交流中心 COMMUNICATION CORE
卫生间 TOILET
办公室 OFFICE

四+五层平面图 THIRD+FOURTH FLOOR PLAN

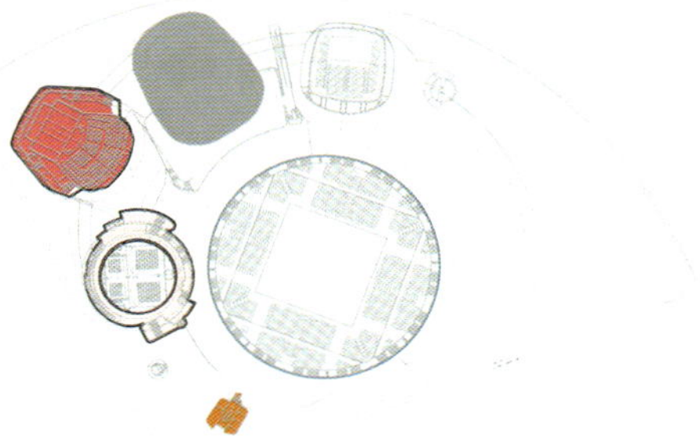

交响乐大厅 SYMPHONY HALL
商会厅 CHAMBER HALL
大会厅 CONGRESS PALACE
展览 EXHIBITIONS
办公室 OFFICES

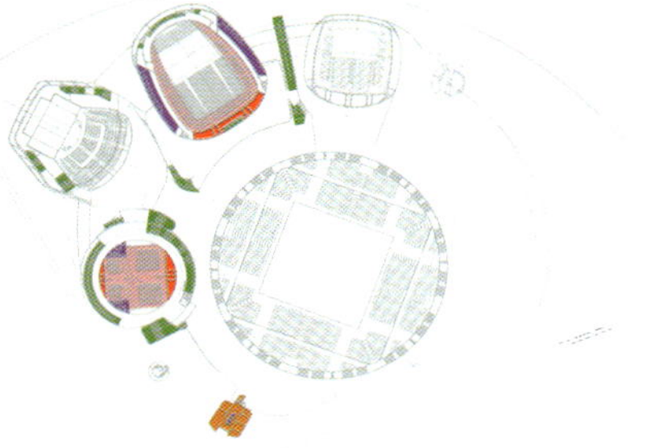

控制中心 CONTROL
商会 CHAMBER
咖啡厅 CAFETERIA
交流中心 COMMUNICATION CORE
卫生间 TOILET
办公室 OFFICE

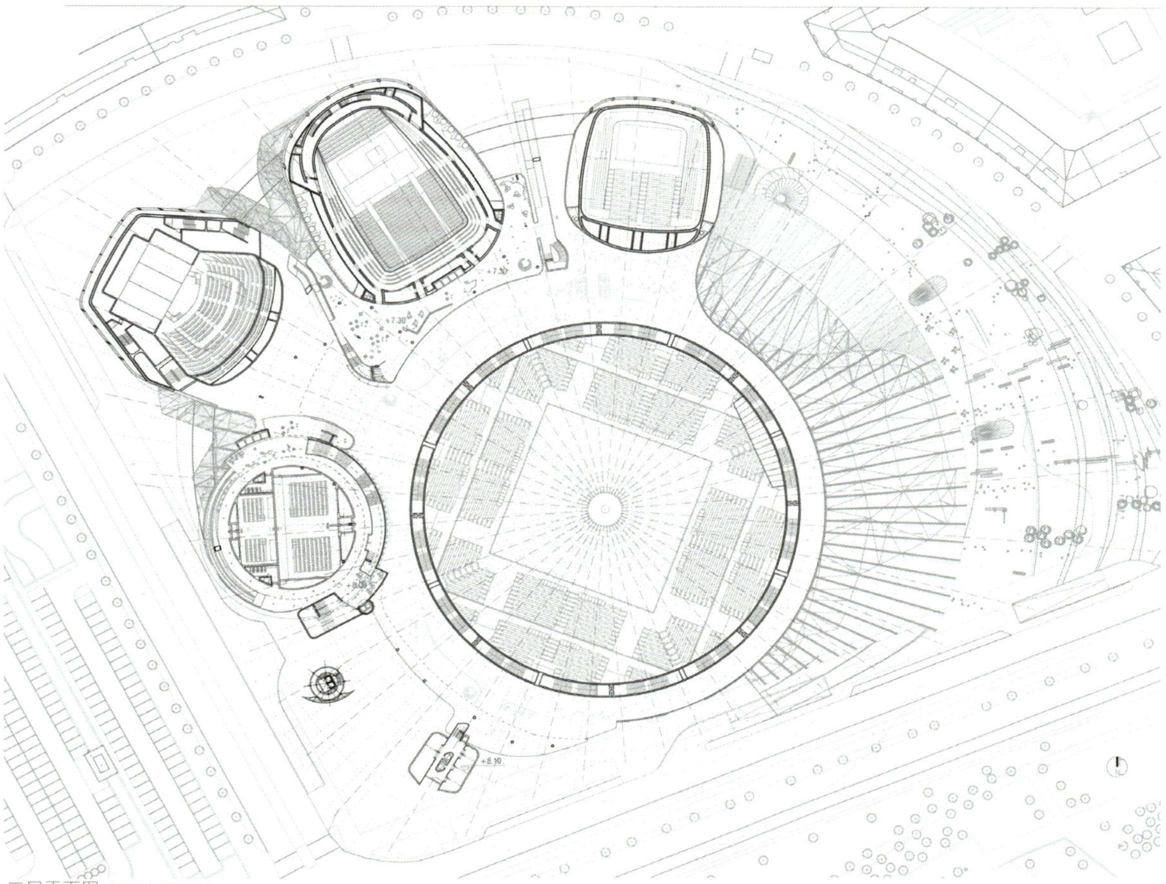

三层平面图 SECOND FLOOR PLAN

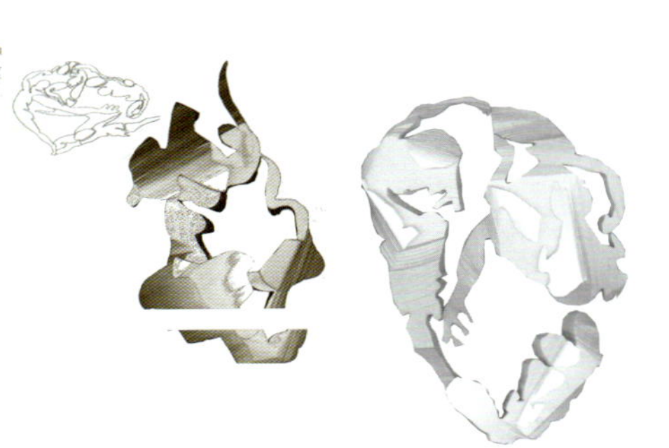

音乐会展中心 · 维多利亚加斯特斯

Music, Congress and Exhibitions Palace · Spain

第二提名奖 · Second Accesit

S&Aa (建筑师事务所) · IDOM (建筑师)

Federico Soriano · Dolores Palacios · Esteban Rodríguez

林间
空地

我们的创意简单明了：由五个主体构成，环一块空地——一个既私密又可共享的露台；五个主体之间职能各异，却又以一种美妙延绵的方式交相辉映。

A GAP BETWEEN THE TREES

Our concept is simple and straightforward; five pieces, five bodies, with different programs and autonomy in their use, which are linked, in a melodious and endless dance around an empty place, a private and shared patio.

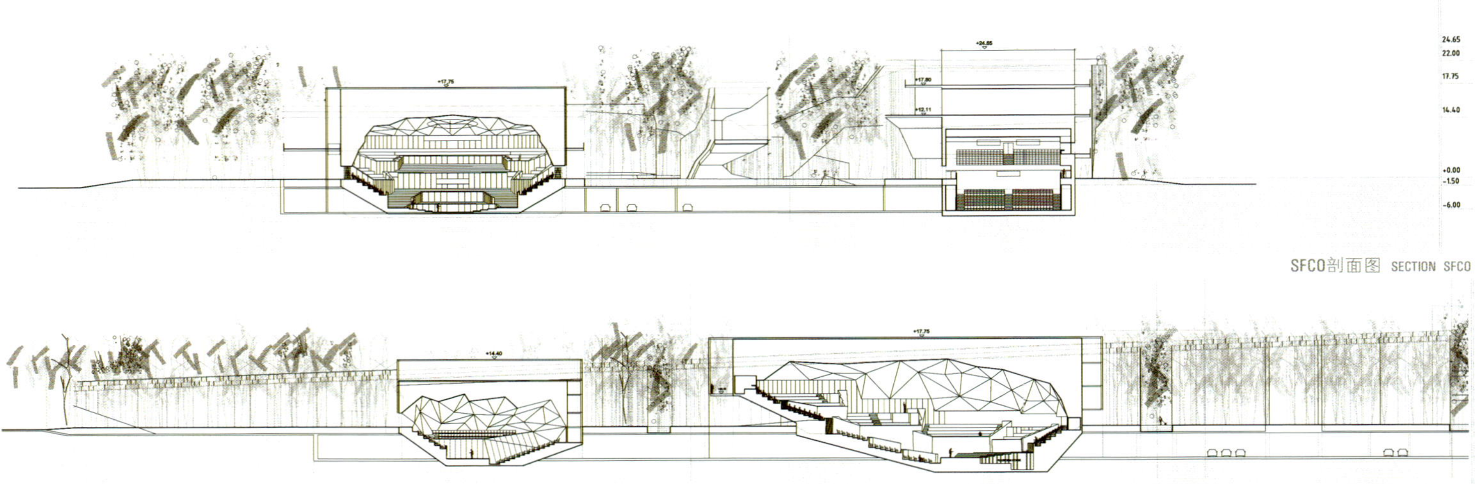

SFCO剖面图 SECTION SFCO

SCSF剖面图 SECTION SCSF

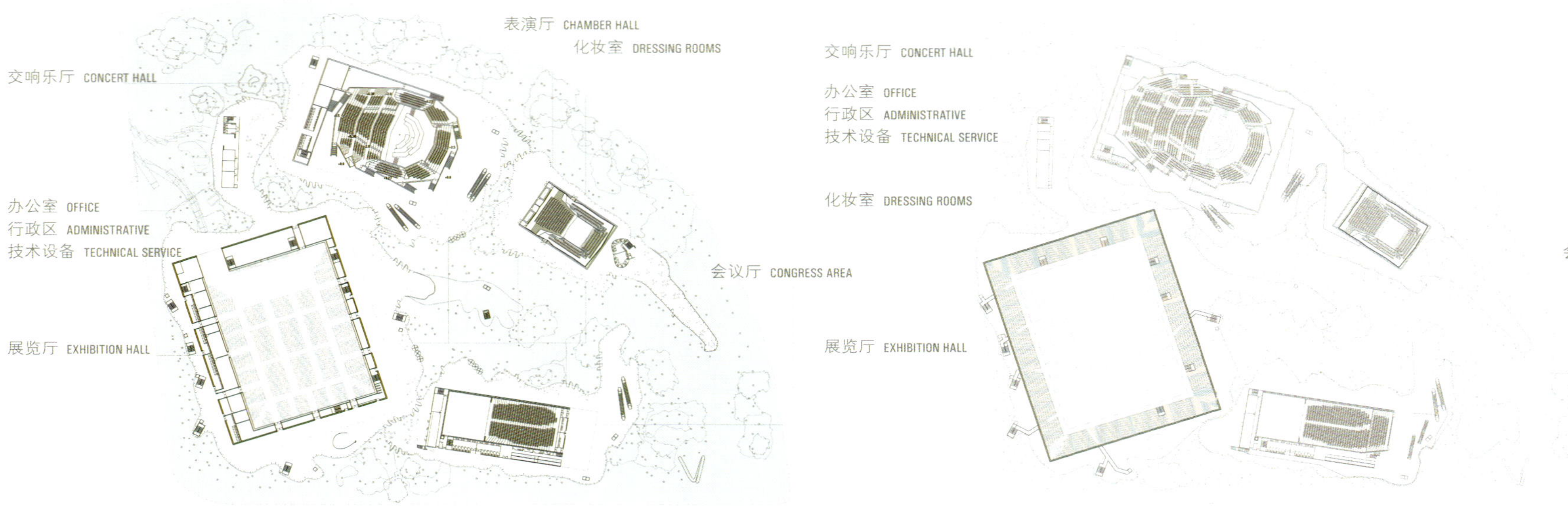

入口层平面图 ACCESS FLOOR PLAN

二层平面图 FIRST FLOOR PLAN

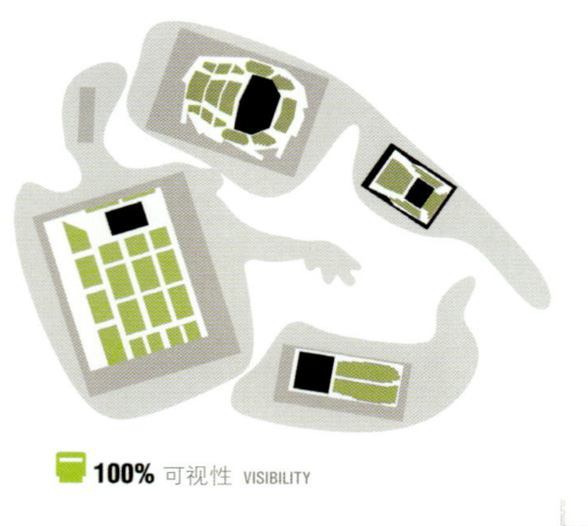

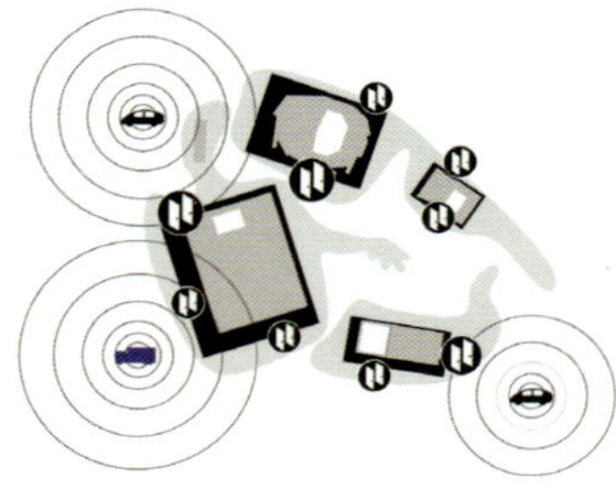

通向大厅的入口拥有声学前厅
THE ACCESS TO THE HALLS HAVE ACOUSTIC LOBBIES

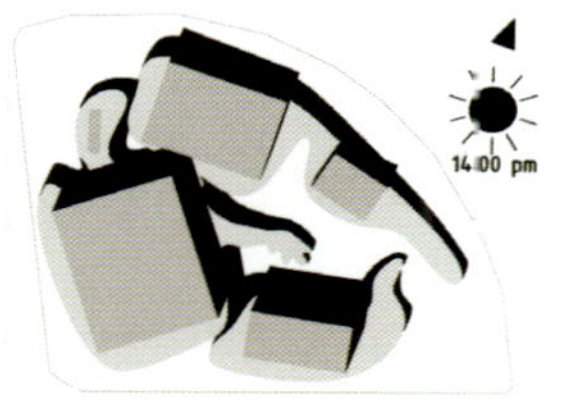

围绕周边建筑的阴影最小化
SHADOW OVER SURROUNDING BUILDINGS IS MINIMUM

建筑促进景观
THE BUILDING PROMOTES A LANDSCAPE

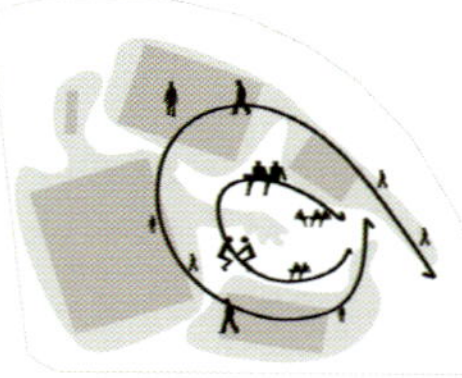

内部流通是连续的
THE INTERIOR CIRCULATION IS CONTINUOUS

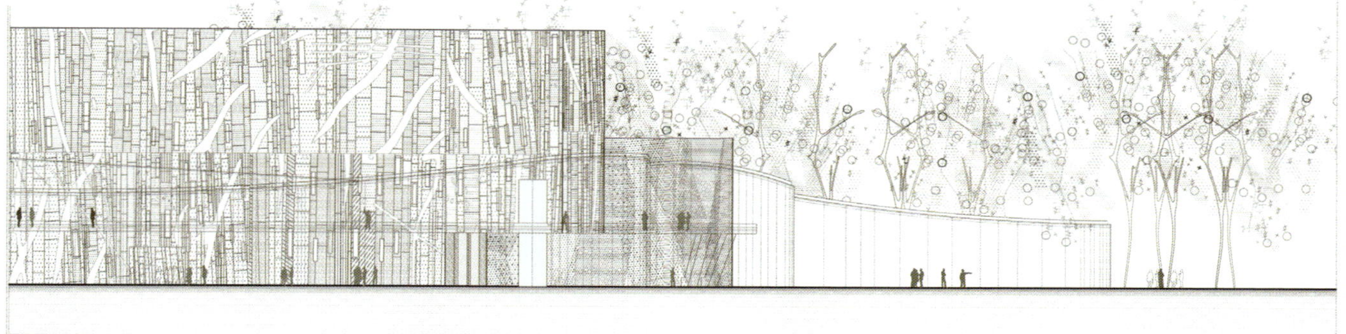

立面图B1-B2 ELEVATION B1-B2

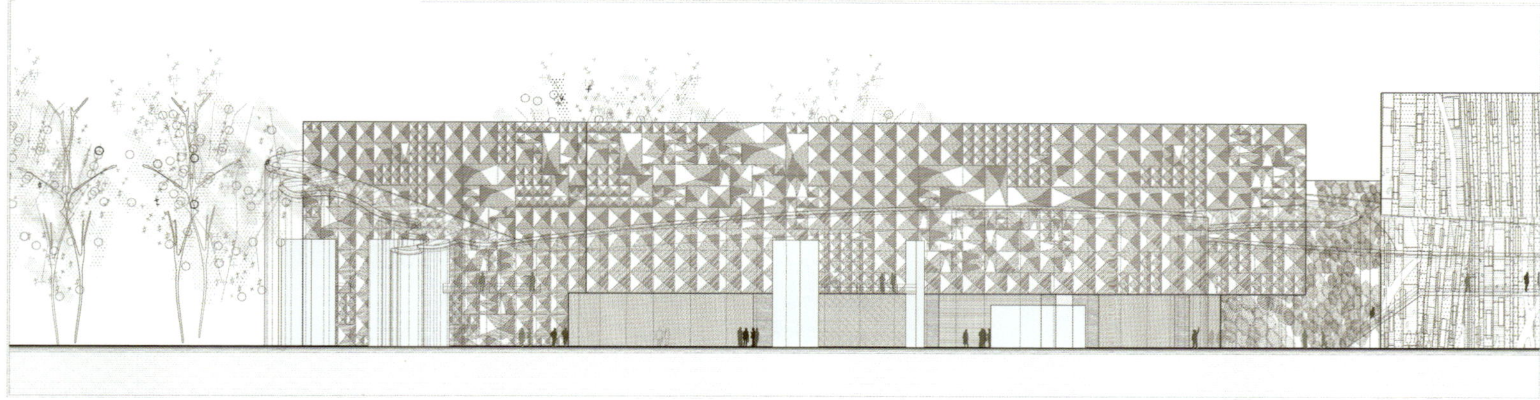

立面图B0-B1 ELEVATION B0-B1

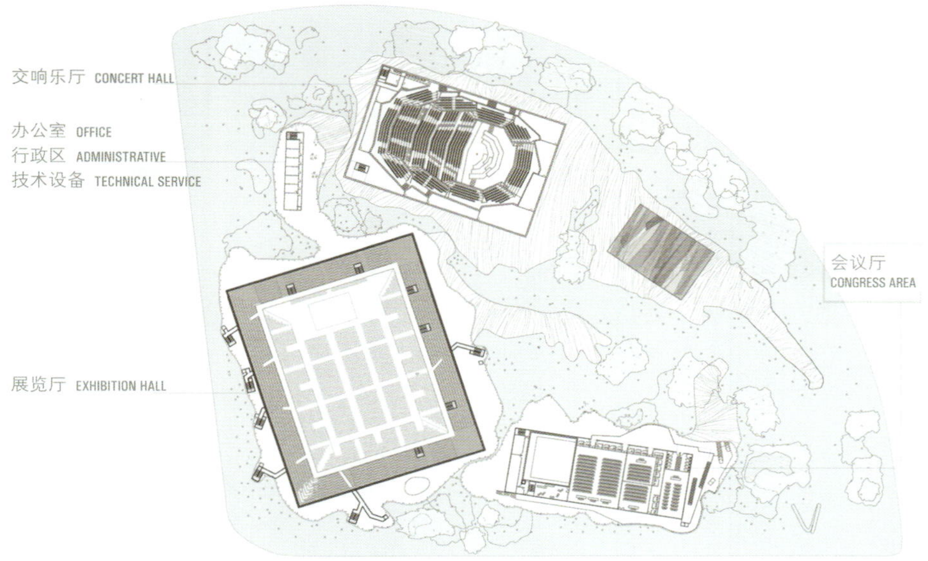

三层平面图 SECOND FLOOR PLAN

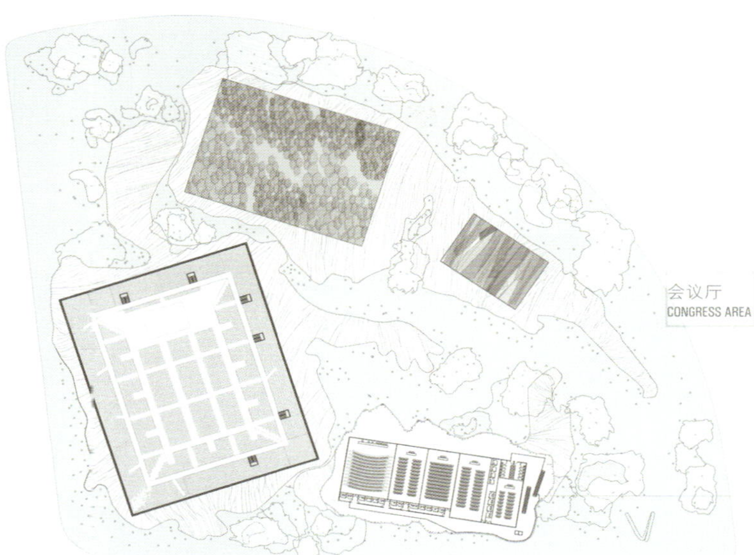

四层平面图 THIRD FLOOF PLAN

音乐会展中心 · 维多利亚加斯特斯

Music, Congress and Exhibitions Palace · Spain

第三提名奖 · Third Accesit

cmArquitectos (建筑师事务所)

Javier Camacho · Mª Eugenia Maciá (建筑师)

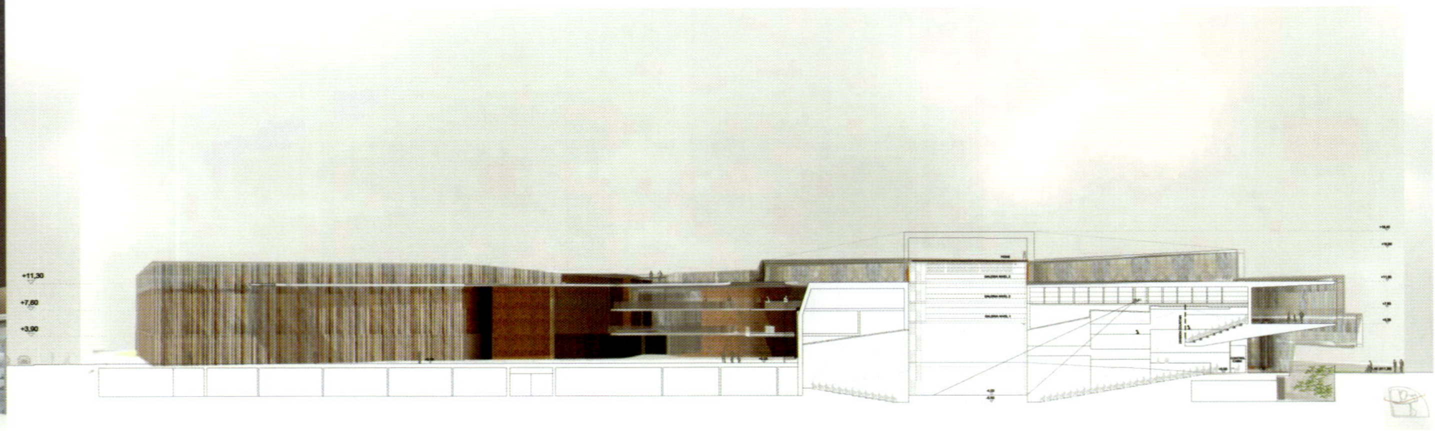

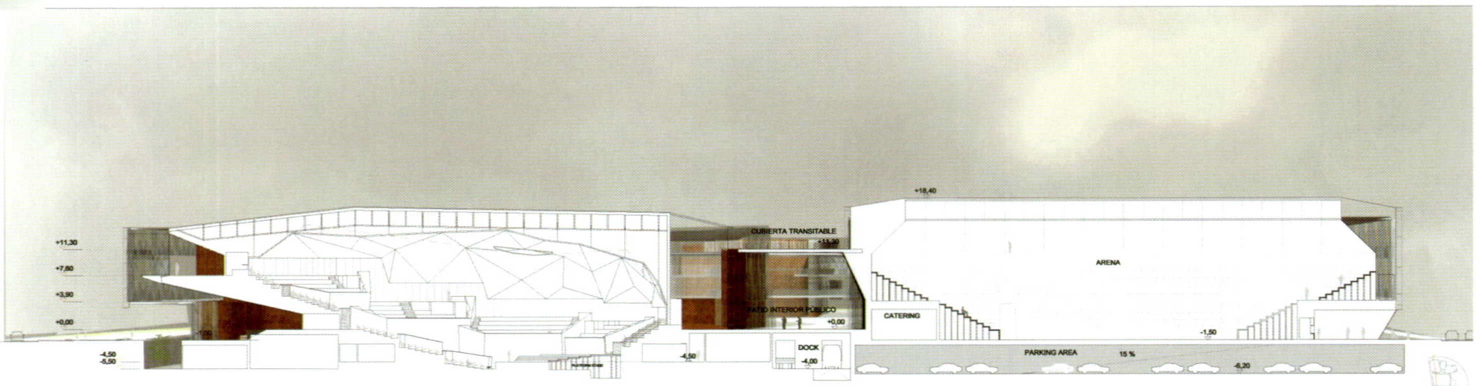

剖面图 SECTIONS

二层平面图 FIRST FLOOR PLAN

四层平面图 THIRD FLOOR PLAN

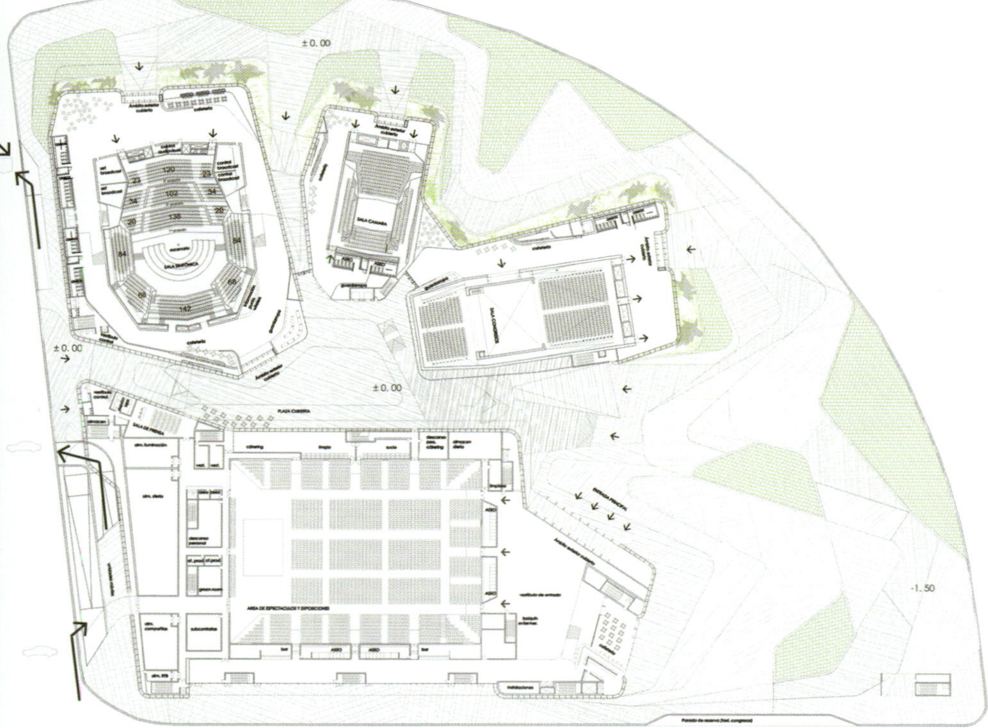

底层平面图 GROUND FLOOR PLAN

三层平面图 SECOND FLOOR PLAN

四个结构体

FOUR VOLUMES

我们刻意避免将新礼堂打造成一个传统意义上的封闭式建筑，相反，竭力将其打造成一个向城市和公众开放的建筑。因此，建筑物的外立面呈蜿蜒曲折状，其间形状各异的空间被设计为建筑的入口和出口，整幢建筑形似城市中一只张开的大手。此外，我们还尝试将这个大空间作为一个带顶的大型广场对外开放，力求在新的室外广场范围内实现空间的流畅性。该建筑拟设计为四个极具雕塑感的大型结构体，由多个平面结构组成，其外立面由精致的玻璃和钢架结构构建而成。

We deliberately avoid the usual picture of the new auditoriums as opaque and closed buildings and we bet on a building open to the city and its inhabitants. Thus the building shapes with a sinuous perimeter, with different voids which invite us to access and scour the building, like a giant open hand to the city. We also propose the possibility to open the big space as a large covered plaza, creating continuity within the new outdoor square. The building is conceived as four large sculptural volumes, formed as faceted masses and wrapped in a delicate skin of glass and steel lattice.

音乐会展中心·维多利亚加斯特斯

Music, Congress and Exhibitions Palace · Spain

JAAM + ARRANZ (建筑师事务所)

入围 · Finalist

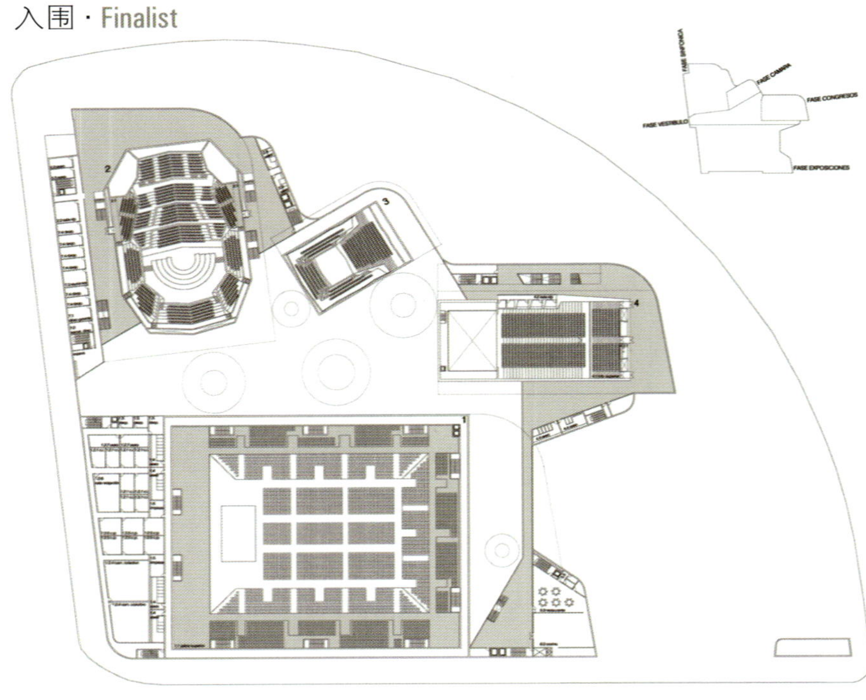

1.展览厅 EXHIBITION HALL
2.交响乐厅 CONCERT HALL
3.表演厅 CHAMBER HALL
4.会议厅 CONGRESS HALL
5.公共设施 COMMON USE
6.设备 INSTALLATION
7.交响乐厅 ADMINISTRATIVE AREA
8.表演者区 ARTIST AREA
9.技术人员区 TECHNICAL AREA
10.雇用迎宾人员区 HOSTESS AREA
11.停车场 PARKING

二层平面图 FIRST FLOOR PLAN

独立建筑

闪亮、透明、洁白的玻璃立面反射出柔和的自然光线和日出、日落时的霞光。夜间，建筑外观灯火璀璨，并且可借助低能耗的LED技术变幻出千变万化的图案。大殿中央是一个大厅，通过天幕采光；大厅里有许多直线元素，赋予了该建筑鲜明的特色。天幕采光符合生物气候学原理，起到了节能和能量收集的作用。

INDEPENDENT CONSTRUCTIONS

The facade of reflecting, translucent and white glass, softly reflects the natural light, the sunrises and sunsets. At night, the skin of the building is evenly lit, enabling countless variations in its appearance using low-power LED technology. The skylights that illuminate the lobbies, in the heart of the Palace, are crowned with vertical elements which as well as completing the identity of the building, are linked to bioclimatic approaches as energy efficiency and collection.

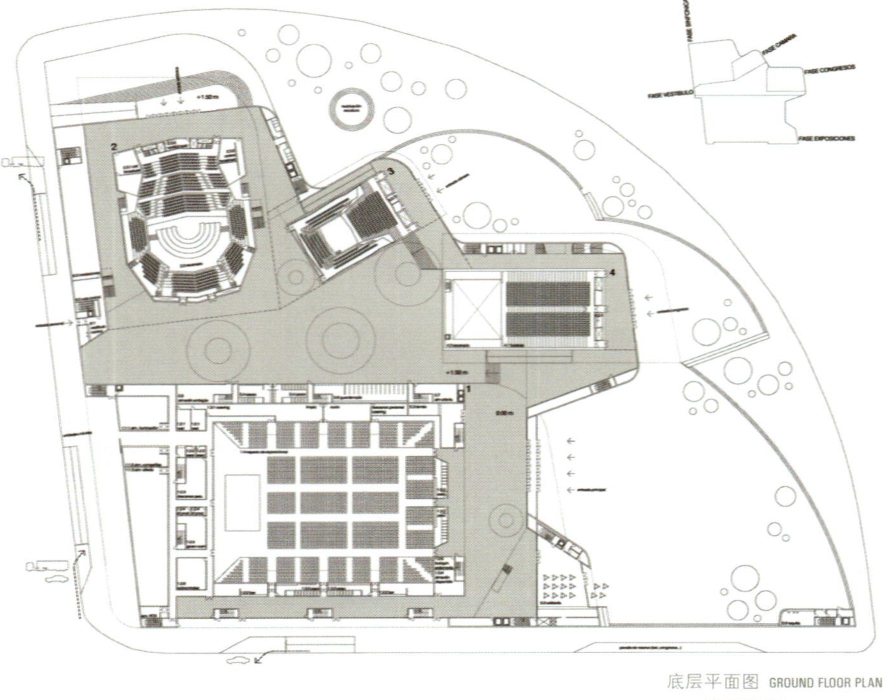

底层平面图 GROUND FLOOR PLAN

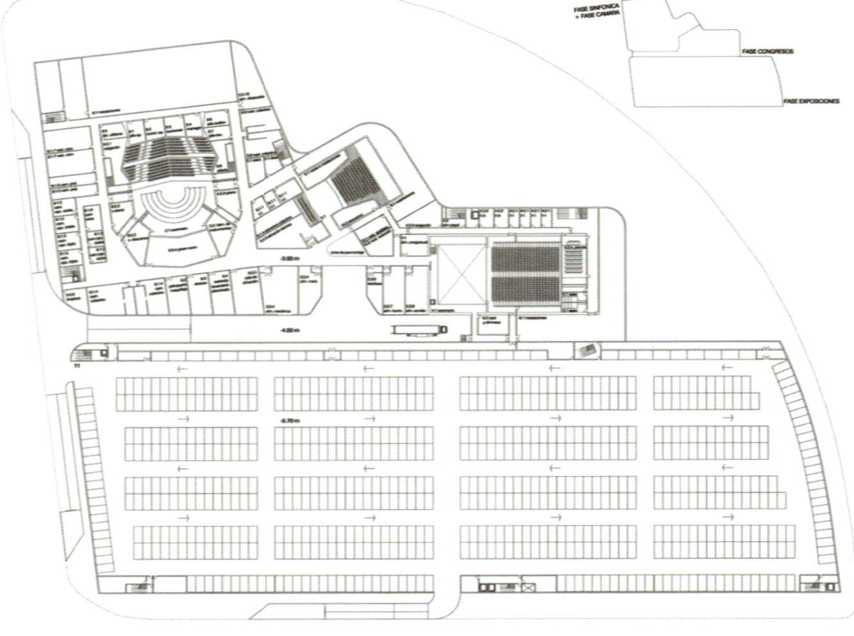

地下一层平面图 BASEMENT FLOOR PLAN

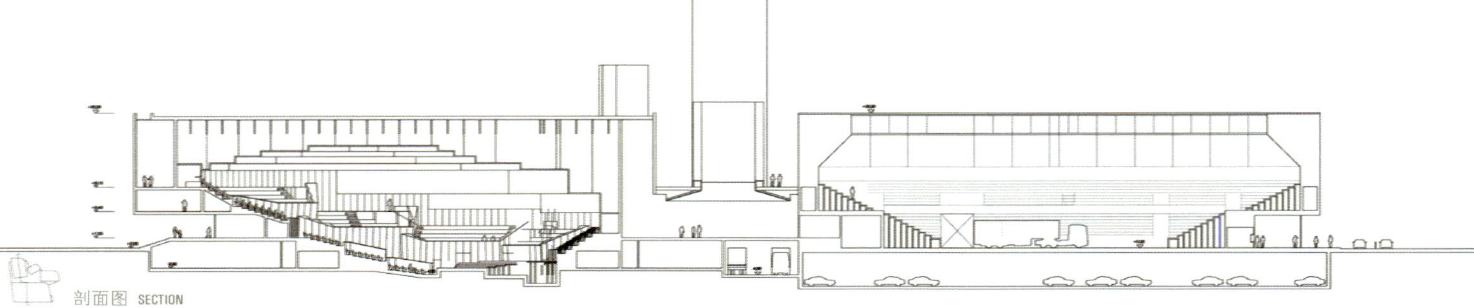

剖面图 SECTION

顶层平面图 ROOF PLAN

三层平面图 SECOND FLOOR PLAN

四层平面图 THIRD FLOOR PLAN

剖面图 SECTION

音乐会展中心 · 维多利亚加斯特斯

Music, Congress and Exhibitions Palace · Spain

selgascano-fhecor (建筑师事务所)

短名单 · Shortlisted

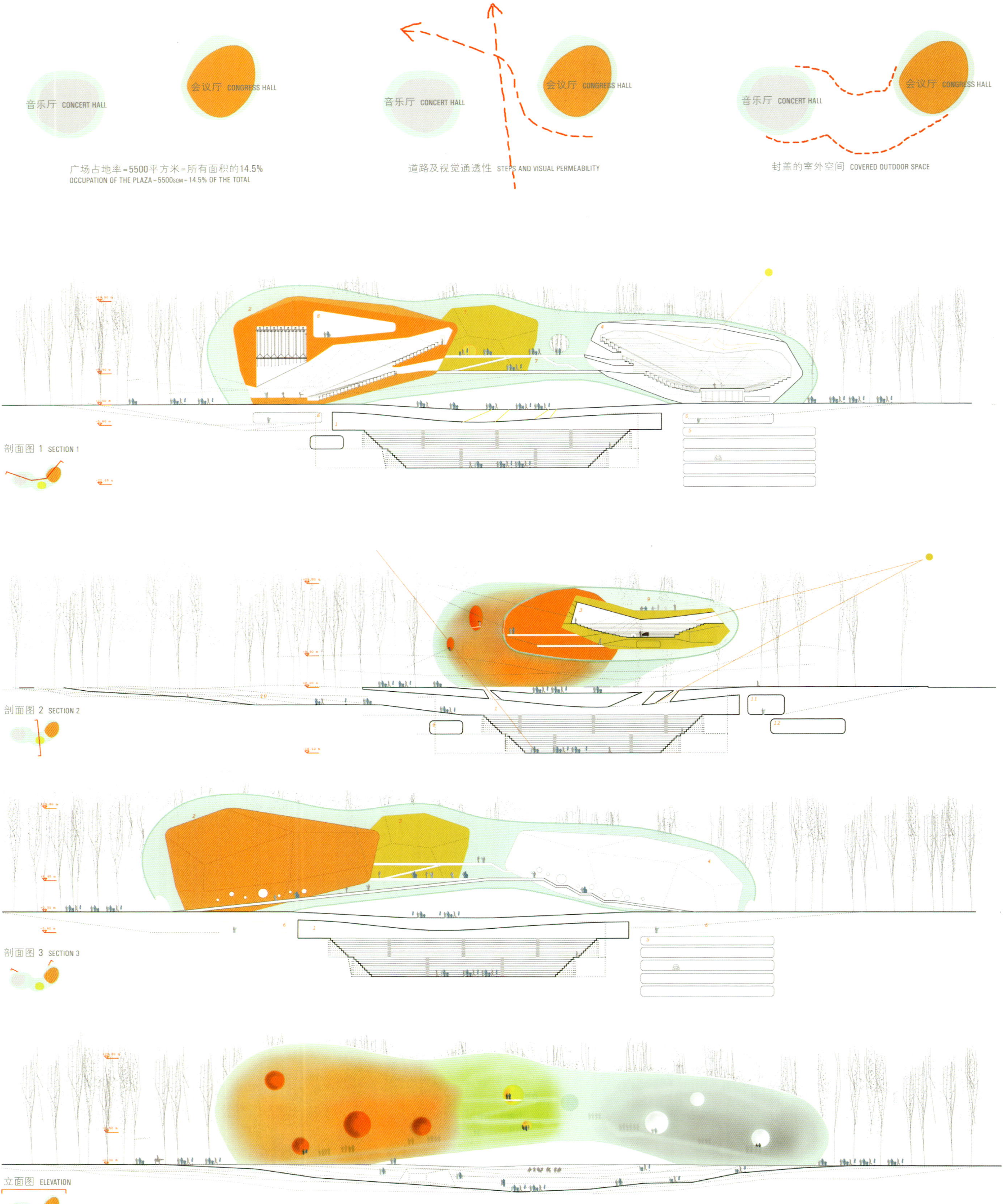

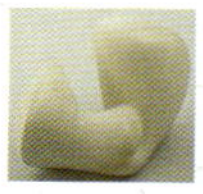

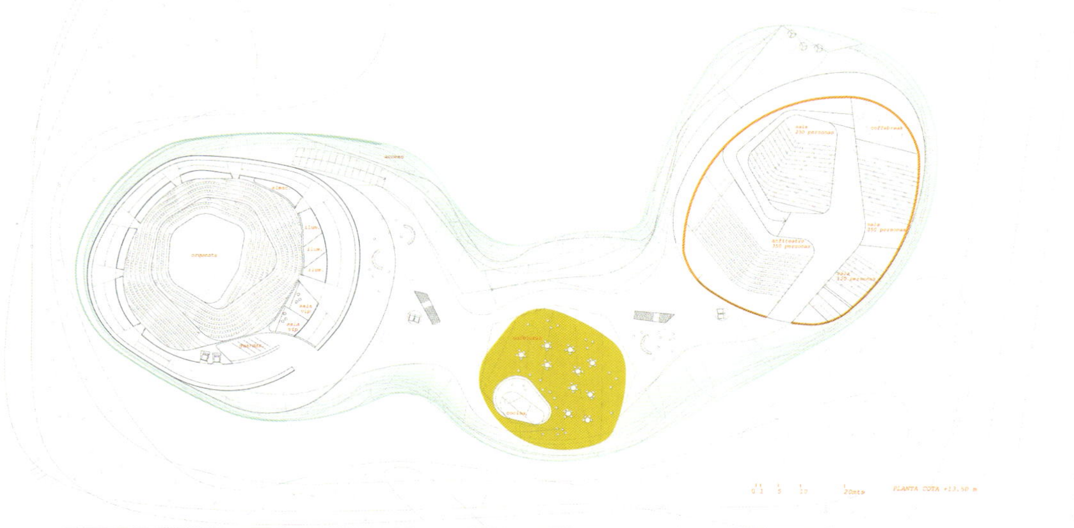

上层平面图标高+13.50 UPPER FLOOR PLAN LEVEL +13.50

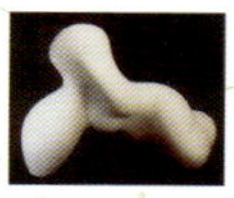

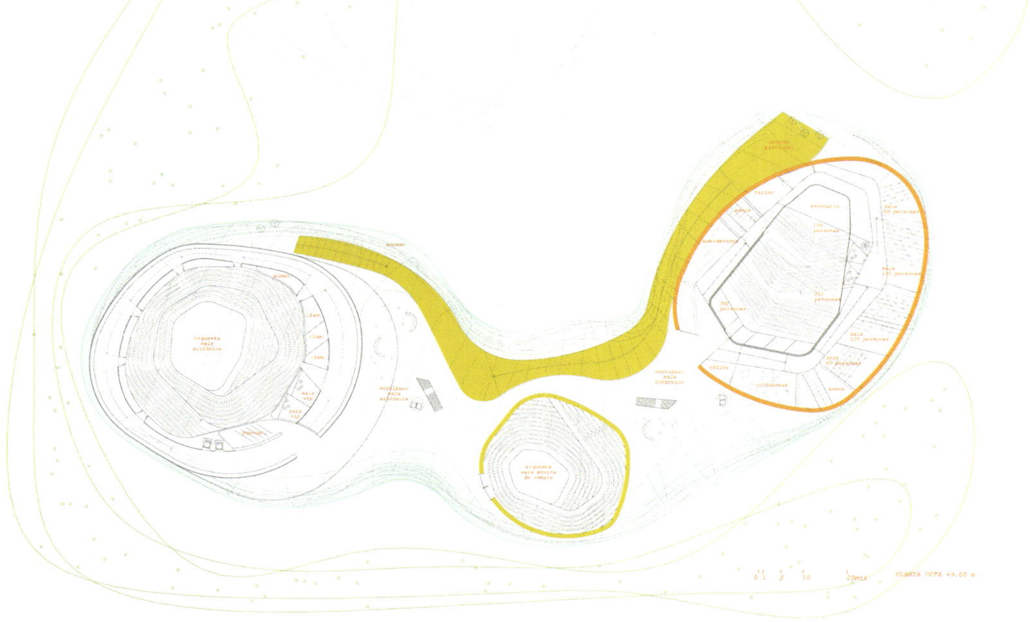

门厅层平面图标高+9.00 FLOOR PLAN FOYER LEVEL +9.00

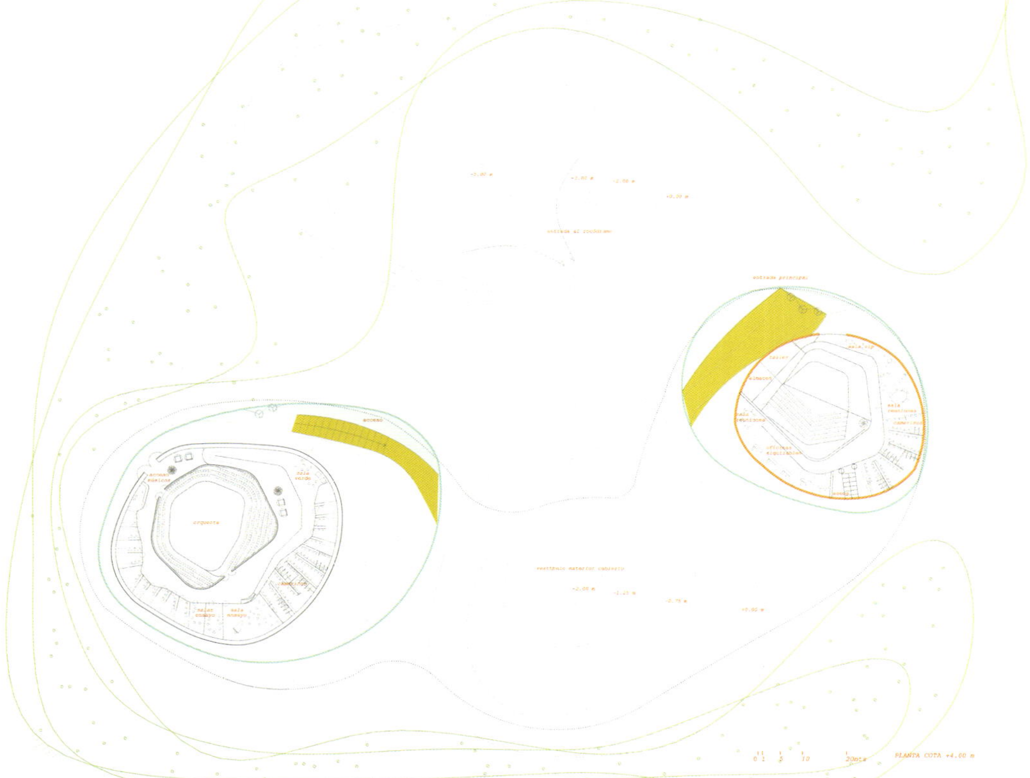

入口层平面图标高+4.00 FLOOR PLAN ACCESS LEVEL +4.00

一片新森林

我们在维多利亚加斯特斯的居民中做了一项调查，调查结果不出所料：绝大多数居民支持修建可穿行的花园、供孩子们玩耍的游乐场地以及方便自由进出的空间。我们的设计区域占地**5,500**平方米，仅占整个公共空间面积（**3.3**万平方米）的**14.5%**，其中**5,000**平方米是带顶的户外空间，即使在雨天也能举行沙龙式音乐会、体育运动及游戏项目；剩下的**2.8**万平方米的面积全部用来栽种杨树。

A NEW FOREST

A survey among residents within the area is predictable: there is priority over the gardens to walk through or to cycle, over playing areas for children and especially over the maintenance of their freedom to move in and through this space which belongs to them. Our proposal covers only 5,500m2 of this public space (14.5%) releasing 33,000m2, where 5,000m2 are outdoor covered exterior spaces to house concerts or sports and games without getting wet. The rest, 28,000m2, are colonized by poplars on a forest.

10 欧罗潘 EUROPAN 10

青年建筑师竞赛
YOUNG ARCHITECTS COMPETITION

欧罗潘的目的在于发挥欧洲年轻建筑师和城市设计专家的重大作用，宣传和进一步探索他们的构想，并为提供地点寻找城市位置迁移的创新建筑和城市方案的各城市和开发者提供帮助。所有参赛者在作品递交截止日时的年龄必须小于40岁。

The objective of Europan is to bring to the fore Europe's young architecture and urban design professionals, and to publicise and develop their ideas. Its objective is also to help cities and developers which have provided sites to find innovative architectural and urban solutions for the transformation of urban locations. All candidates must be under 40 years old on the closing date for submission of entries.

复兴

复兴是一项深思熟虑、组织严密且有意识进行的行为，需要小组成员的共同努力来创造新的文化体系。就“欧罗潘10”而言，里面就提及了选址的问题。这些地点存在的最大转变在于：**对于那些缺乏空间感和社会感的区域来说，如何能使其城市生活变得紧凑？**

城市针灸

针灸疗法通过用针刺入与全身系统关联的一些特定穴位来解决健康方面的大问题。在“欧罗潘10”中，一些选址往往超出了其实际尺寸。从地域的角度看，怎样适时介入才能对城市也产生类似的效用呢？

德绍 · 德国
慕尼黑 · 德国
特隆赫姆 · 挪威
埃曼 · 荷兰

-

公共线性空间

线性空间在现有城市区域结构中发挥着举足轻重的作用，**但是如何基于周边环境开发线性空间使其相互融合，从而彰显线性空间作为公共区域的功能呢？**

卡塞雷斯 · 西班牙

磁极

磁铁是产生磁场、吸引或排斥其他磁铁的材料或物体。在“欧罗潘10”中，一些选址可为打造城市磁铁创造条件。**某一地点需要具备哪些条件（如更改和加强空间的使用、提升公共空间和私有空间的动态感等）来吸引市民以增加人气呢？**

特鲁埃尔 · 西班牙
萨格勒布 · 克罗地亚

revitalization

Revitalization is a deliberate, organized, conscious effort by members of a group to create a new culture. Regarding Europan 10, it relates to the sites where the main question of mutation is: **in spatially and socially disqualified areas, how can urban life be intensified?**

URBAN ACUPUNCTURE

Acupuncture treats large health problems by interventions on some specific points that interact with the global system. Regarding Europan 10, some sites are larger than their real dimensions. **In which way can punctual interventions have an urban effect at the territorial level?**

Dessau · Germany
München · Germany
Trondheim · Norway
Emmen · Netherlands

PUBLIC LINES

Linear spaces can play a strong role in the structure of existing urban areas, but **how can their development be linked to their surrounding, thus reinforcing their role as public spaces?**

Cáceres · Spain

MAGNETIC POLE

A magnet is a material or an object that produces a magnetic field and attracts or repels other magnets. Regarding Europan 10, some sites can provide the opportunity to create urban magnets. **What are the necessary conditions in a specific location - change and reinforcement of uses, increasing of public and private dynamics... - to reactivate a space by attracting citizens?**

Teruel · Spain
Zagreb · Croatia

特隆赫姆 · 挪威
Trondheim · Norway
人口 population: 78.060

德绍 · 德国
Dessau · Germany
人口 population: 77.394

慕尼黑 · 德国
Munich · Germany
人口 population: 1.330.440

埃曼 · 荷兰
Emmen · Netherlands
人口 population:108.863

萨格勒布 · 克罗地亚
Zagreb · Croatia
人口 population: 804.200

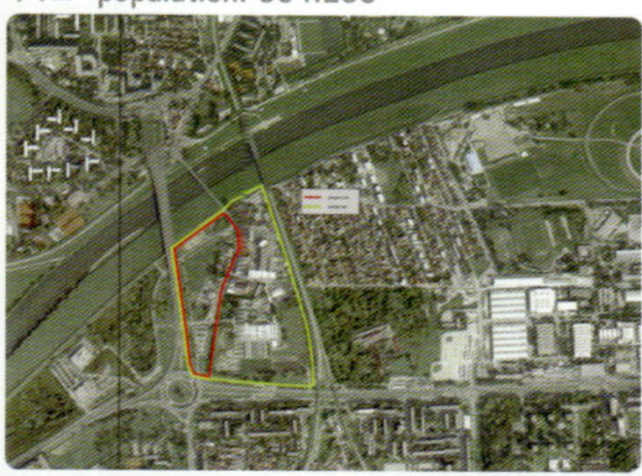

卡塞雷斯 · 西班牙
Cáceres · Spain
人口 population: 93.131

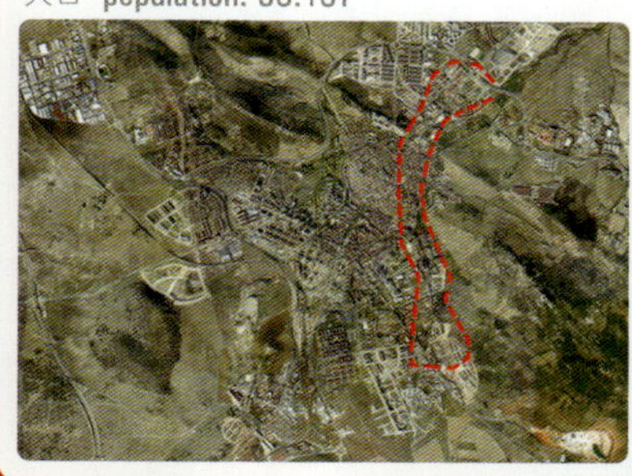

特鲁埃尔 · 西班牙
Teruel · Spain
人口 population: 35.396

德绍·德国 (竞标代码) Roll in! Dessau · Germany

zectorarchitects
(建筑师事务所)
Norbert Kling · Carsten Jungfer (建筑师)
中标 winner

全新联系和关联

按照传统，在正式场合人们总会铺上红地毯来给重要人物引路。同样在举行活动时，红地毯也是个适配工具，它能将任何指定的地点转变为一个特殊场所。

本次大赛将贯穿南北的交通干道Kavalierstraße发展成了一个主要活动场所。设计方案提出了一个策略，但并未做出任何确定的计划。它为各种事件、活动和情境提供了平台，能使这个城市变得朝气蓬勃。该理念是基于“活动格局”的城市概念形成的，城市概念的重点是“红地毯策略”。城市中心的格局不再由Kavalierstraße上的车水马龙来划分，取而代之的是“地毯”界限。“地毯”既包括暂时元素也包括永恒元素。“地毯”及附近的定位点将有助于各项活动的开展、积聚和宣传，是人们消磨时间、交谈或放松的好地方。

NEW LINKS AND CONNECTIONS

A red carpet is traditionally used to mark the route for important people on formal occasions. In combination with the event, the red carpet acts as an adaptive device that transforms any given location into a special place.

The competition brief identifies Kavalierstraße, currently a major North-South running traffic artery, as the main zone for interventions.The proposed scheme provides a strategy but does not make fixed propositions. It offers platforms for events, activities and situations which could become energy generators for the city themselves. This concept is based on the notion of the city as "pattern of events". The concept key is the red "carpet strategy". Instead of heavy traffic rolling through Kavalierstraße dividing the core of the city, "carpets" are now rolled into and across this former boundary. The "carpets" consist of both, temporary and permanent elements. The "carpets" and adjacent anchor points will help to enable, accumulate and radiate activities, thus become a place where people spend time, talk, or just relax.

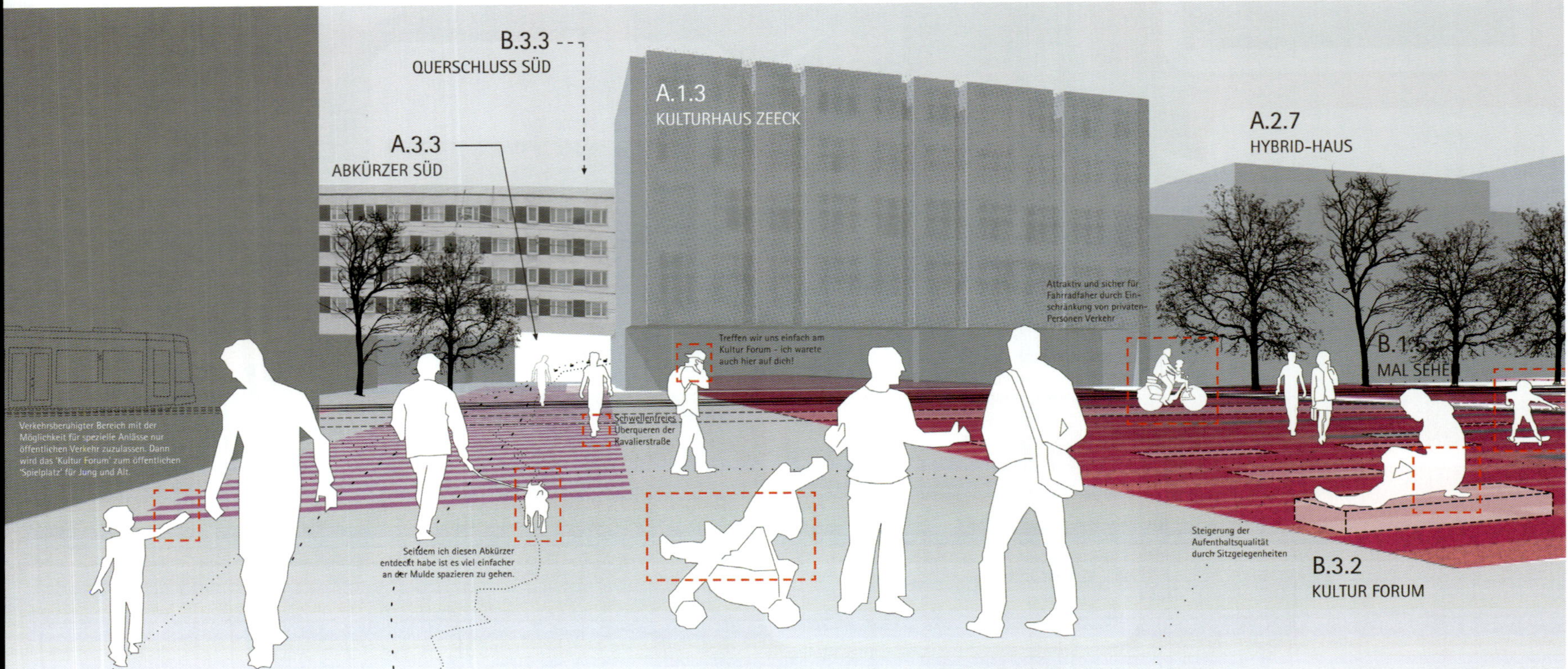

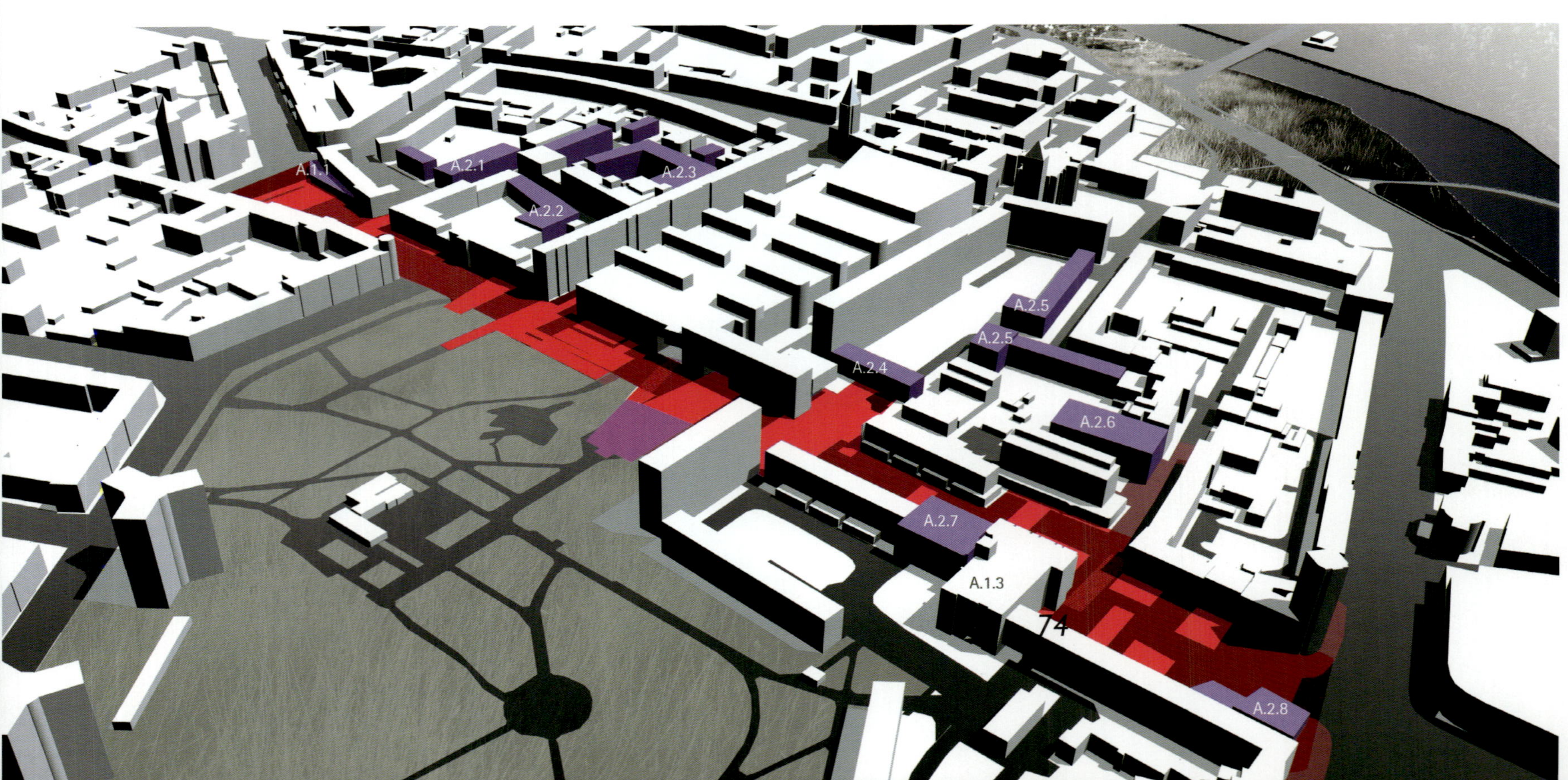

A.2 现有建筑的改善 IMPROVEMENT OF EXISTING STOCK
A.1.1 前沿 PRELUDE
A.1.2 啤酒厂 BREWERY QUARTER
A.1.3 文化·仓储区域 CULTURE-STORE ZEECK

A.2 致密化 DENSIFICATION
A.2.1 开放区块 OPEN PERIMETER BLOCK
A.2.2/A.2.3 新啤酒厂 NEW BREW
A.2.4 边缘空间 EDGE SPACE
A.2.5 齐格扎格 ZIG ZAG
A.2.6 运动中心 SPORTS HUB
A.2.7 混合建筑 HYBRID BUILDING
A.2.8 前沿 PRELUDE

A.3 连接性 CONNECTIVITY
A.3.1 北面捷径 SHORT-CUT NORTH
A.3.2 穿孔 PERFORATION
A.3.3 南面捷径 SHORT-CUT SOUTH

B.1 地面的高架 SURFACE UPGRADE
B.1.1 街角公园 PARK
B.1.2 通向公园的通道 STEP INTO THE PARK
B.1.3 交通的中央核心 CENTRAL TRANSPORT HUB
B.1.4 公园捷径 SHORT-CUT PARK
B.1.5 让我们看！ LET'S SEE!
B.1.6 公寓庭院 RESIDENTS YARD

B.2 致密化 DENSIFICATION
B.2.1 新集市 NEW MARKET
B.2.2 透明公园 TRANSPARENT PARK

B.3 连接性 CONNECTIVITY
B.3.1 公众舞台 OPEN STAGE
B.3.2 文化论坛 CULTURAL FORUM
B.3.3 南面横向连接 CROSS LINK SOUTH

C.1 高架 UPGRADE
C.1.1 临时用户 TEMPORARY USER
C.1.2 体育·保健 SPORT · HEALTH

C.2 致密化 DENSIFICATION
C.2.1 露天影院 OPEN-AIR CINEMA

C.3 连接性 CONNECTIVITY

...可卷起的地毯！ ...ROLL-IN! CARPET
更多...可卷起的地毯！ MORE...ROLL-IN! CARPETS
KAVALIERSTRASSE相关区域 KAVALIERSTRASSE INTERVENTION ZONE
NEUEMARKT-ZEECK相关区域 NEUEMARKT-ZEECK INTERVENTION ZONES
原有和加强后的链接 EXISTING AND STRENGHTENED LINKS
公共绿地 PUBLIC GREEN SPACE
计划的填充发展 PROPOSED INFILL DEVELOPMENTS
项目区域 PROJECT AREA

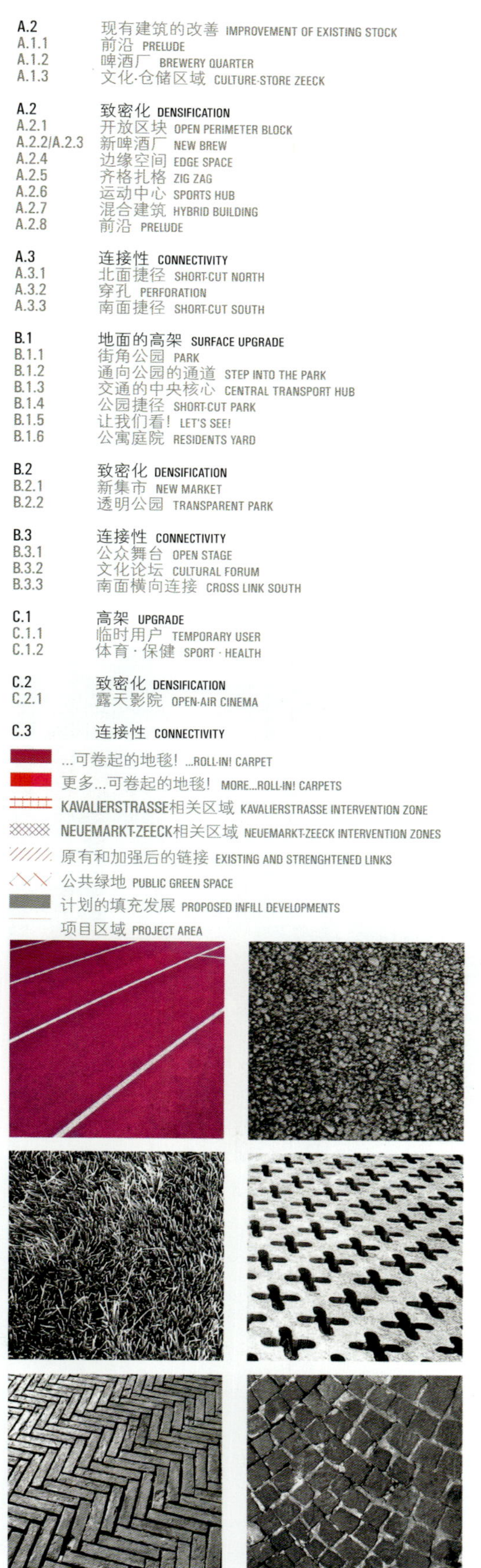

按照分期开发和公众参与的策略，“地毯”的精确设计将取决于所有入住这里的居民。对那些永久性固定的外观元素来说，在材料上的选择余地更大。坚硬而陈旧的铺砌面将和绿化景观、街道设施以及照明相得益彰。临时的“地毯”将采用一些非材料元素打造，如照明或自动清除的涂料表面。

Following the strategy of a phased development and of public participation, the precise design of the carpets will be determined by the input of all house-holders involved. For the more permanent surfaces there is a wide choice of materials. Hard wearing paving could be combined with planting, street furniture and lighting. A temporary carpet can be also defined by means that are non-material, like lighting or self removing paint surfaces.

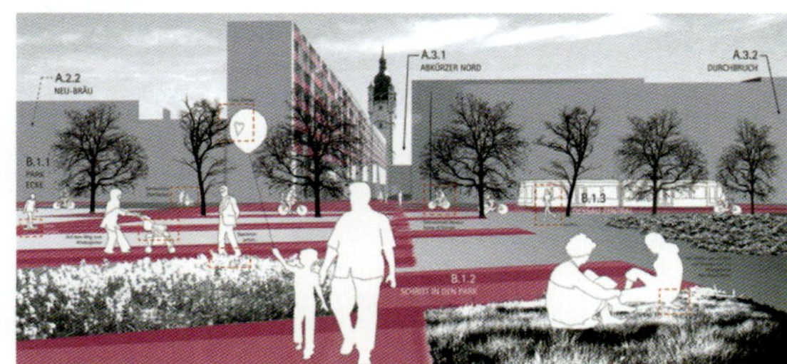

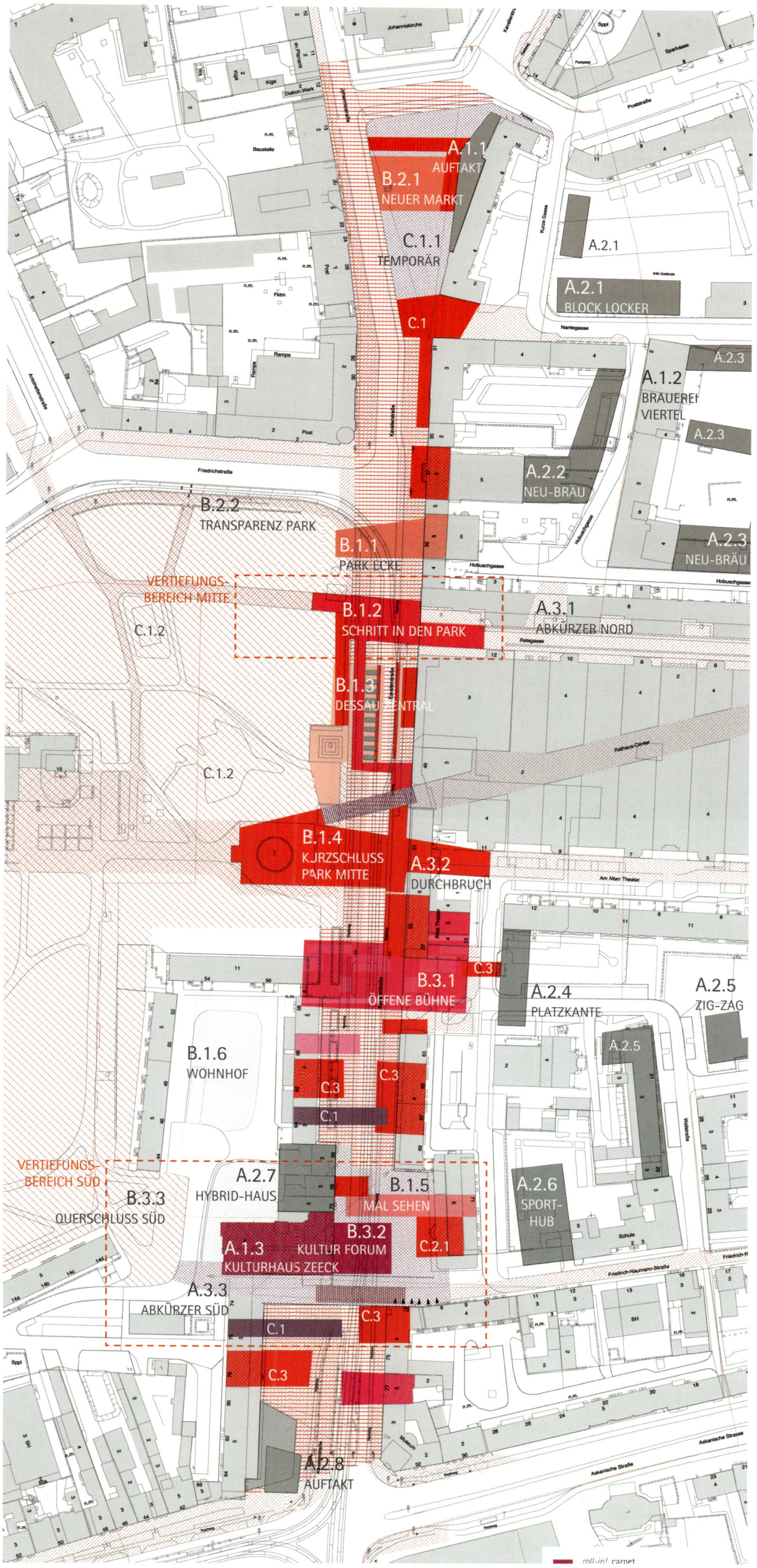

(竞标代码) Combined Worlds

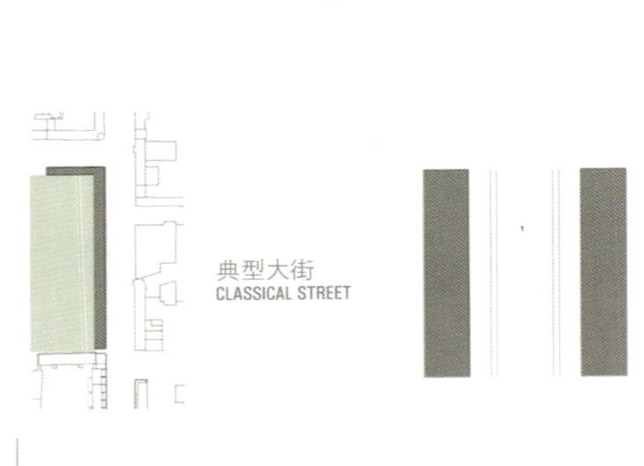

Tiago Cardoso Tomás
(建筑师)
合作 (c) Nuno Mesquita · Paul Roy

二等奖 runner-up

复兴的都市生活方式

RENEWED URBAN LIFESTYLE

通过利用现有元素和新元素对街道加以处理来实现Kavalierstraße的复兴。从改造后的Neumarkt广场开始，街道将微微曲折地向前延伸开去，构造出大小各异的人行道，从而形成社交活动集中的指定场所。街道视角也各不相同，大到宽阔的都市街道，小到仅两米宽、如步行道一般的神秘森林。我们提议翻新街道两边的建筑物并增加建筑密度，以便清楚地重新划定界限。

The Kavalierstraße revitalization is achieved through the manipulation of the street's perception using existing and new elements. Departing from the reformulated Neumarkt square, the street will be developed in a gentle curve succession, generating variable sized sidewalks to form selected poles of concentrated social activity. The street view will vary from a wide urban street, to a 2m-wide enigmatic forest like footpath. We propose the progressive building's refurbishment and densification along the street in order to redefine clearly the limits.

典型大街
CLASSICAL STREET

方案-复合世界(城市+城市公园)
PROPOSAL-COMBINED WORLDS (CITY+URBAN PARK)

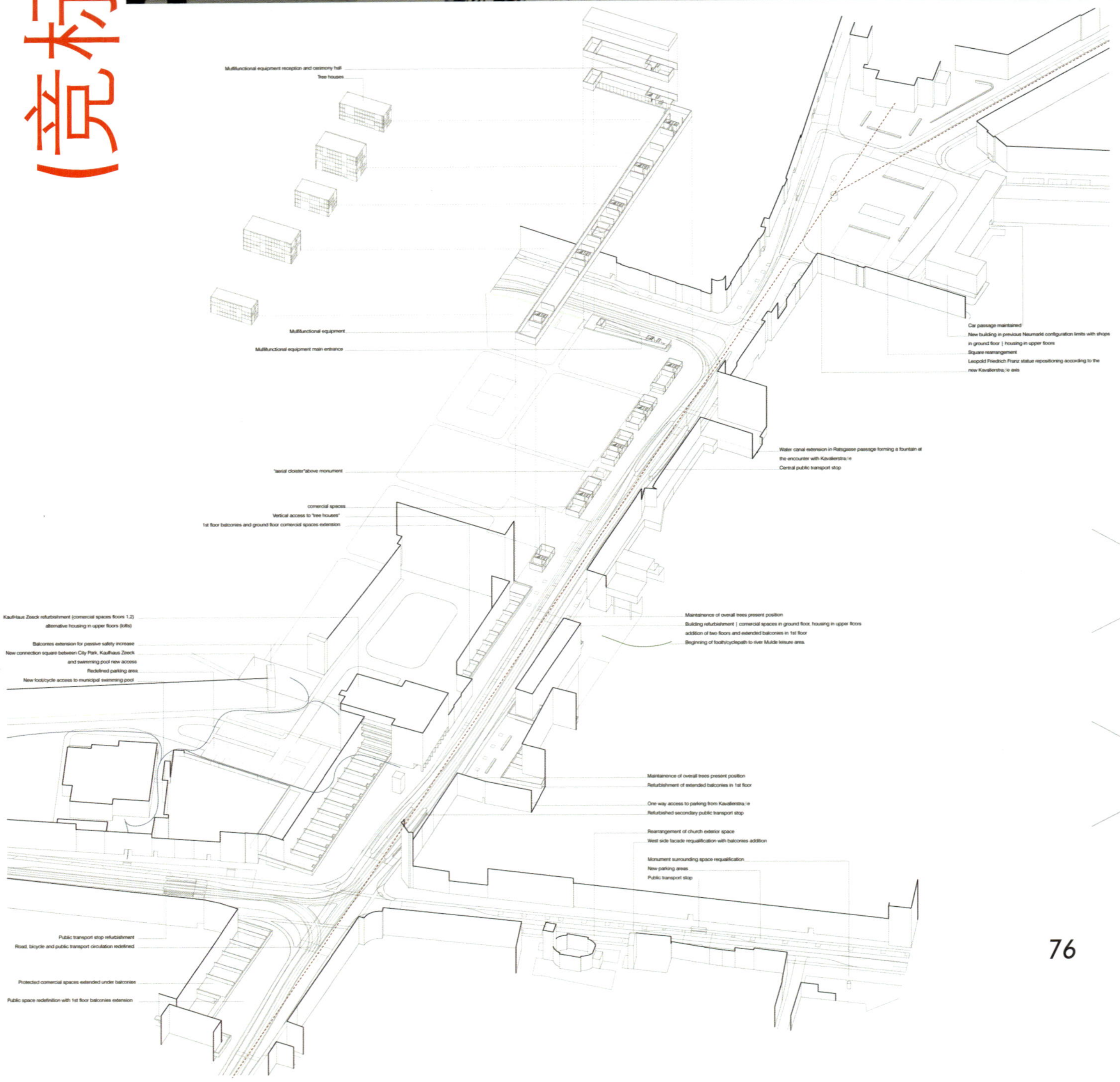

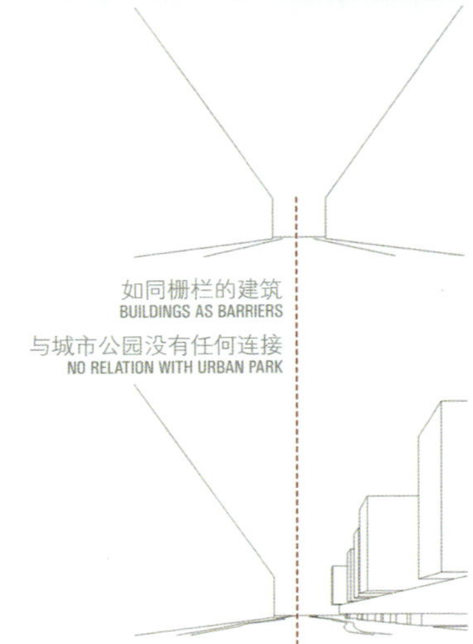

如同栅栏的建筑
BUILDINGS AS BARRIERS
与城市公园没有任何连接
NO RELATION WITH URBAN PARK

可穿透建筑
PERMEABLE BUILDINGS
与城市公园敞开连接
OPEN RELATION WITH URBAN PARK

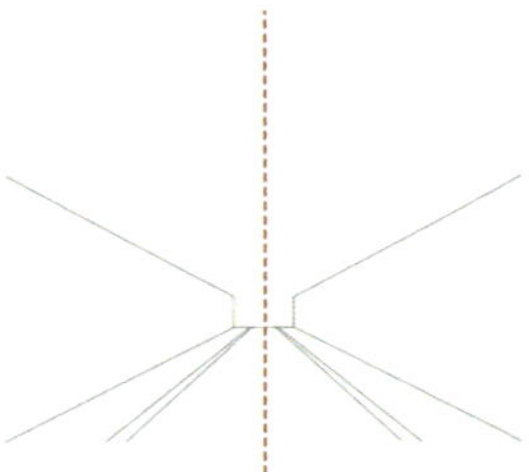

限制的使用可能
LIMITED USE POSSIBILITIES
城市环境不可变+无等级
INVARIABLE+NON-HIERARCHIZED URBAN ENVIRONMENT

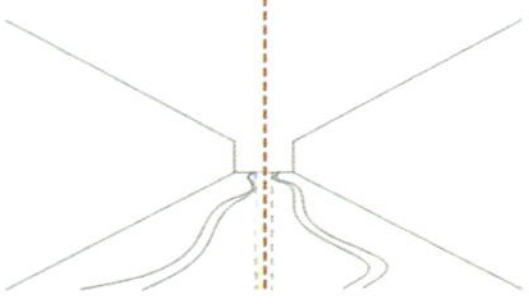

附加的使用可能
ADDITIONAL USE POSSIBILITIES
公共空间的可变
VARIABLE SIZE OF PUBLIC SPACE

剖面图 AA · 新集市
SECTION AA · NEW MARKET
剖面图 BB · 文化场所的主入口
SECTION BB · CULTURAL SPACES MAIN ENTRANCE
剖面图 CC · “植物房” · 瑞兹盖斯通道
SECTION CC · "TREE HOUSES" · RATSGASSE PASSAGE
剖面图 DD · 纪念碑 · 市政厅
SECTION DD · MONUMENT · CITY HALL
剖面图 EE · 城市公园 · 休闲·文化
SECTION EE · URBAN PARK · LEISURE-CULTURE
剖面图 FF · 城市公园 · 重修后的泽克中心
SECTION FF · CITY PARK · REFURBISHED KAUFHAUS ZEECK
剖面图 GG · 通往泳池道路 · 重修后的住宅楼
SECTION GG · PATH TO SWIMMING POOL · REFURBISHED HOUSING BUILDINGS
重修后的楼房里的住宅和商铺 HOUSING AND SHOPS IN REFURBISHED BUILDINGS
重修后的墙面 · 添加的阳台 REFURBISHED FAÇADE · BALCONY ADDITIONS
新住宅楼 NEW HOUSING BUILDINGS
多功能建筑 · 公共活动 MULTIFUNCTIONAL BUILDING · PUBLIC EVENTS
码头和私人花园 DECK AND PRIVATE GARDEN
新停车场 NEW PARKING SPACES
新公共空间 NEW PUBLIC SPACES
喷泉 FOUNTAIN
自行车道 CYCLEWAY
自行车停车区域 BICYCLE NEW PARKING AREA
公共交通站 STOP PUBLIC TRANSPORT
电车 TRAMWAY
工作室-项目区域 STUDY-PROJECT AREA
重新放置的雕塑 REPOSITIONED STATUE-SCULPTURE

慕尼黑·德国 Munich · Germany

(竞标代码) Fresh ideas

Markus Rudolph · Alexander Kneer

(建筑师)

并列二等奖 runner-up ex-aequo

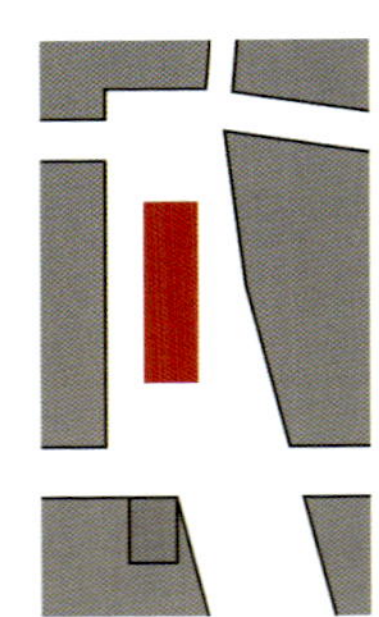

单独建筑体 SOLITARY CONSTRUCTION BODY

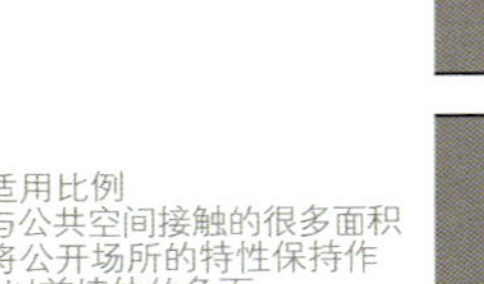

*适用比例
*与公共空间接触的很多面积
*将公开场所的特性保持作为以前墙体的负面

*APPROPIATE SCALE
*A LOT OF CONTACT SURFACE WITH THE PUBLIC SPACE
*KEEPS THE IDENTITY OF THE OPEN SPACE AS A NEGATIVE OF THE FORMER CITY WALL

分割 SEGMENTATION

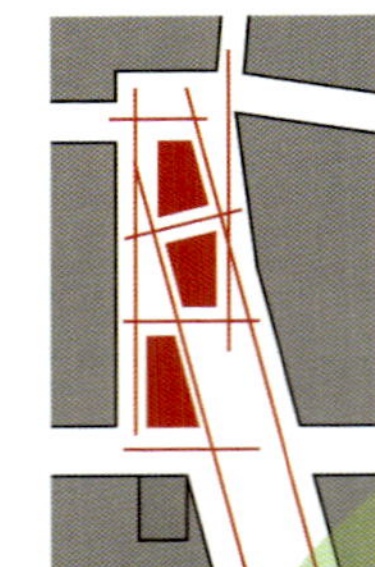

*公开场所方向的归总(以前墙体)
*交叉连接

*INCLUSION OF THE DIRECTIONS OF THE OPEN SPACE (FORMER CITY WALL)
*CROSS CONNECTIONS

渗透性 PERMEABILITY

*作为公共磁场
*二层平面功能布局: 学生+教师的展示平台
*可选择的·改变的: 商业用途及活动、餐饮、及商店、展览

*SERVES AS PUBLIC MAGNET
*PROGRAM FIRST FLOOR: CENTRAL PRESENTATION PLATFORM FOR ALL STUDENTS+FACULTIES
*ALTERNATIVELY · ALTERNATING: COMMERCIAL USES-EVENTS, GASTRONOMY, SHOP, EXHIBITION

MUC学生论坛 MUC STUDENT FORUM

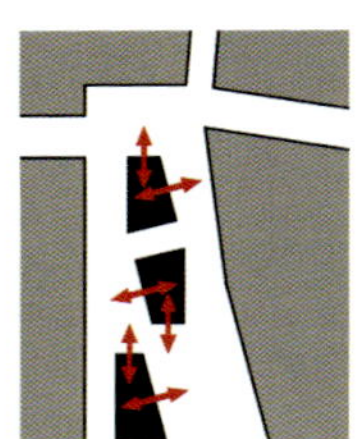

*强力的指导及开放面向公共场所的功能空间
*为所有行人提供的服务

*STRONG ORIENTATION AND OPENING OF SPECIAL USES TOWARDS THE OPEN SPACE
*OFFER FOR ALL PEDESTRIANS

与外部空间的联系 OUTDOOR SPACE RELATIONSHIPS

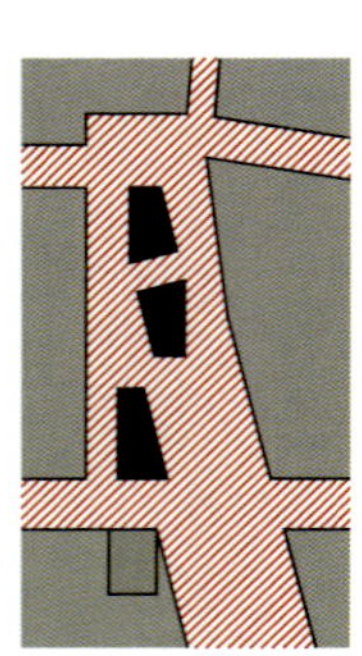

复兴 REVITALIZATION

界面

INTERFACE

该项目提出将MUC学生论坛作为该地区的一个磁体和催化剂。这将是该城市所有大学和学院全体师生的新兴中心演讲论坛。之前旧城墙的位置被打造成了一个开放的空间，其中，新开发的独立空间使得这个独立空间仍可被视作一个连续的开放空间。建筑位于视角主轴线上，外立面上覆有植被，是绿化走廊的一部分，达到了通透的效果。

The project proposes the MUC Student Forum as a magnet and a catalyst for the area. It shall be a new, central presentation forum for the different faculties of all universities and academies spread over the city. The new solitary volumes ensure that the open space, formerly the place where the old city wall stood, can still be perceived as a continuous open space. The buildings are related to main view axes and provide permeability while becoming part of the green corridor through its plant-clad facades.

利用底层平面的改变来促进城市空间 IT USES THE CHANGE OF THE GROUND FLOOR PLAN TO LIVEN THE URBAN SPACE UP

场景 1 · 学生—联合国活动展览 SCENARIO 1 · STUDENTS-EXHIBITIONS OF UNITED NATIONS EVENTS

场景 2 · 学生—烹饪活动展览 SCENARIO 2 · STUDENTS-EXHIBITIONS OF GASTRONOMIC EVENTS

场景 3 · 公司及餐饮商业用途 SCENARIO 3 · COMMERCIAL USE OF FIRMS AND GASTRONOMY

一个更有力的连接 A STRONGER CONNECTION

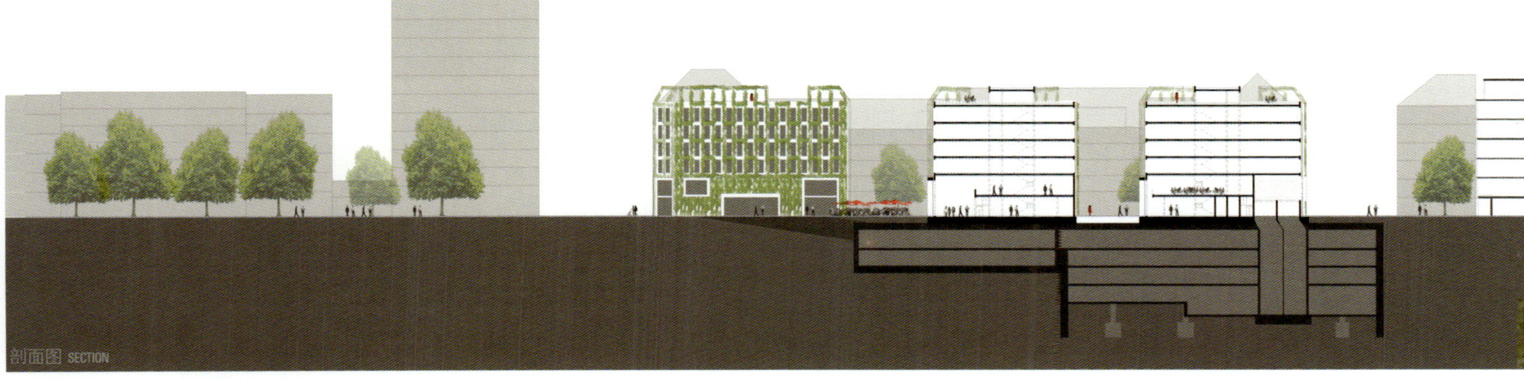

剖面图 SECTION

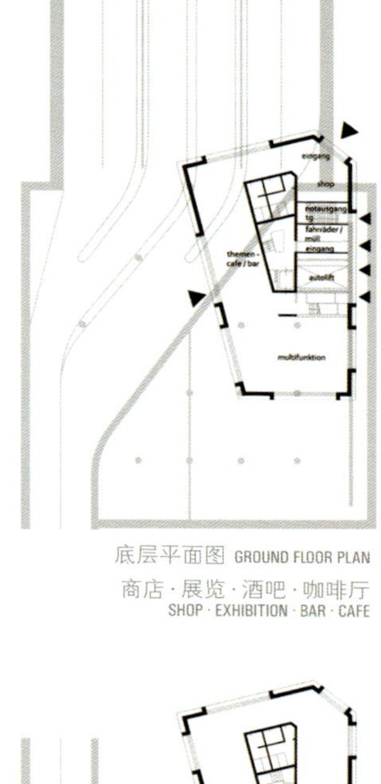

底层平面图 GROUND FLOOR PLAN

商店 · 展览 · 酒吧 · 咖啡厅

SHOP · EXHIBITION · BAR · CAFE

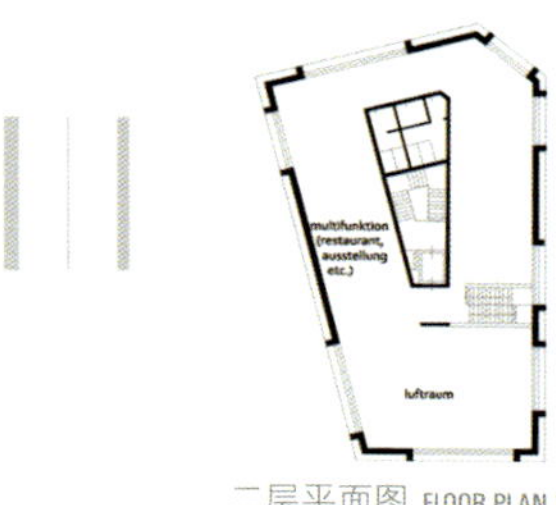

二层平面图 FLOOR PLAN 1

商店 · 展览 · 餐厅

SHOP · EXHIBITION · RESTAURANT

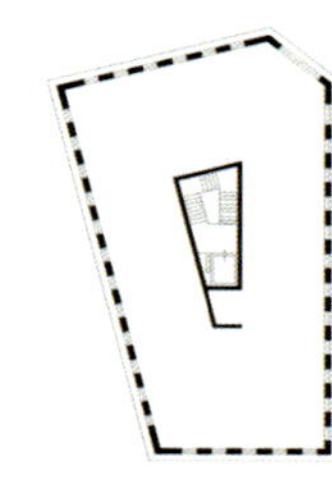

三至六层平面图 FLOOR PLAN 2-5

机动楼层用于住宅 · 学生公寓 · 办公室

FLEXIBLE FLOOR PLAN FOR HOUSING · RESIDENCE FOR STUDENTS · OFFICES

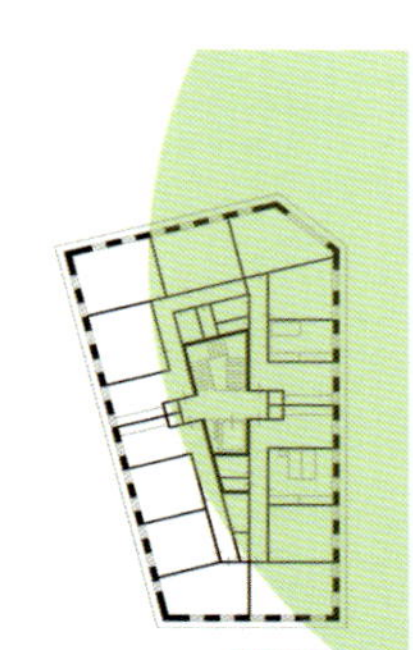

住宅层 HOUSING FLOOR PLAN

1–4个卧室机动布置

FLEXIBLE OF 1-4 BEDROOMS

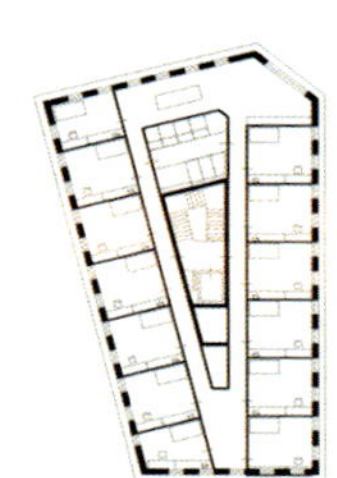

学生公寓 RESIDENCE FOR STUDENTS

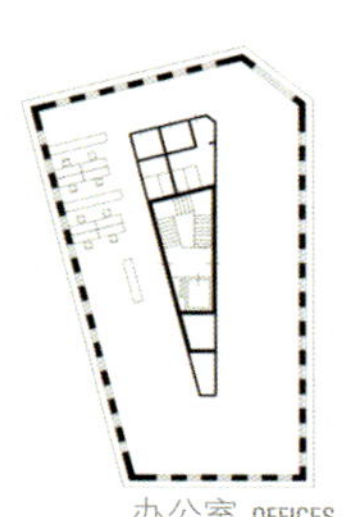

办公室 OFFICES

(竞标代码) The House

Wolfgang Zeh · Moritz Kaiser · Kai Beck
(建筑师)

并列二等奖 runner-up ex-aequo

城市空间整合 SPATIAL URBAN INTEGRATION

概念图 CONCEPT IMAGE

Stachus / Karlstor
STACHUS广场 · KARLSTOR STACHUS PLAZA · KARLSTOR

Herzog-Wilhelm-Straße
HERZOG-WILHELM大街 HERZOG-WILHELM STREET

Grünraum / Sendlinger Tor
GRÜNRAUM · SENDLINGER门 GRÜNRAUM · SENDLINGER DOOR

Grünraum / Blumenstraße
GRÜNRAUM · BLUMEN大街 GRÜNRAUM · BLUMEN STREET

Blumenstraße
BLUMEN大街 BLUMEN STREET

Getreidemarkthalle
玉米市场 CORN MARKET HALL

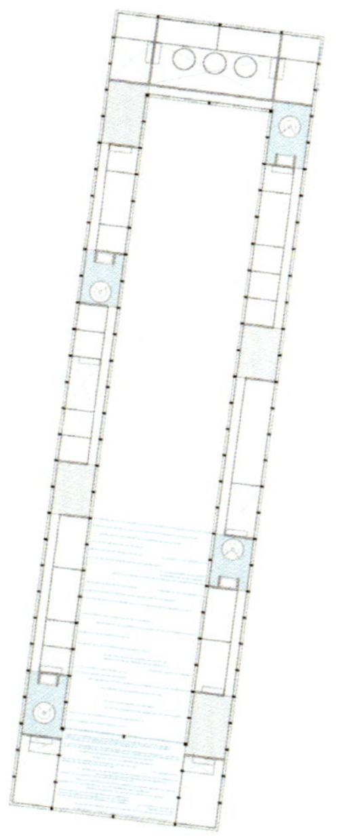
三层平面图 办公 FLOOR PLAN 2 WORK

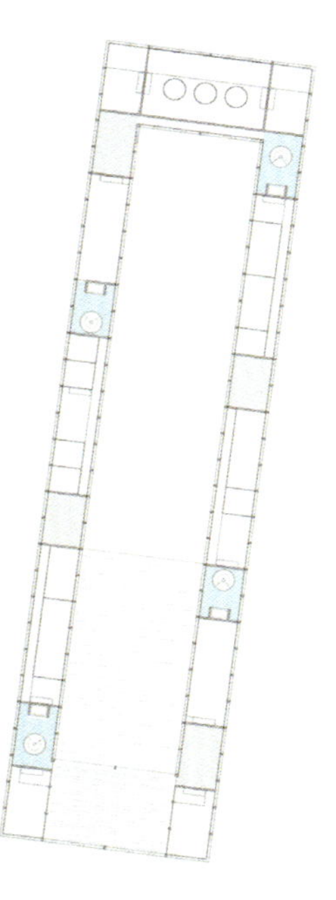
四层平面图 办公 FLOOR PLAN 3 WORK

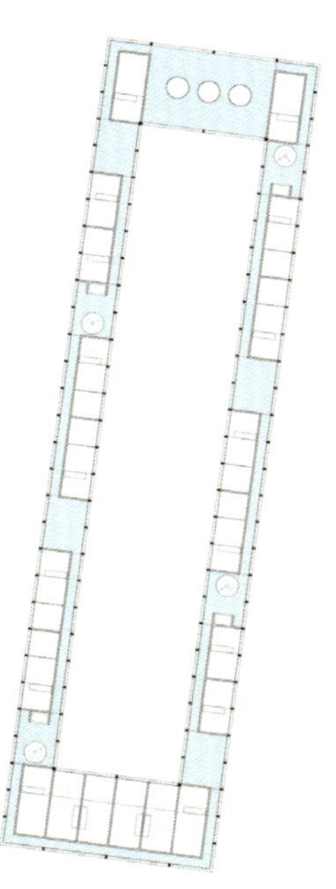
五层平面图 住宅 FLOOR PLAN 4 HOUSING

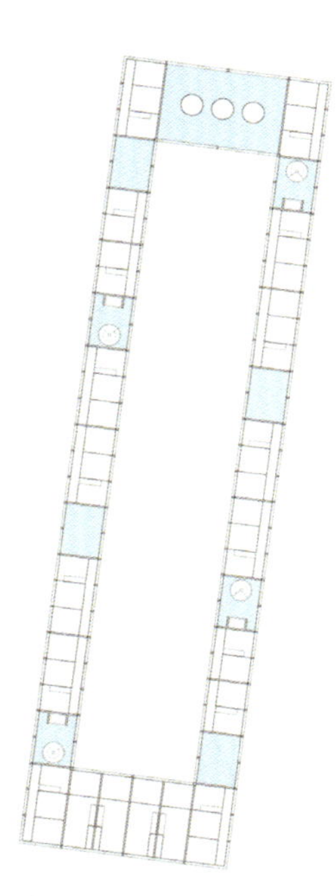
六层平面图 住宅 FLOOR PLAN 5 HOUSING

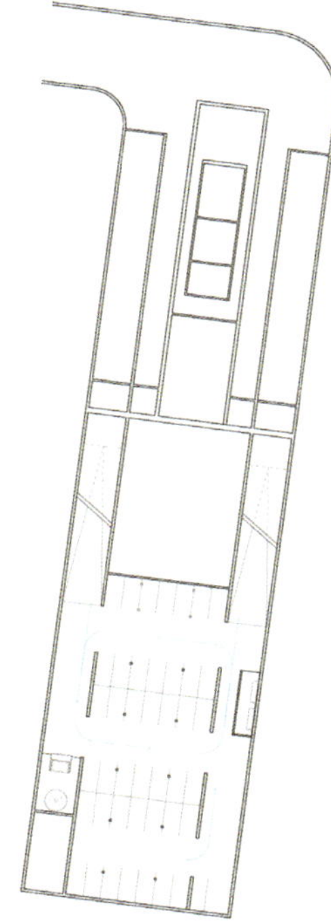
地下一层平面图 停车场 FLOOR PLAN -1 PARKING

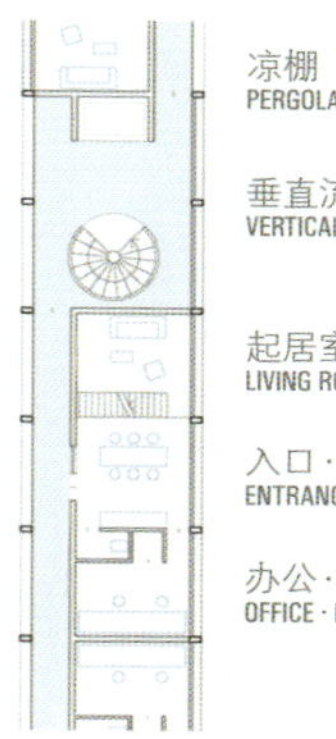

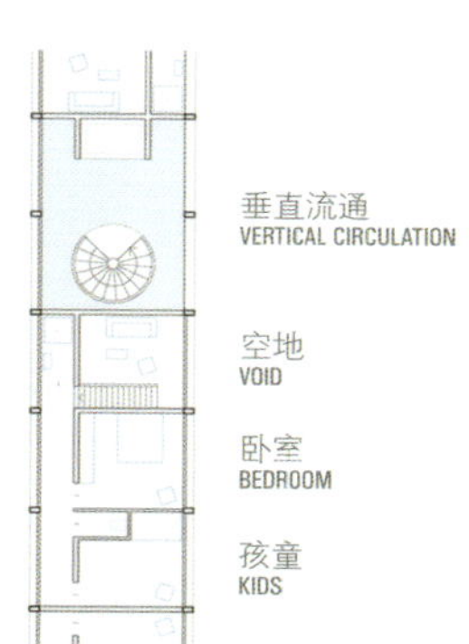

复式 DUPLEX

更新的都市生活方式

该项目是设防严密的中世纪城市与慕尼黑在威廉大帝风格时期新扩展地区之间的过渡，它看上去像一颗闪耀在空旷绿化空间里的独粒宝石，也像人口密集城市的周边环境中的一个街区。它在现有结构——地下停车通道的基础之上修建了一处引人注目的地上扩建物，这个地下停车通道几十年来一直是该地区的地标性建筑。美术馆就是这个公共绿化空间的地面凉廊。三层层高加倍的楼层没有任何垂直层次，却满足了其功能需求。

RENEWED URBAN LIFESTYLE

As an intermediator between the fortified, medieval city and Munich's new extension in the period of Wilhelminian style, the project appears as a solitaire in the open green space and as a block in the dense urban surrounding. It is also a conscious, aboveground continuation of an existing structure, rising from the underground parking gateway which characterized the site for decades. The gallery acts as a loggia in the public green on the street level. Three double height floors meet functional demands without any vertical hierarchy.

(竞标代码) Elevating Munich

XML Architecture Research Urbanisms

(建筑师事务所)
Max Cohen de Lara and David Mulder
(建筑师)

荣誉提名奖 honourable mention

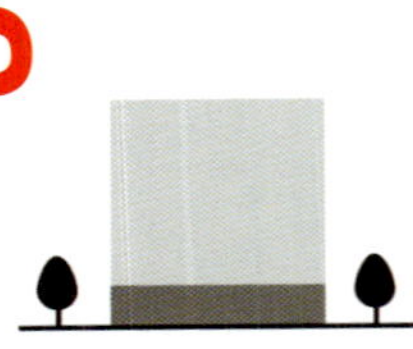

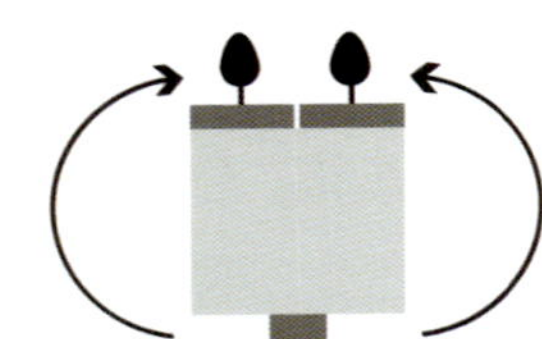

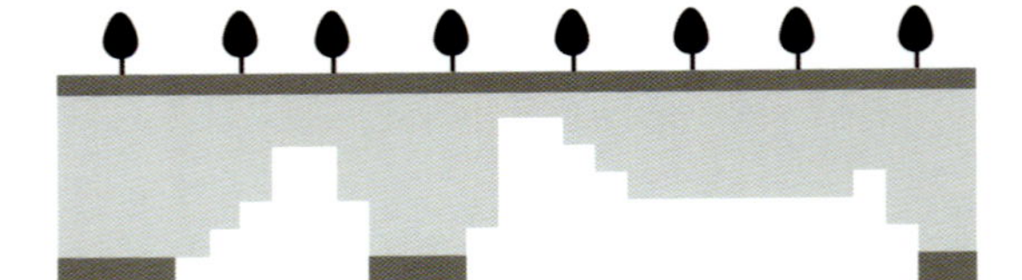

概念 CONCEPT

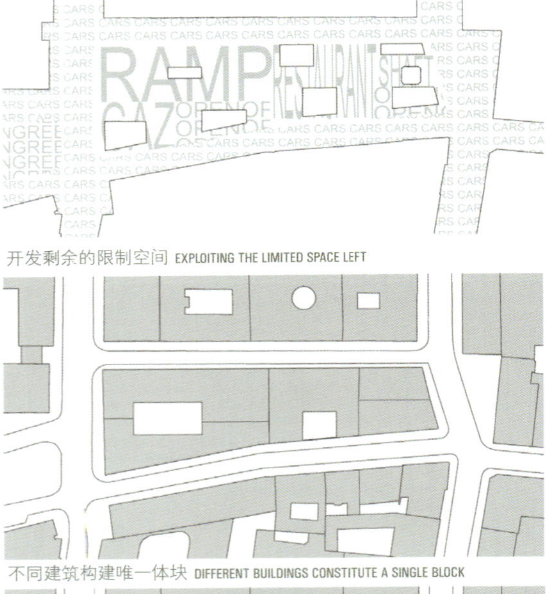

开发剩余的限制空间 EXPLOITING THE LIMITED SPACE LEFT

不同建筑构建唯一体块 DIFFERENT BUILDINGS CONSTITUTE A SINGLE BLOCK

“大腿” 激活公共空间 "LEGS" ACTIVATING PUBLIC SPACE

视线构成“大腿”的布局 SIGHTLINES ORGANIZE ARRANGEMENT OF "LEGS"

街道被不同的方式激活

如今，Herzog Wilhelm Strasse是一个位于两个独立购物区之间的过渡区域，不引人注目也不紧凑。场所中心有个通道通向地下停车场。剩余的有限空间囊括了许多地下停车场的通风井、一个多少有点被弃置的餐厅以及边上的一个加油站。我们在新开发过程中所面临的挑战是在完成项目要求的同时拓宽公共空间。

该项目旨在将工程建造在充分利用此处有限空间的建筑物之间。每幢建筑均有各自的具体功能——酒店、办公、零售和住宅，与其周边环境相辅相成。屋顶区域被建造成一个新的公共空间，大家可以搭乘位于该建筑“腿部”的电梯通往该区域。

THE STREET IS ACTIVATED IN DIFFERENT WAYS

Today, the Herzog Wilhelm Strasse location is an unattractive, incoherent transit zone positioned between two separate shopping areas. The heart of the site is a hole that provides access to an underground parking facility. The limited space that is left contains a number of ventilation shafts for the underground parking, a somewhat forlorn restaurant, and a gas station to one side. The challenge for the new development is to achieve the requested program and at the same time enlarge the public space.

This project proposes to accommodate the requested program in a series of buildings that fully exploit the limited space available on this location. Built on a scale that creates a dialogue with the surrounding area, each building has its own specific program: hotel, office, retail and housing. The roof area becomes a new public space, which the public can access by the elevators housed in the "legs" of the building.

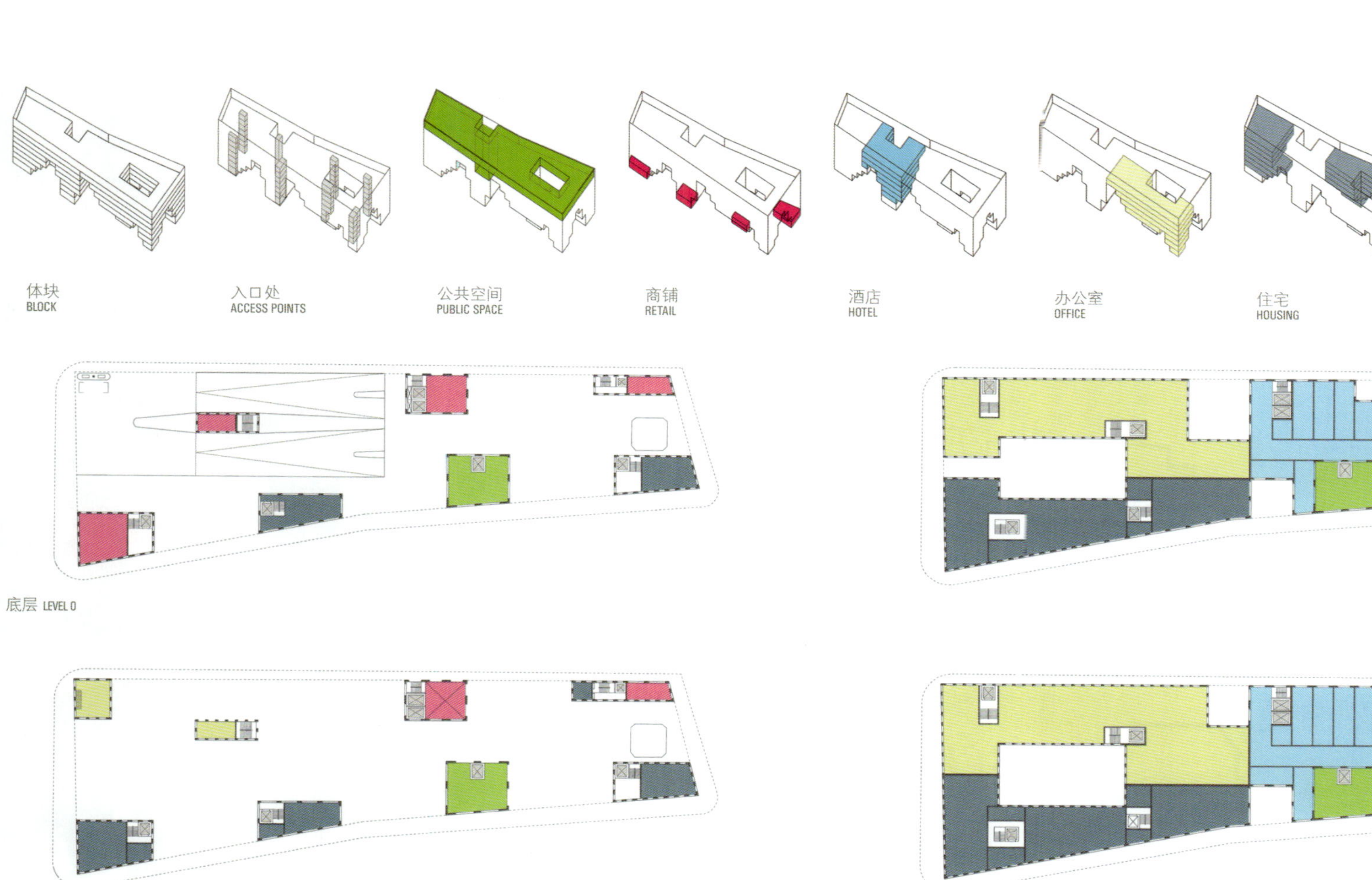

体块 BLOCK　入口处 ACCESS POINTS　公共空间 PUBLIC SPACE　商铺 RETAIL　酒店 HOTEL　办公室 OFFICE　住宅 HOUSING　功能复合 COMBINED PROGRAM

底层 LEVEL 0

五层 LEVEL +4

二层 LEVEL +1

六层 LEVEL +5

三层 LEVEL +2

七层 LEVEL +6

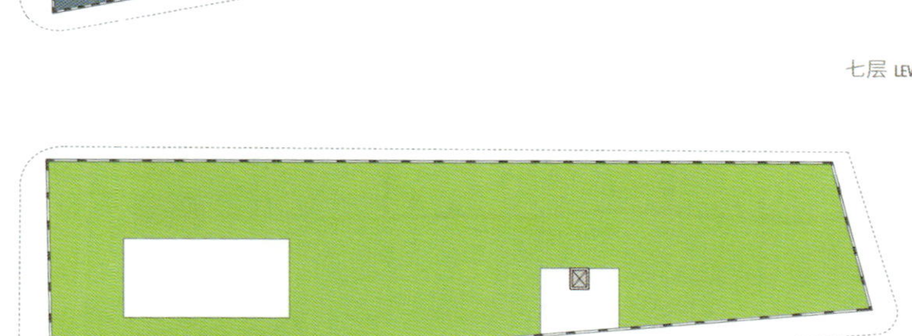

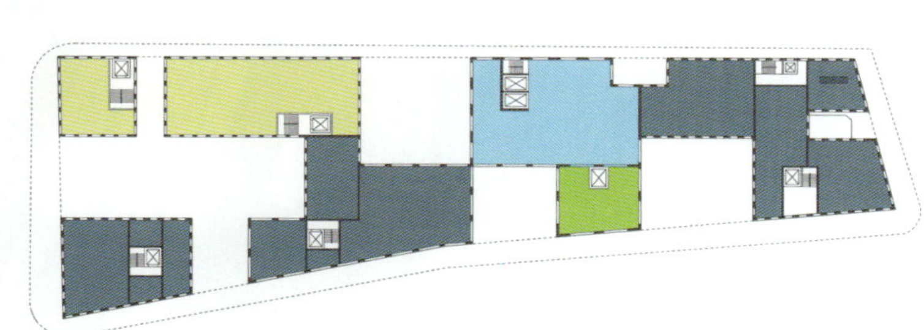

四层 LEVEL +3

八层 LEVEL +7

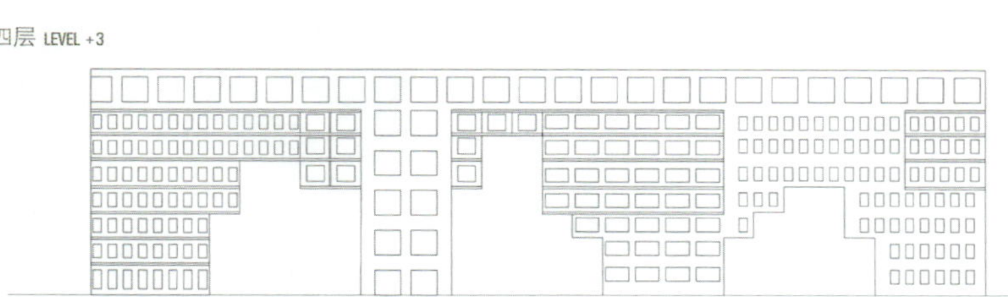

纵向剖面图 LONGITUDINAL SECTION

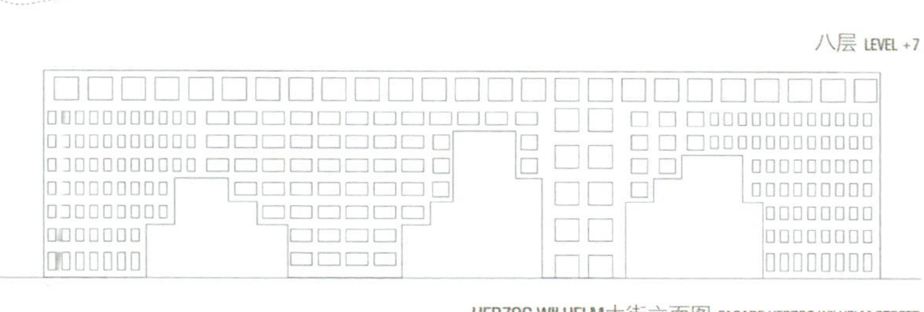

HERZOG-WILHELM大街立面图 FAÇADE HERZOG-WILHELM STREET

(竞标代码) HWS:Re-Enactment

Elisa Stellacci · Andrea Verdecchia
(建筑师事务所)

荣誉提名奖 honourable mention

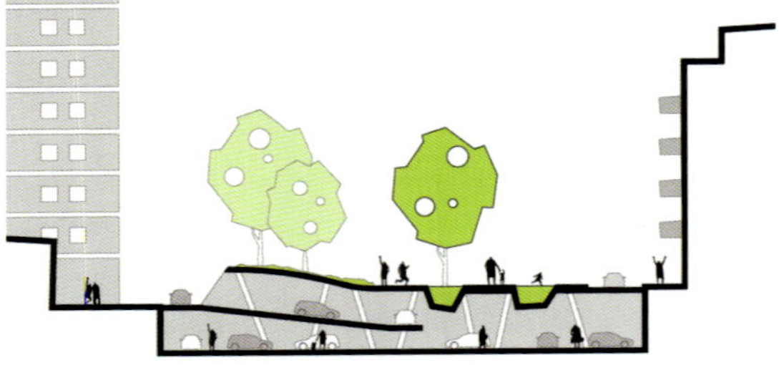

剖面图 A-A´ 新停车场 SECTION A-A´ NEW CARK PARK

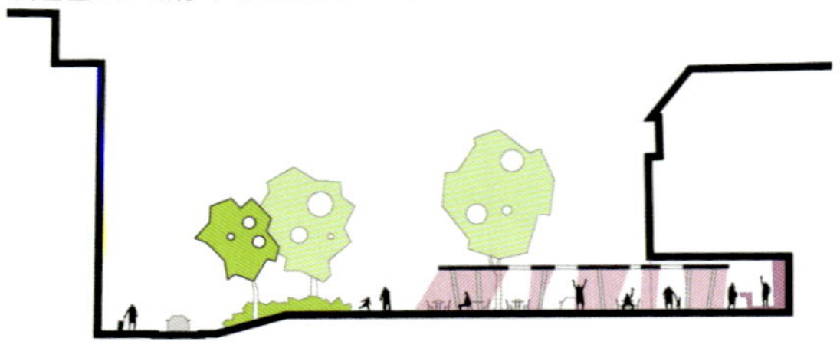

剖面图 B-B´ 主食 · 饮料 SECTION B-B´ FOOD · DRINK

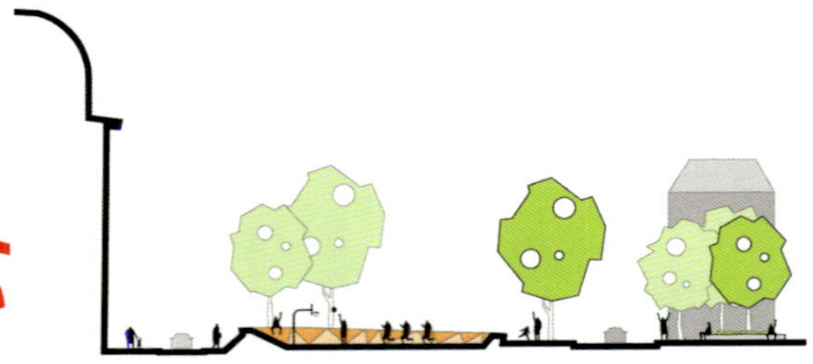

剖面图 C-C´ 游戏 SECTION C-C´ PLAYGROUND

剖面图 D-D´ 艺术 · 文化 SECTION D-D´ ART · CULTURE

剖面图 E-E´ 溜冰场 · 真冰场 SECTION E-E´ SKATEPARK · ICE SKATING

非连续性

该项目旨在通过在操场上配置休闲设施，并在艺术平台周边打造自由的绿化空间，将功能单一的Herzog-Wilhelm Strasse边界打造成一个兼具多项功能的中心地带。该项目是一个特征鲜明的大型结构体，占地7,000平方米，用于建造空间、开阔视野和打造现有地下停车场的坡道。该设计与城市规划的原则保持一致：利用项目的非连续性来推动社会交际。

DISCONTINUITY

The aim is to replace the actual mono-functionality at the borders of Herzog-Wilhelm Strasse with a central great MIX. The way is to make playgrounds cohabiting with leisure facilities, and art platforms with green free spaces. The 7000 m2 required floor space has been concentrated in a unique mass volume, used to structure the space, canalize views and over-pass the floor cut of the existing underground carpark's ramp. The design follows the same principles adopted for the urban planning: use discontinuity of programs to produce social intercourse.

自然 NATURE
商铺 SHOP
主食 · 饮料 FOOD · DRINK
住宅 HOUSING
艺术 · 文化 ART · CULTURE
游戏 PLAYGROUND
办公室 OFFICE
独立办公室 (PR) OFFICE (PR)

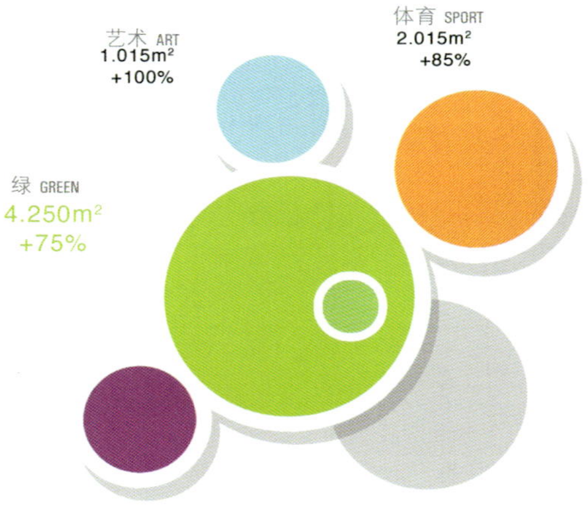

艺术 ART 1.015m² +100%
体育 SPORT 2.015m² +85%
绿 GREEN 4.250m² +75%
主食+饮料 FOOD+DRINK 1.005m² +80%

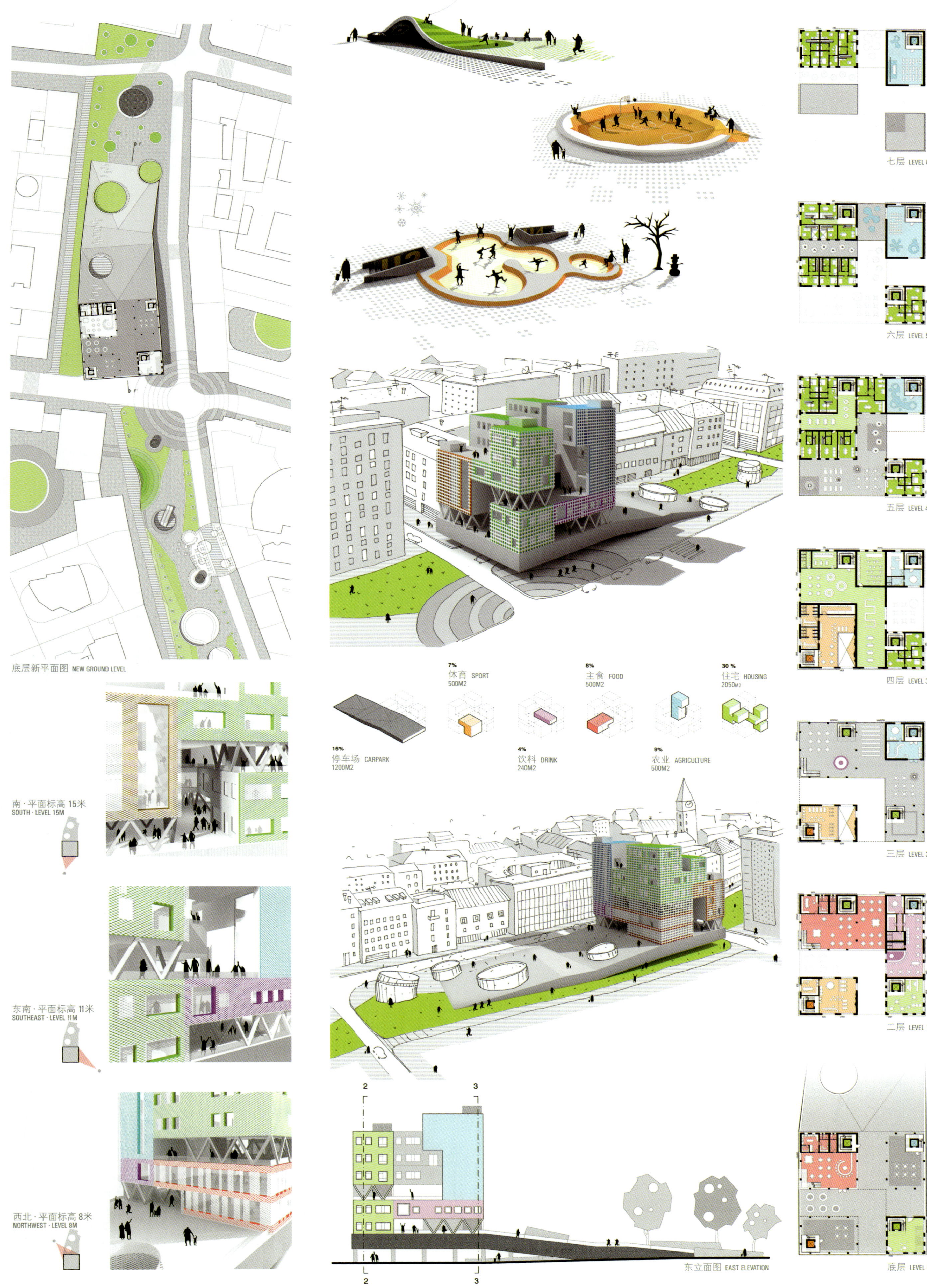
七层 LEVEL 6
六层 LEVEL 5
五层 LEVEL 4
四层 LEVEL 3
三层 LEVEL 2
二层 LEVEL 1
底层 LEVEL 0
底层新平面图 NEW GROUND LEVEL
7% 体育 SPORT 500M2
8% 主食 FOOD 500M2
30% 住宅 HOUSING 2050M2
16% 停车场 CARPARK 1200M2
4% 饮料 DRINK 240M2
9% 农业 AGRICULTURE 500M2
南·平面标高 15米
SOUTH · LEVEL 15M
东南·平面标高 11米
SOUTHEAST · LEVEL 11M
西北·平面标高 8米
NORTHWEST · LEVEL 8M
东立面图 EAST ELEVATION

特隆赫姆 · 挪威 Trondheim · Norway

(竞标代码) Proscenium

Point Supreme Architects and Alexandros Gerousis

(建筑师事务所及建筑师)

中标 winner

地块位置 SITE PLAN

0m

平面图 +12.00 352平方米 FLOOR PLAN +12.00 352M2

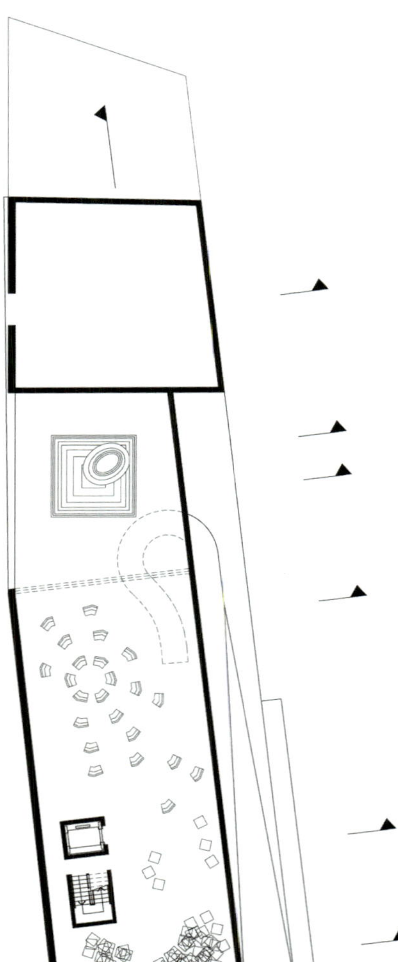

平面图 +0.00 550平方米 FLOOR PLAN +0.00 550M2

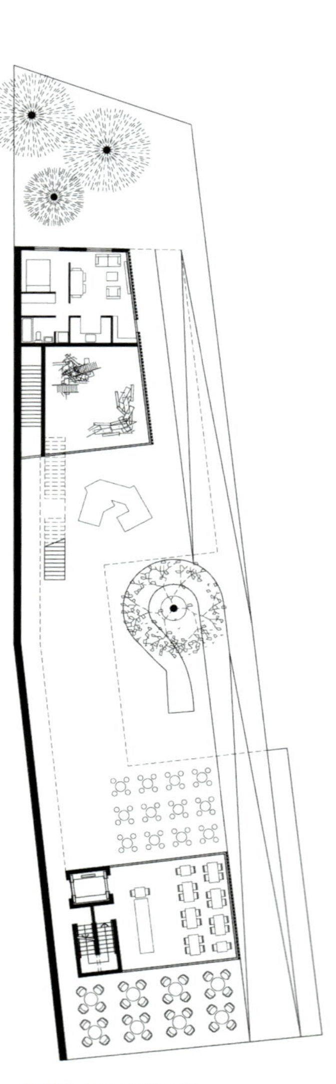

平面图 +3.00 168平方米 FLOOR PLAN +3.00 168M2

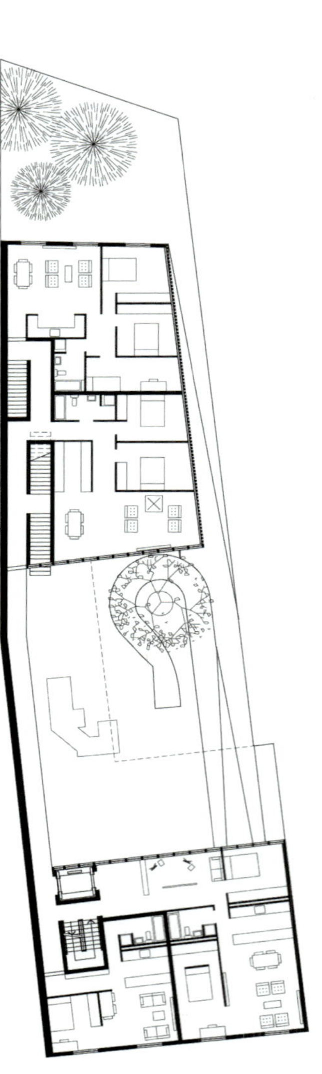

平面图 +6.00 416平方米 FLOOR PLAN +6.00 416M2

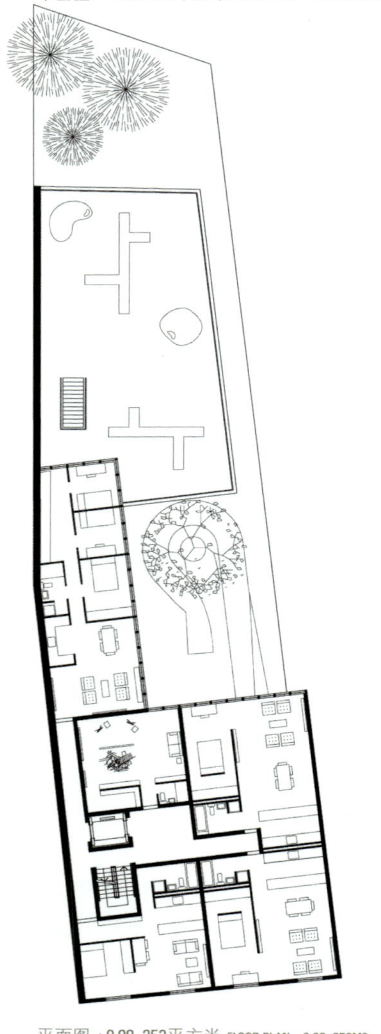

平面图 +9.00 352平方米 FLOOR PLAN +9.00 352M2

表达敬意

该设计蓝图提出了一系列仔细布置的结构，打造了以前水下地堡“Dora 1”的附加物：一个位于屋顶的驱动式电影院和露天多功能空间；一个穿越铁道下方的下沉式广场，构成了主入口；行政大楼和送货楼，构成了次入口；一个螺旋形的圆形剧场面积翻倍，形成了一个市场和集会空间，并扩展成为一个港口；一个长方形的水池面积翻倍，形成了一个冬日里的溜冰场。这些附加物定位且打造了Dora 1的周边空间，并在战略上扩大了其对周边环境的影响。为了满足日益增长的住房、办公空间和停车场方面的需求，该设计蓝图在Dora 1的周边还设计了另外三个场所。

RESPECTFUL

The masterplan suggests a series of carefully placed interventions that act as appendages to the former submarine bunker 'Dora 1': a drive-in cinema and open air multi-function space on the roof, a sunken plaza passing under the railway forming the main entrance, administration and drop off building creating a second entrance, a spiral amphitheatre that doubles as market and gathering space, and extended into the harbour, a rectangular pool that doubles as an ice skating rink in the winter. These appendages define and shape the space around Dora1 and strategically expand its influence on the immediate context. To meet growing demands for housing, office space and parking, the masterplan identifies three further sites around Dora 1.

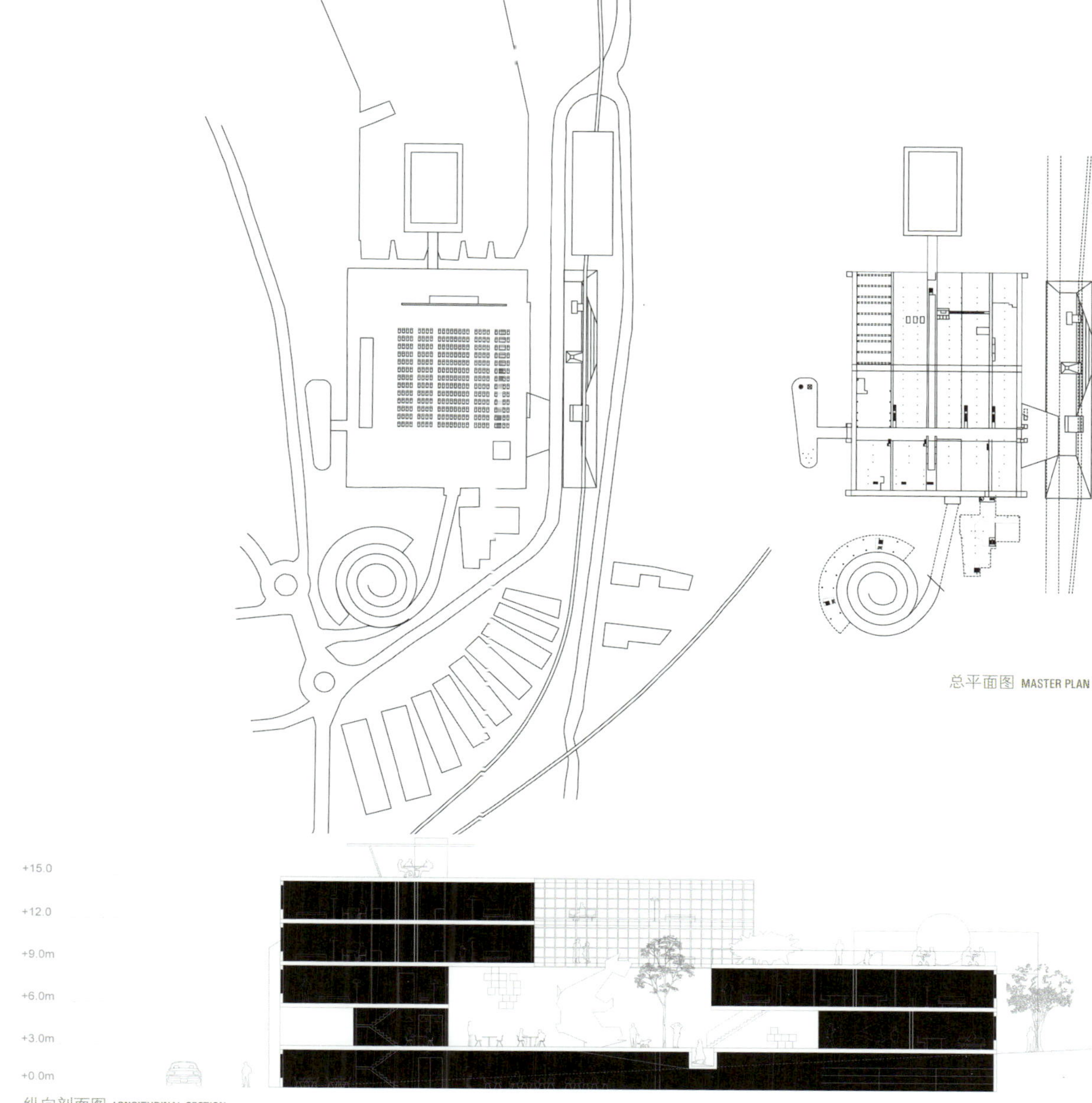

总平面图 MASTER PLAN

纵向剖面图 LONGITUDINAL SECTION

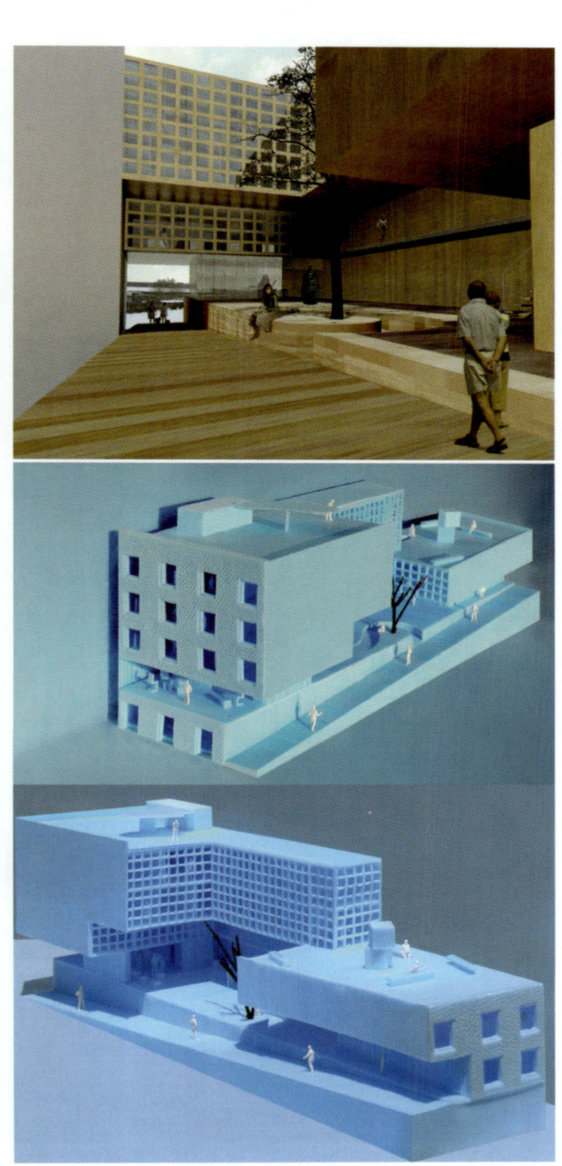

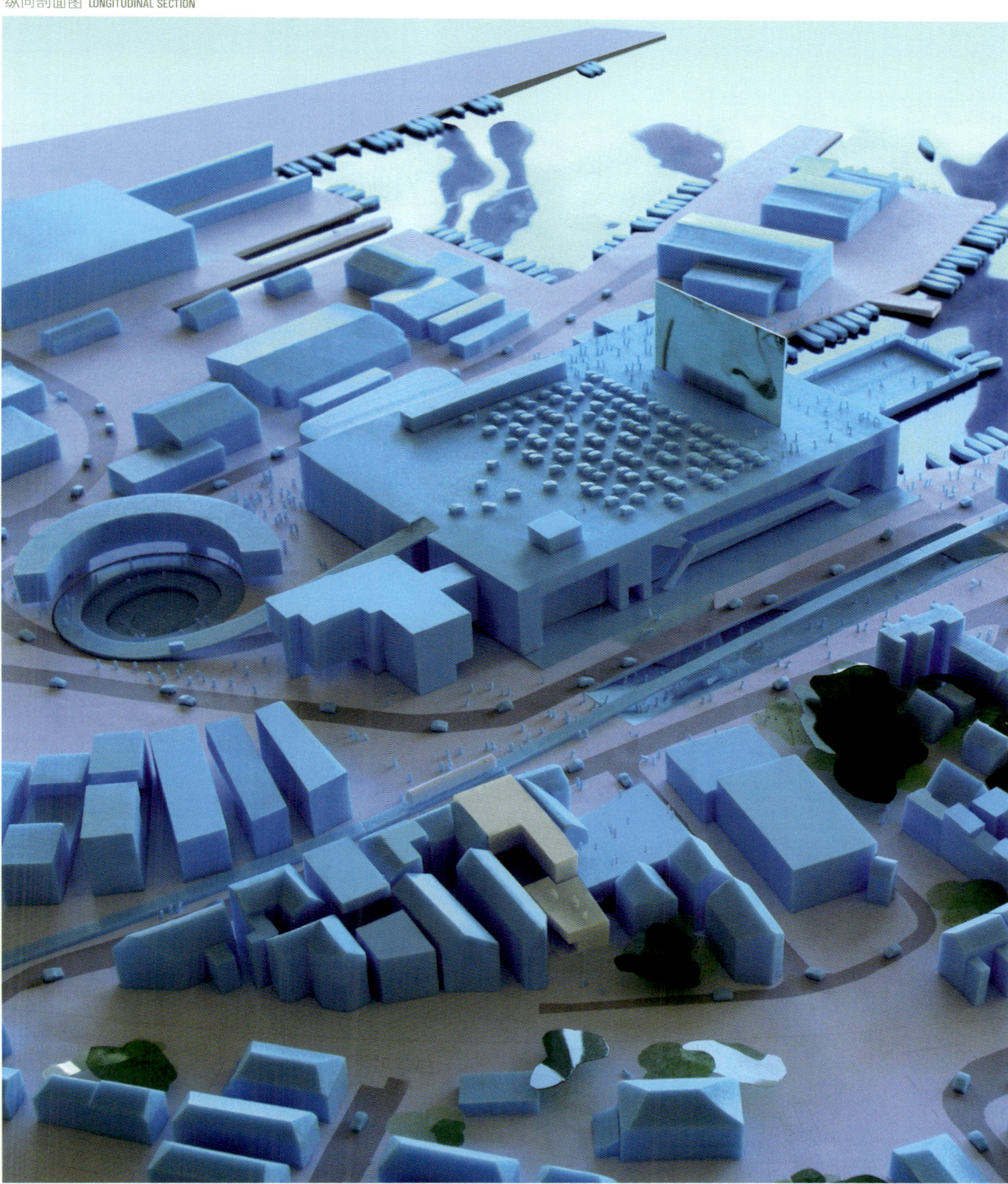

(竞标代码) From Trondheim, with love

Guillaume Jounet · Leo Thafelin · Remy Bardin

(建筑师)

二等奖 runner-up

艺术中心细部 ART CENTRE DETAILS

剖面图 SECTIONS

五层平面图 FLOOR PLAN 4

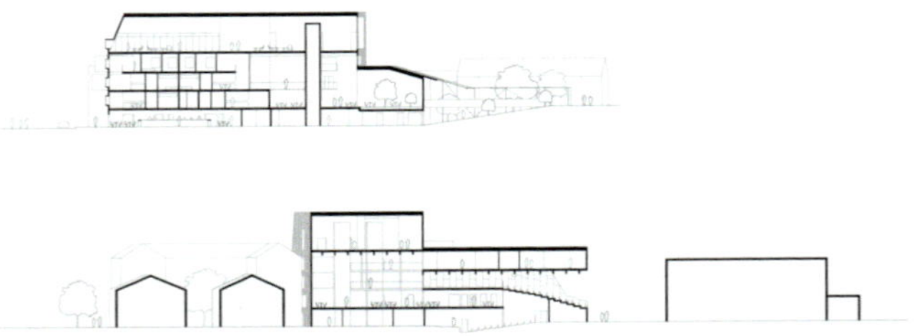

三层平面图 FLOOR PLAN 2

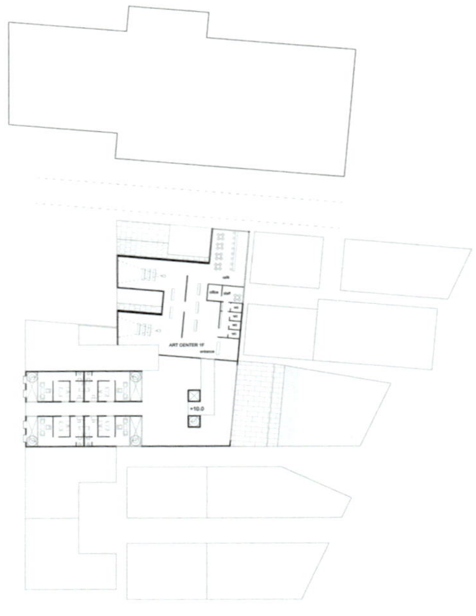

四层平面图 FLOOR PLAN 3

底层平面图 GROUND FLOOR PLAN

二层平面图 FLOOR PLAN 1

艺术中心 1 ART CENTRE 1

艺术中心 2 ART CENTRE 2

集装箱内设备 1 CONTAINER INSTALLATION 1

集装箱内安装 2 CONTAINER INSTALLATION 2

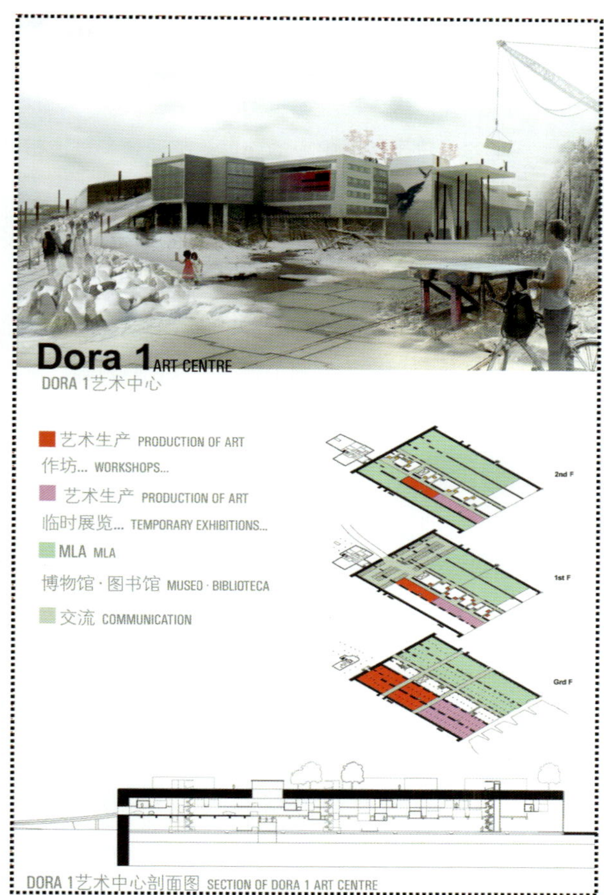

DORA 1艺术中心剖面图 SECTION OF DORA 1 ART CENTRE

集装箱设备剖面图 SECTION OF THE CONTAINER FACILITY

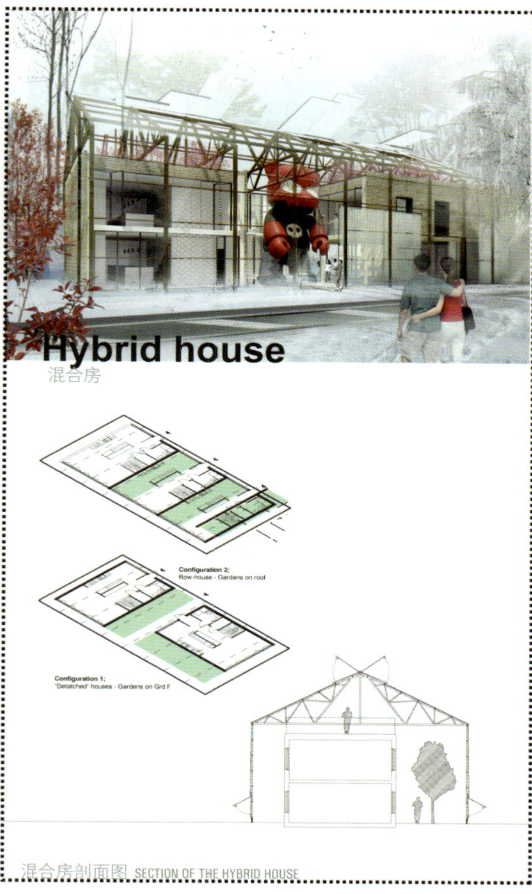

混合房剖面图 SECTION OF THE HYBRID HOUSE

30M3立方米的明信片

若艺术也能像明信片一样在世界上广泛传播会怎么样呢？世界会变成一个由1,001个分散房间组成的博物馆吗？将一个当代艺术馆装在一个20英寸的货柜里，随意发出，就如同给世界上的某人寄一张明信片一样：你不会知道那个人何时何地会以怎样的方式收到这张明信片。那个人甚至可能都不知道明信片上的寄件地址在何方。这就是特隆赫姆变得盛行的原因，同时也是特隆赫姆发展成为一个文化热点的原因。

与其指望全世界的人都来参观特隆赫姆，还不如将其艺术和文化传播到世界各地。Svartlamoen坐落在港口和铁道之间，使得该建筑的地理位置十分有利于艺术的传播。这里制造的任何东西均能传播到世界各地。最后，货柜也会变得像一个个艺术馆，每次有人寄出一个艺术馆，他/她就会在“分散的博物馆”里增加一个新房间。利用该建筑的演艺厅和街对面的另一建筑的优势，一个具有意象特征的文化中心就此形成了，其规模大到让居民们能经常以不同的方式前来造访。

30M3 POSTCARD

What if art could be spread worldwide like a postcard? Could the world become a museum with 1001 scattered rooms? Randomly shipping a contemporary art gallery inserted in a 20' container is almost like sending a postcard to someone across the world: you never know how, when or where the person will read it. This person might not even know the place you write from. That is how Trondheim unassumingly became ubiquitous. That is how Trondheim became a cultural hotspot.

Instead of expecting the whole world to come visit, Trondheim decided to travel anywhere by exporting its art and culture. The plot location could hardly be more convenient to export art, as Svartlamoen is sitting between the harbor and railways. Anything that is produced within the site can be sent anywhere on Earth. In the end of the process, containers would be like galleries, and each time someone send a gallery, he/she also adds a new room to the "scattered museum".By taking advantage of having a performance hall and another project across the street, an iconic cultural centre for the neighborhood is formed. Its scale allows inhabitants to use it regularly and in many ways

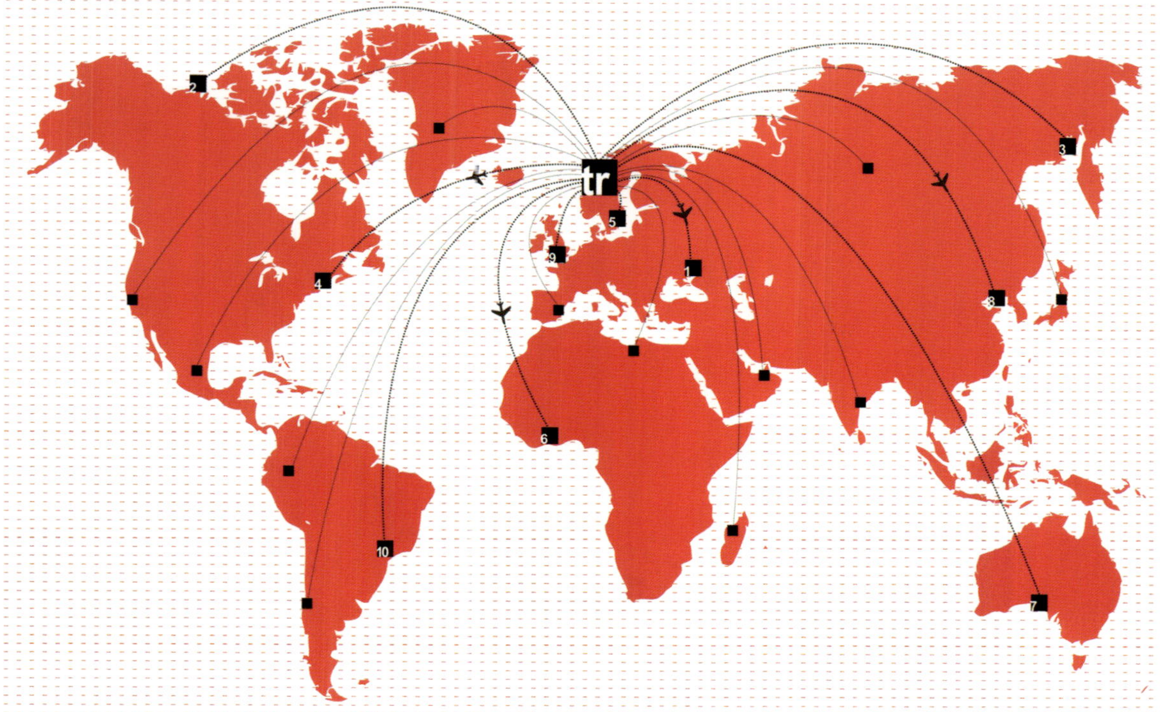

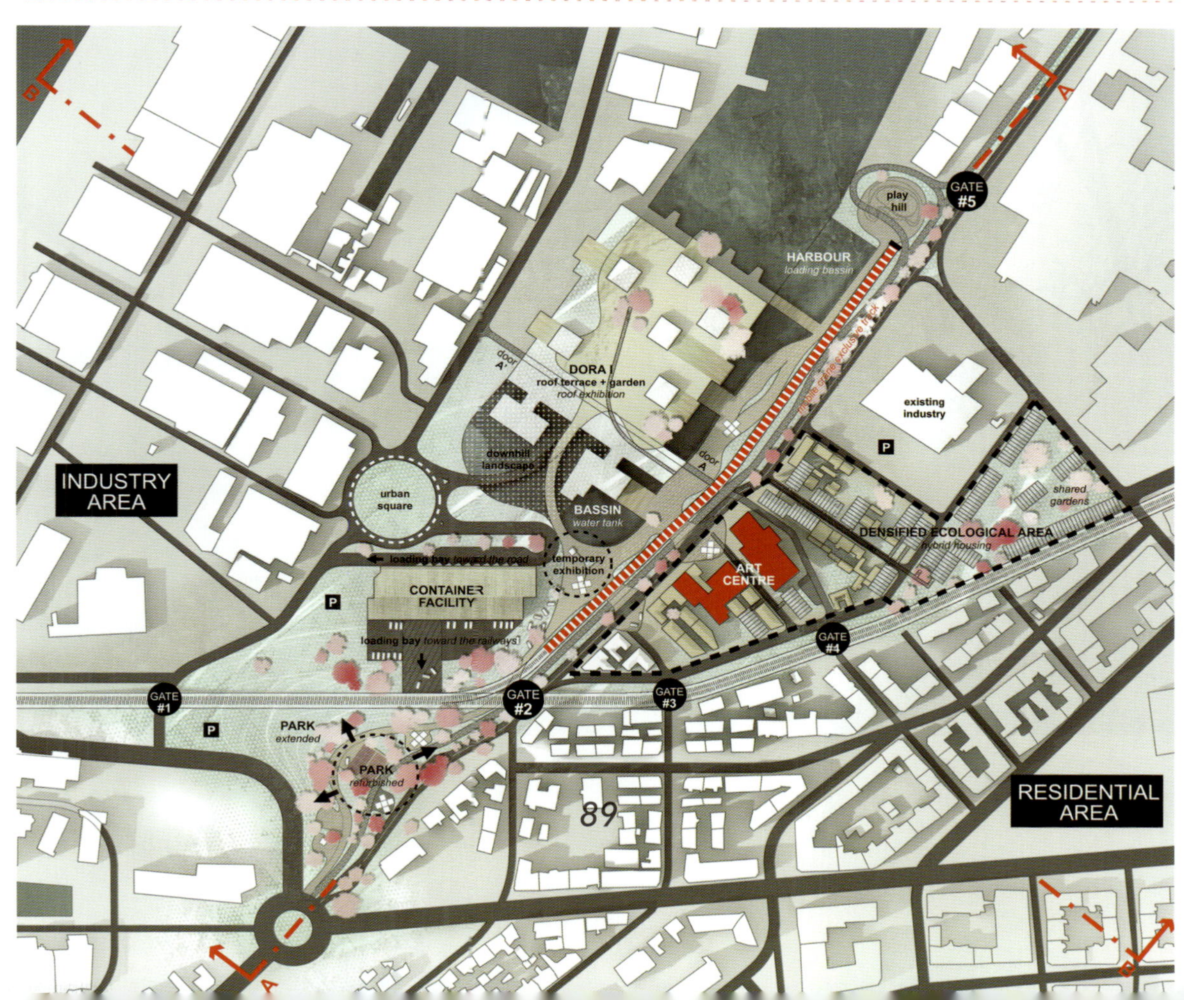

Suture (竞标代码)

Marco Galasso · Timur Shabaev
(建筑师)
中标 winner

A 办公室 OFFICE B 超市 SUPERMARKET C 缝合处 THE SUTURE
D 交流中心儿童学校 COMMUNITY CENTRE-KINDERGARTEN E 联体别墅·隔音墙 ROWHOUSES-NOISE BARRIER

重新规划

学习区位于埃曼中心的南侧，与通道和铁道相邻，是一个隔离的区域。我们将这里打造成了一个多元化的区域：将教堂改造成办公室；将南边的仓库拆除，在原地兴建了住房；将Buitenweg的一幢复式楼改造成一个社区中心和幼儿园。新建楼房继续保留了现有的都市结构。

我们建造了缝合区——社区的公共空间。这些公共空间按照现有不规则小道的格局进行打造，修建通往各主要场所的捷径。

KINTSUGI (金继ぎ)·在日语中：黄金木工·是用粉状树脂漆金修复陶器的日本艺术

KINTSUGI (金继ぎ)·JAPANESE: GOLDEN JOINERY · IS THE JAPANESE ART OF FIXING BROKEN POTTERY WITH A LACQUER RESIN SPRINKLED WITH POWDERED GOLD

RE-PROGRAMMING

The study area is situated on the southern side of the centre of Emmen and it is bordered by access roads and the railway. It creates an isolated location. We bring diversity to the programs of the site: we turn the church into an office; we demolish the warehouses in the south part of the site and replace them with housing; we turn one of the existing duplex on Buitenweg into a community centre and kindergarten. We continue the existing urban tissues with new housing.

We create sutures - public spaces for the community. They follow the pattern of existing informal paths and create shortcuts to the main destinations.

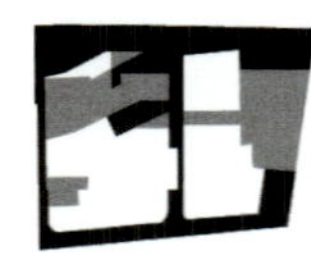

=

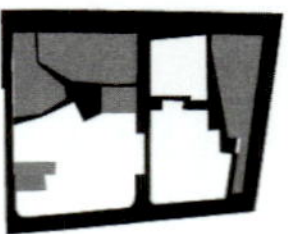

目前 NOW   方案 PROJECT

BUITENWEG景象 VIEW OF BUITENWEG

联体别墅·铁路景象 ROW HOUSES · VIEW FROM THE RAILWAY

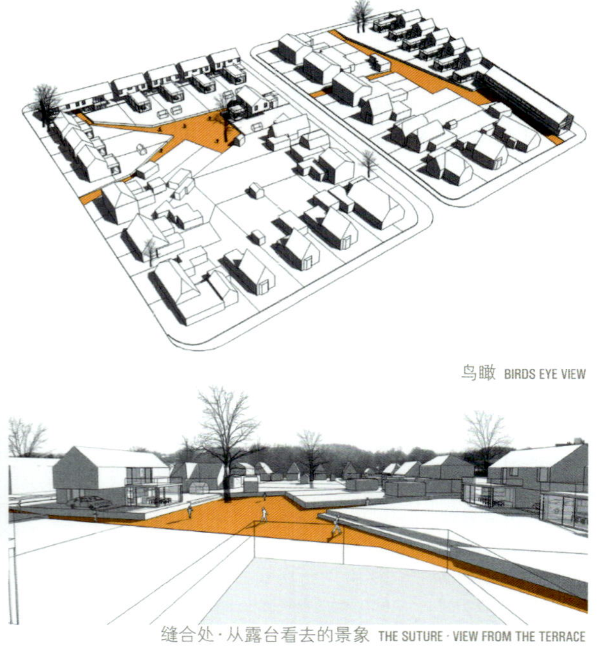

鸟瞰 BIRDS EYE VIEW

缝合处·从露台看去的景象 THE SUTURE · VIEW FROM THE TERRACE

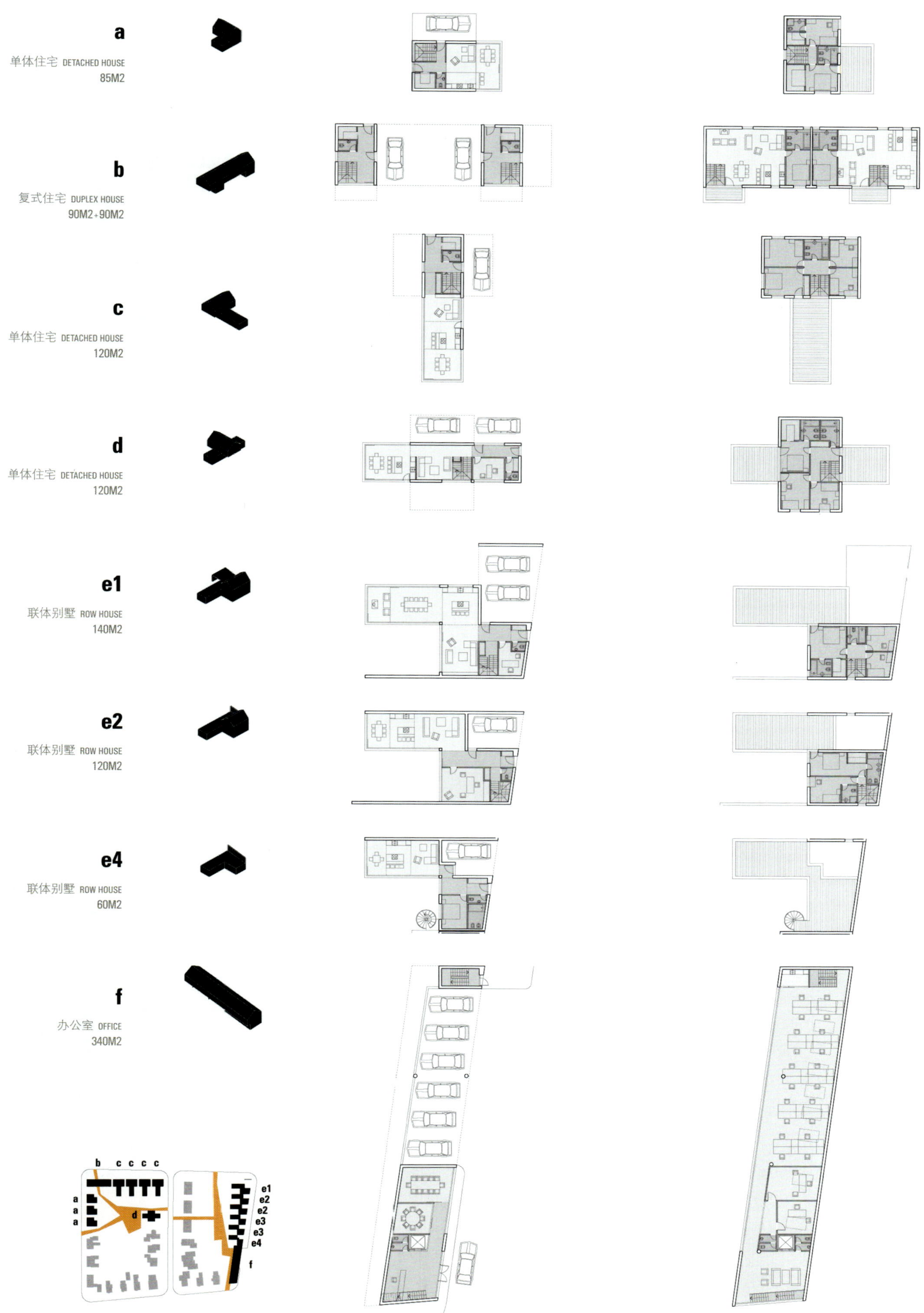
a
单体住宅 DETACHED HOUSE
85M2
b
复式住宅 DUPLEX HOUSE
90M2+90M2
c
单体住宅 DETACHED HOUSE
120M2
d
单体住宅 DETACHED HOUSE
120M2
e1
联体别墅 ROW HOUSE
140M2
e2
联体别墅 ROW HOUSE
120M2
e4
联体别墅 ROW HOUSE
60M2
f
办公室 OFFICE
340M2

(竞标代码) Shaping voids

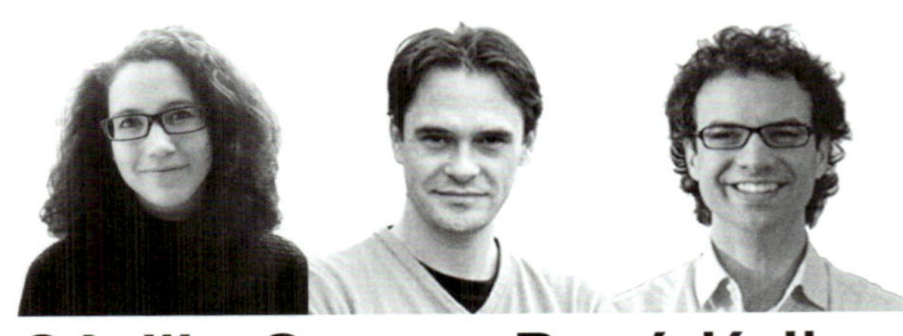

Cécilia Gross · René Kuiken · Matteo Kuijpers

(建筑师)

二等奖 runner-up

0
现有的 EXISTING

1
合并 > 网格 UNIFY > GRID

2
针灸 > 建筑体量 ACUPUNCTURE > BUILDING MASS

3
色彩 > 功能 COLOUR > PROGRAM

4
调整空地 > 类型 SHAPE YOUR VOID > TYPOLOGY

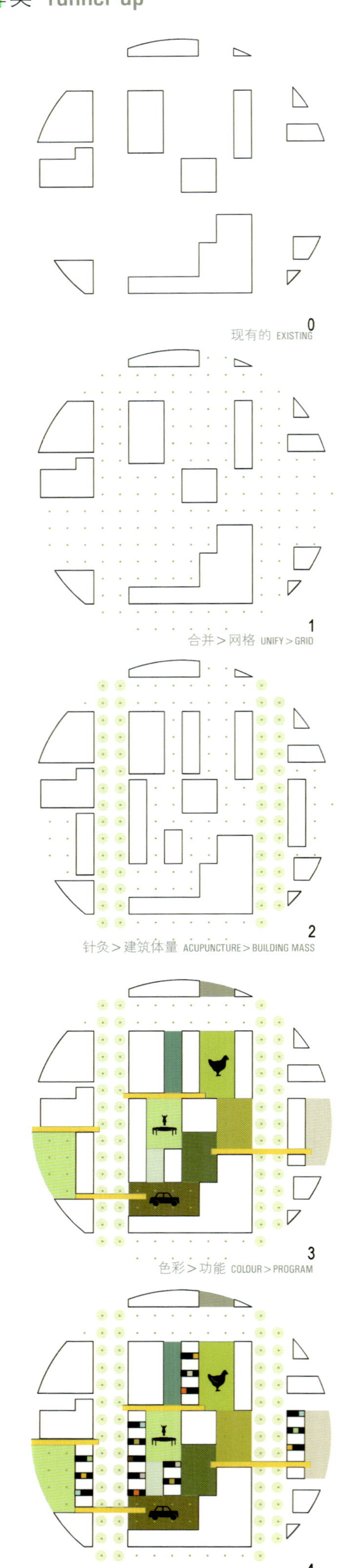

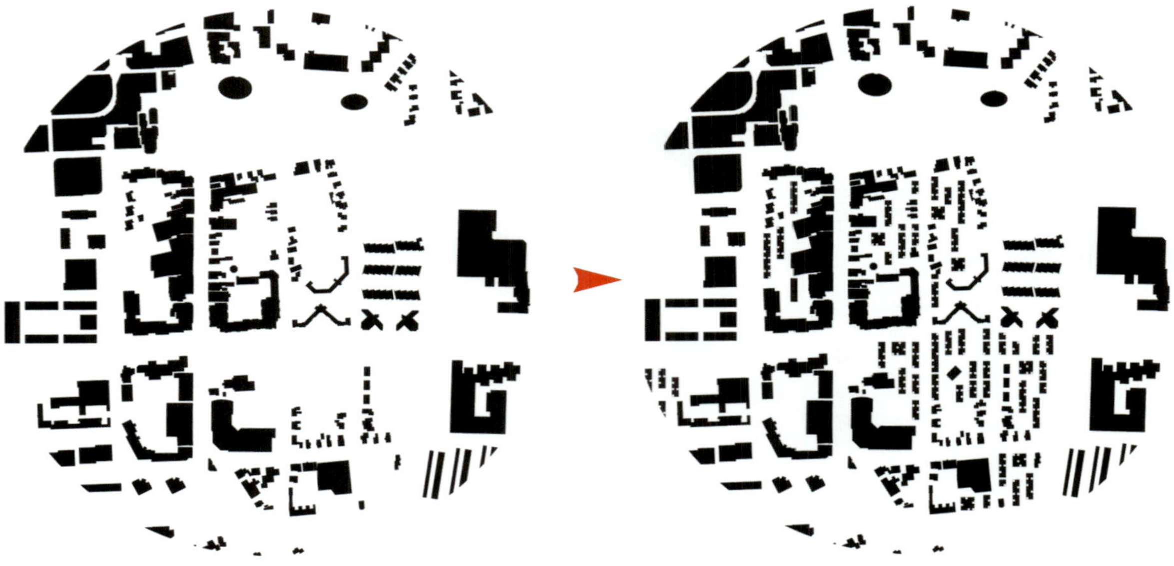

现有的城市结构 EXISTING CITY STRUCTURE

新的城市结构 NEW CITY STRUCTURE

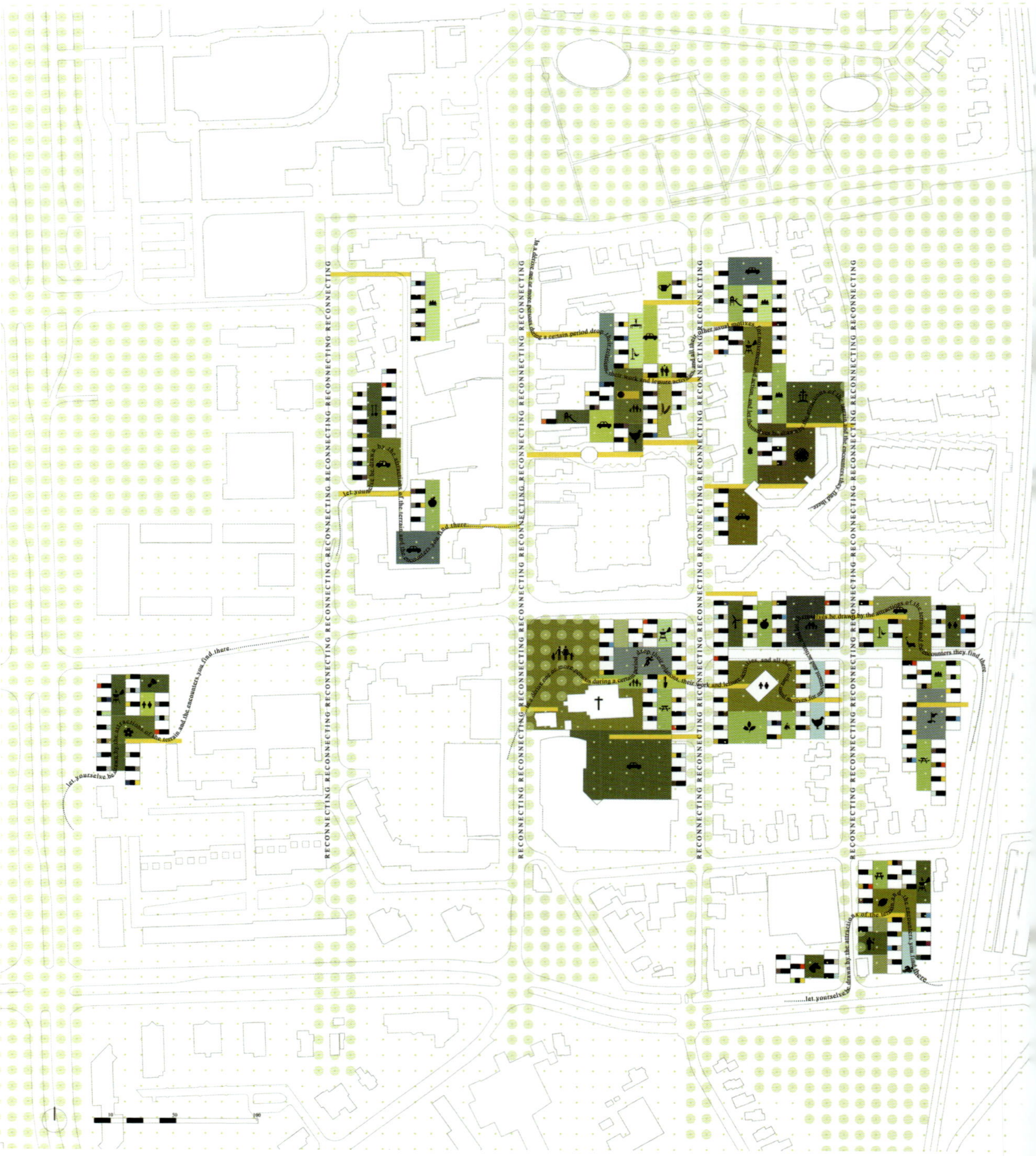

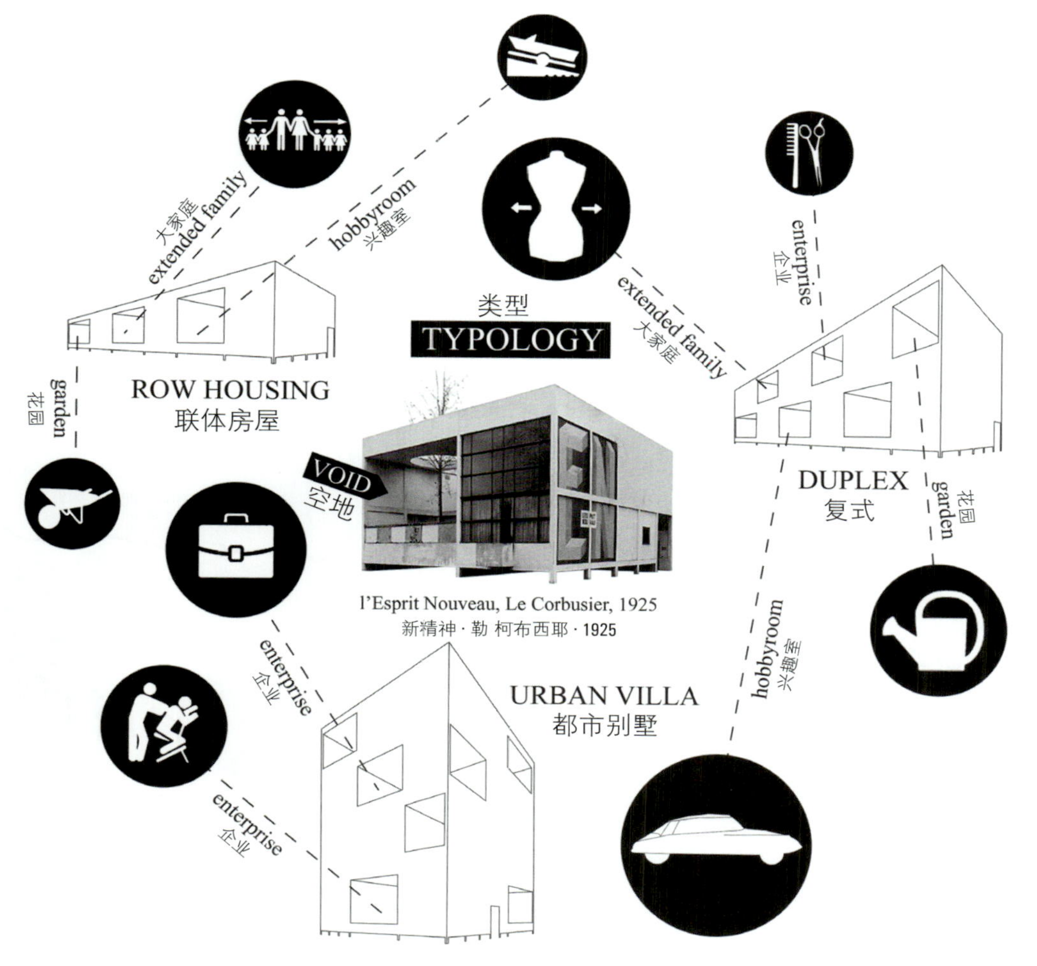

公共空间的特性

打造不同的空间是体现埃曼特性的一项方针策略。该方案在空间上和规划上都将公共空间重新阐释为一个结构清晰、居住区丰富多彩、住房特点鲜明的区域，旨在吸引年轻居住者。排列有致的松树种植于埃曼中心地带，将街道、停车位和各个场所连接为一个整体，并使得森林与城市中心紧密相连。

这个常常被人忽略的不起眼的公共空间将被打造成一个明朗的都市格局。不同场所的建筑空间打造了多种多样的居住区，这里有广场、绿色花园、儿童操场、雕塑公园，等等。这些公共场所体现了公共空间的特性，而私人场所则赋予了年轻的居住者成长并张扬个性的机会。该方案未设计后院或阳台，取而代之的是每家每户都有一个3 x 9的自由区域。

IDENTITY TO PUBLIC SPACE

Shaping Voids is a strategy which operates within the unique character of Emmen. It is about redefining spatially and programmatically the public space into a clear pattern with colourful living areas and unique housing typologies to attract younger inhabitants. A grid of pine trees is positioned within the centre of Emmen to unify streets, parking lots and voids. It will connect the forest even more with the city centre.

The unidentified and neglected public space is shaped into a clear city pattern. A language of different voids creates a colourful living area with squares, green gardens, children´s playgrounds, sculpture gardens... Whereas the public voids give identity to the public space, private voids give the younger inhabitants the opportunity to grow and be unique. A 3x9 free zone is given to every dwelling instead of backyards or terraces.

(竞标代码) It's my forest

Alberto Porras

(建筑师)
合作 (c) Luciano Pablos · Leticia Rodriguez

荣誉提名奖 honourable mention

每个年轻人或每个家庭都将拥有一棵树。这是一个与众不同的花园，一次独具匠心的发展。

Every youngster or family will be in possession of one tree. It is a particular garden, it is a particular growth.

现在的年轻人 CURRENT YOUNG PEOPLE

埃曼的发展理念是“开放城市和绿色城市”。

Emmen has pursued the concept of the open and green town.

都市针灸。我们转化城市的弱点并予以改造。

Urban acupuncture.
We transform the weak points and connect them.

我们创造了就业机会。我们有农业园。

We have generated employment.
We have agricultural gardens.

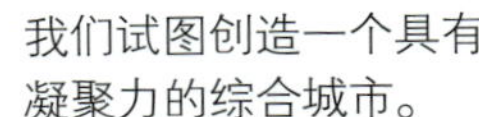

我们试图创造一个具有凝聚力的综合城市。

We tried to create a cohesive and integrate city.

2035 2060 2085 2110 2135

现在的森林区域 CURRENT FOREST AREAS

在埃曼，森林只有市郊才有，而年轻人也居住在市郊。因此埃曼的发展必须追随这一现状。

In Emmen, young people and forests are in in the outskirts. Emmen must grow towards them.

树木的生长伴随着居民和住房的发展。我们开始打造埃曼大花园。

The growth of the trees is parallel to the growth of inhabitants and housing. We begin the big garden of Emmen.

埃曼采用的是一个空间等级结构，形成了一个由大大小小子中心补充的中心。

Emmen is the centre which, following a spatial hierarchical structure is complemented by a number of larger and smaller sub-centres.

……埃曼森林。树木是我们的里程标。我们可以根据树的种类、高度和健康状态清楚地知道我们处于何方、正与谁共处。

... Emmen forest. We use the tree as a milestone. The type, height and health of the tree will give us a clear reference where and with whom we are.

我们试图将城市的发展与森林的田园式的发展交织在一起。

We tried to interlace the growth of a city with the idyllic growth of a forest.

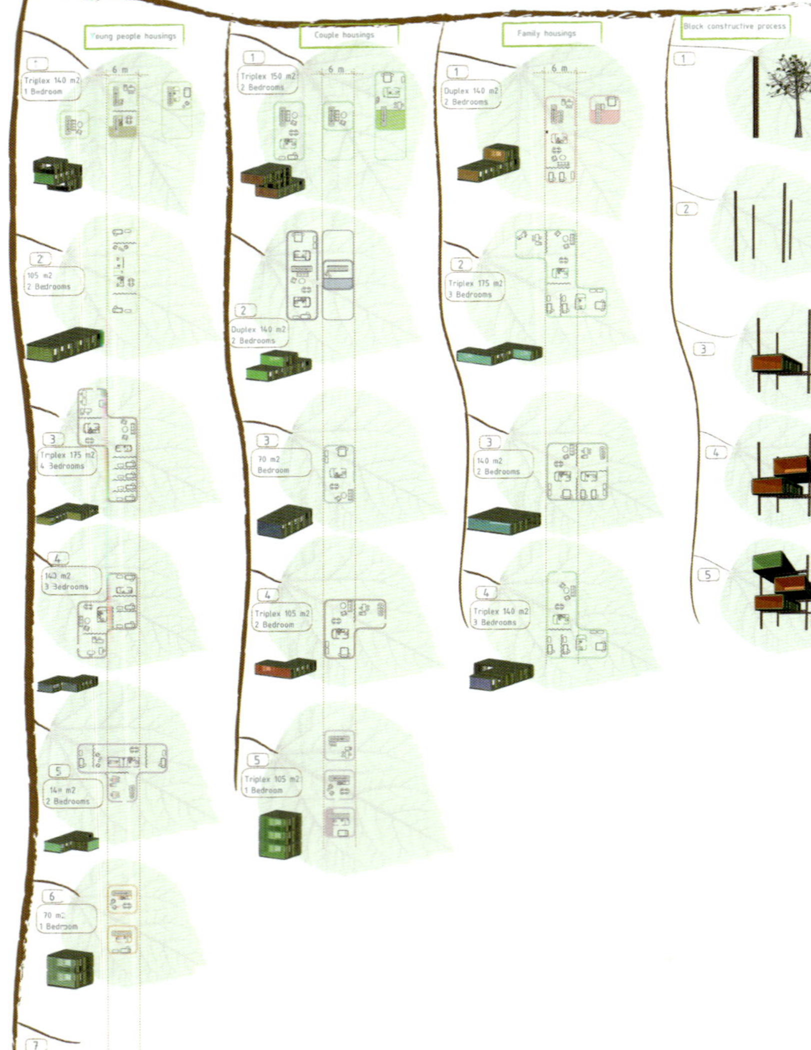

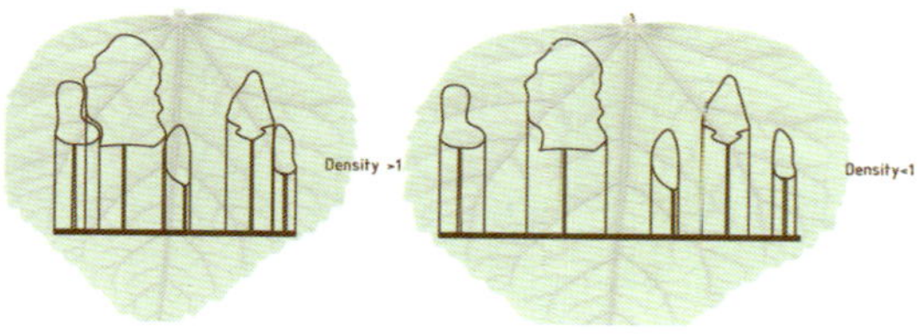

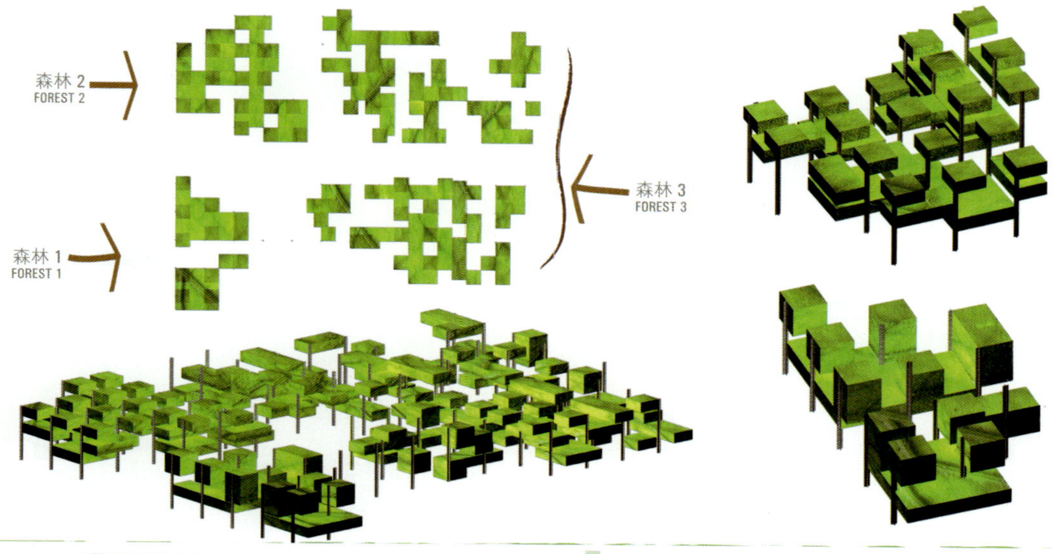

Overstory
Midstory
Understory
Density +1
Density +1
Clustered
Intimate

森林 2
FOREST 2
森林 1
FOREST 1
森林 3
FOREST 3

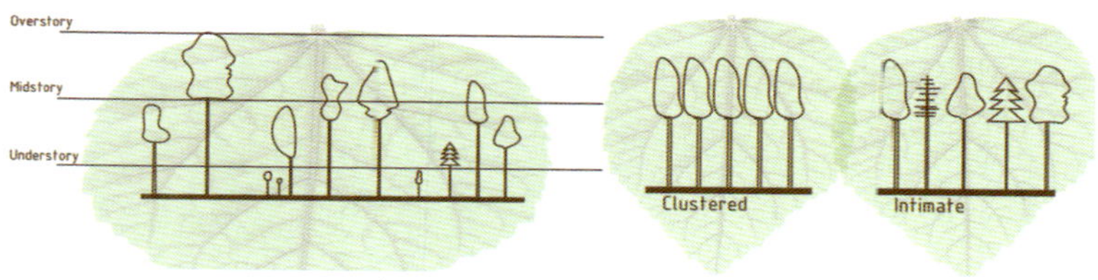

四层平面图 FLOOR PLAN 3

三层平面图 FLOOR PLAN 2

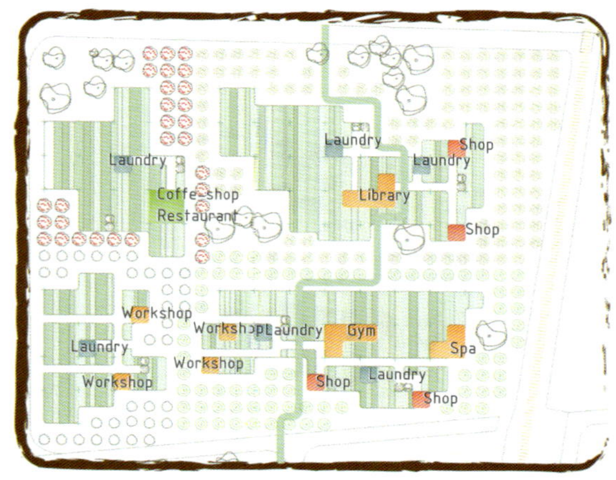

底层平面图 GROUND FLOOR PLAN

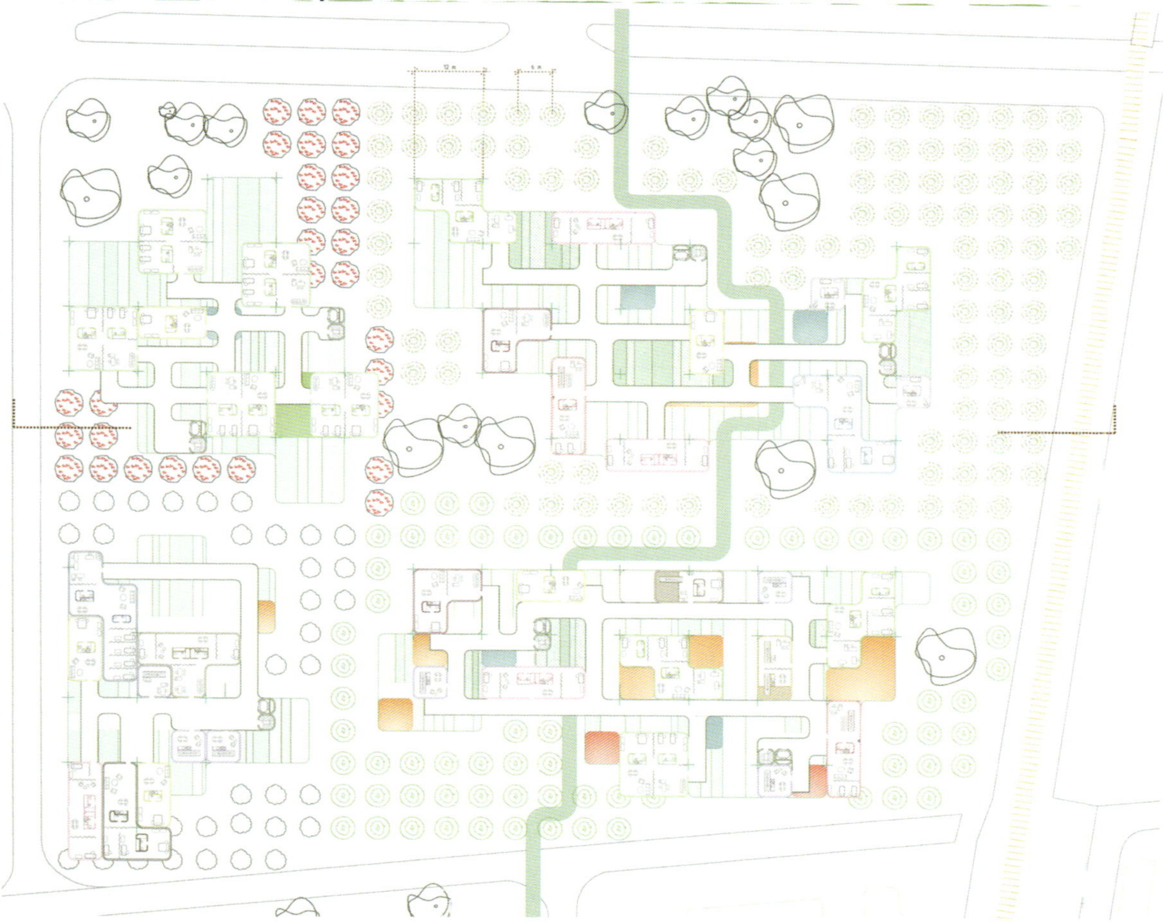

二层平面图 FLOOR PLAN 1

增长
INCREASING

冬天 WINTER

绿地
GREEN AREAS

商铺和购物
COMMERCE & SHOPPING

火车轨道
TRAIN RAIL

自行车道
BICYCLE RAIL

休闲
LEISURE

(竞标代码) Daily Zoo

Enrique Krahe
(建筑师)

荣誉提名奖 honourable mention

第纳尔的过渡比例

Van Schaikweg北部的公寓楼有6层，而这里独栋式的住宅有3层。第纳尔清楚地体现了这一比例上的过渡。

DZD TRANSITION SCALE

Block apartments in the north of Van Schaikweg reach up to 6 floors, while detached houses in the project site exceptionally reach 3. DZD articulates the transition of scales.

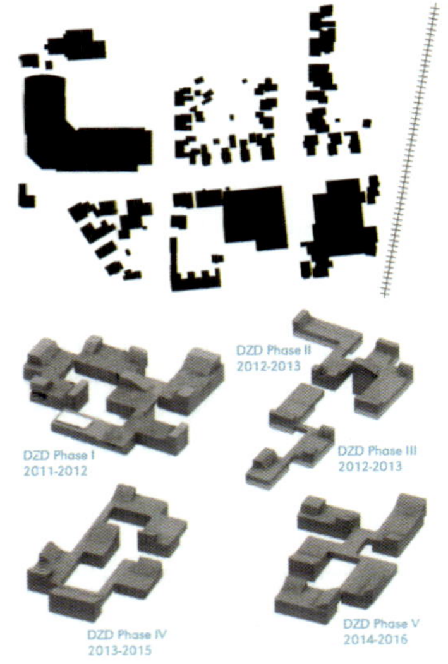

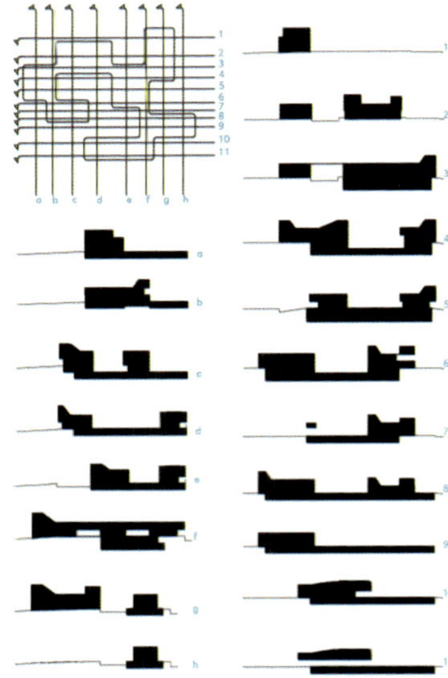

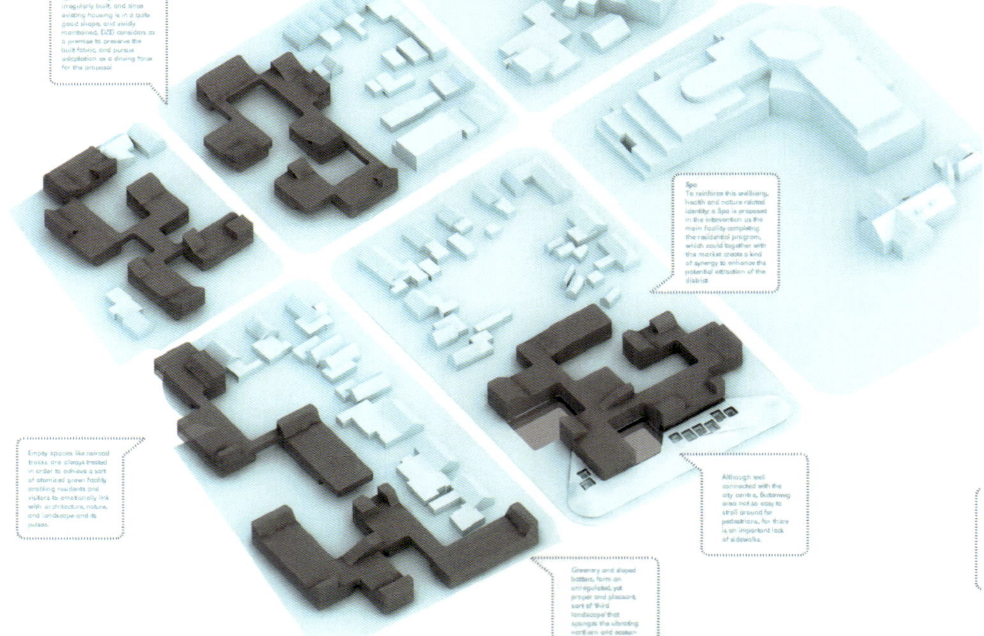

发展

南部的购物中心将在短期内被拆除。这将是一次增加住房密度的好机会。

GROWTH

Malls in the south of the site of intervention are intended to be demolished in a short period. That would be a wise opportunity to increase housing density.

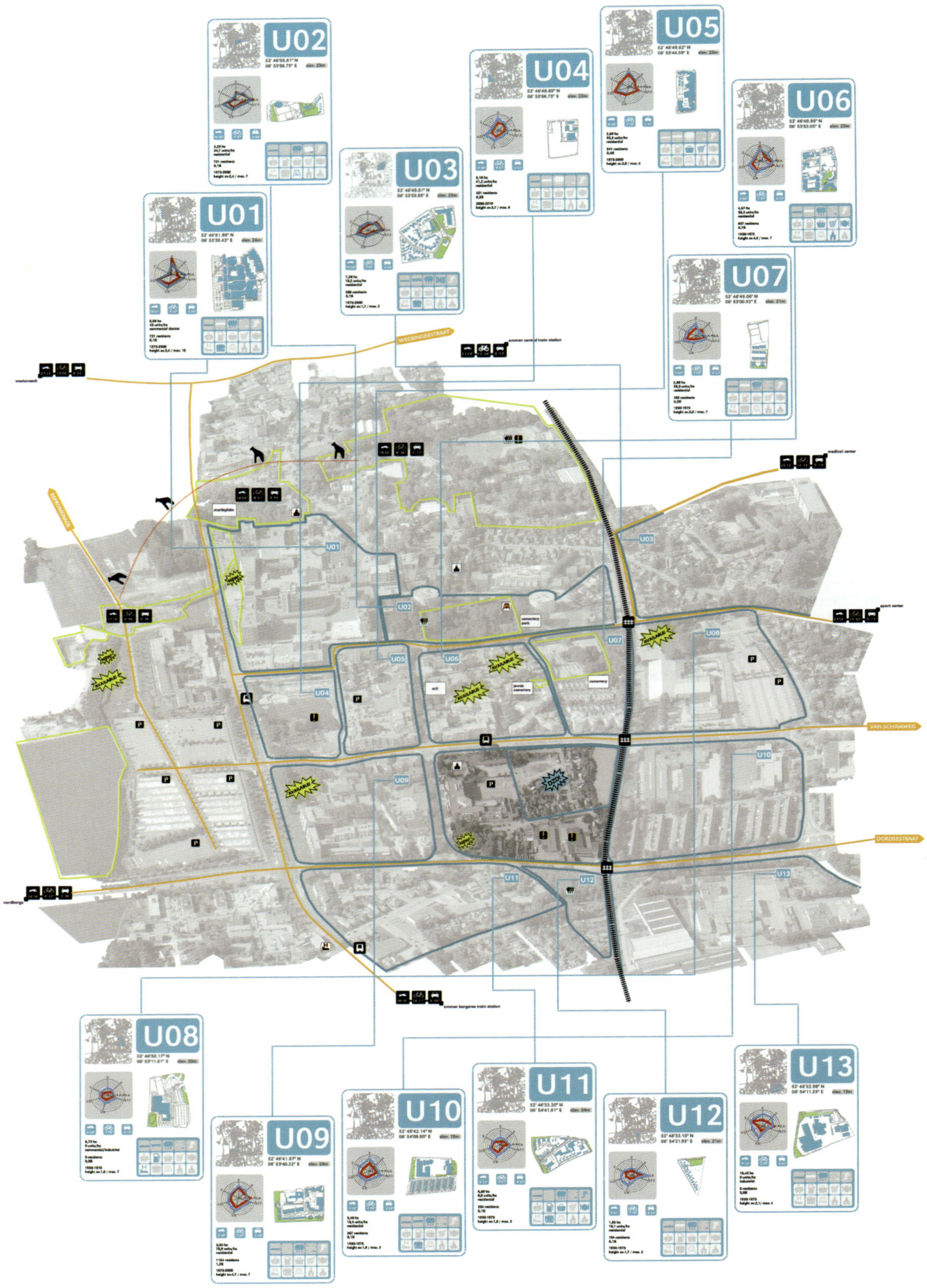

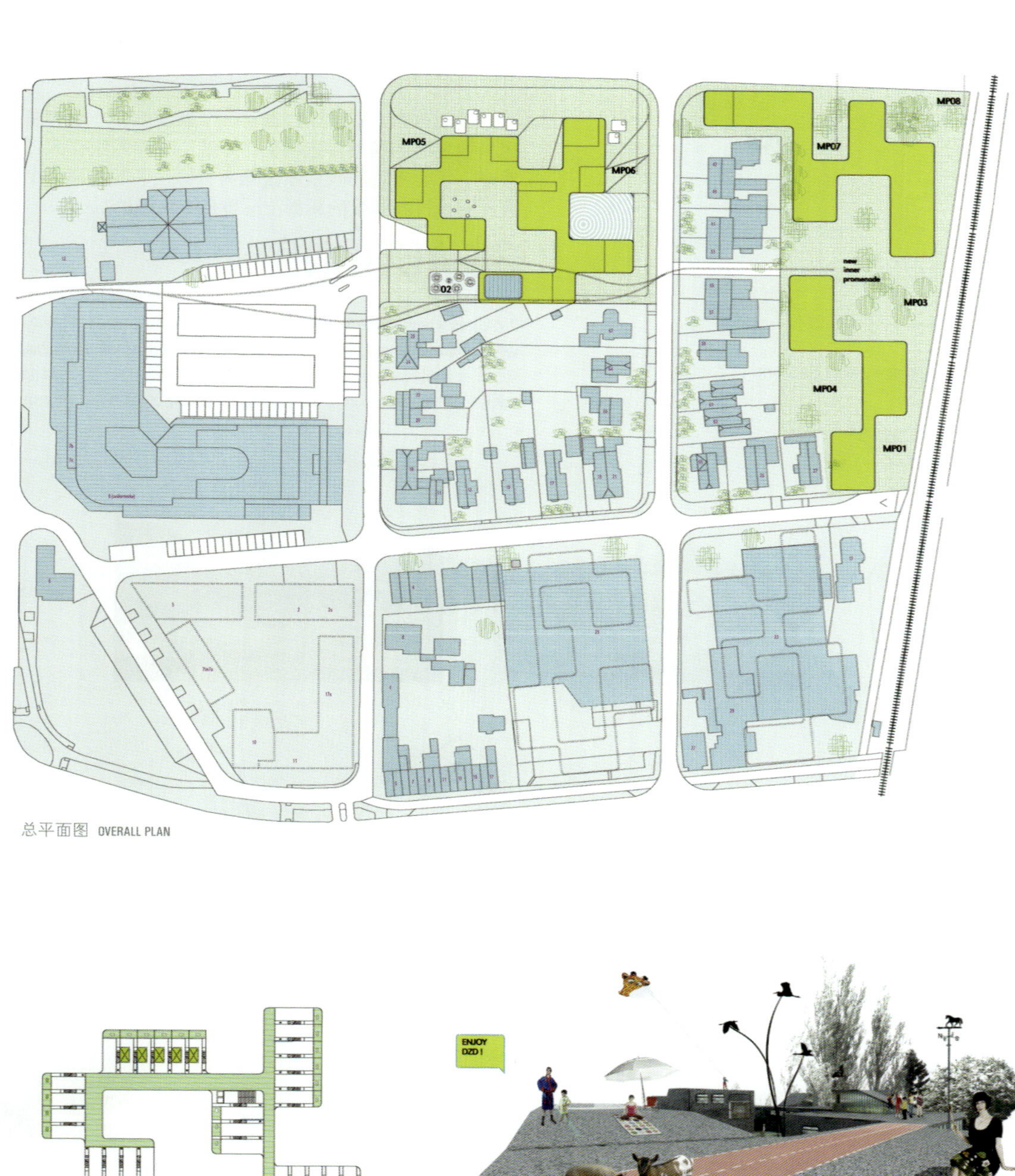

总平面图 OVERALL PLAN

二层平面图 FLOOR PLAN 1

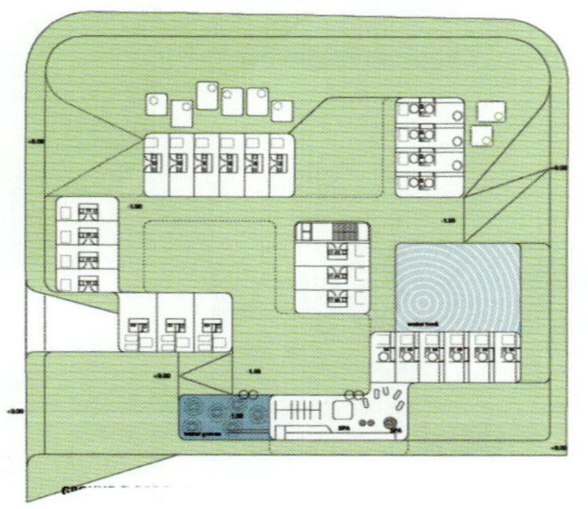

底层平面图 GROUND FLOOR PLAN

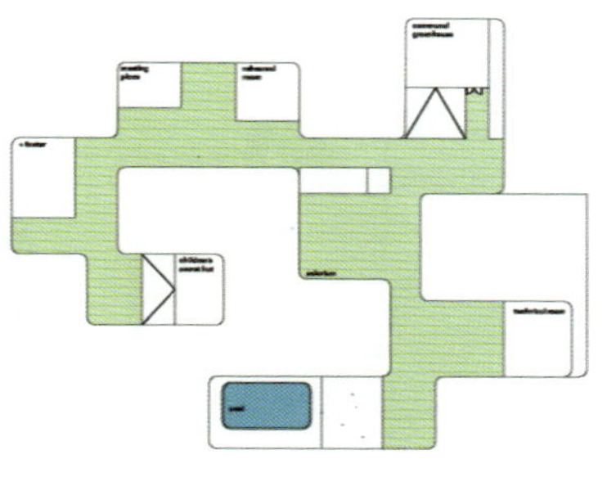

露天层平面图 TERRACE FLOOR PLAN

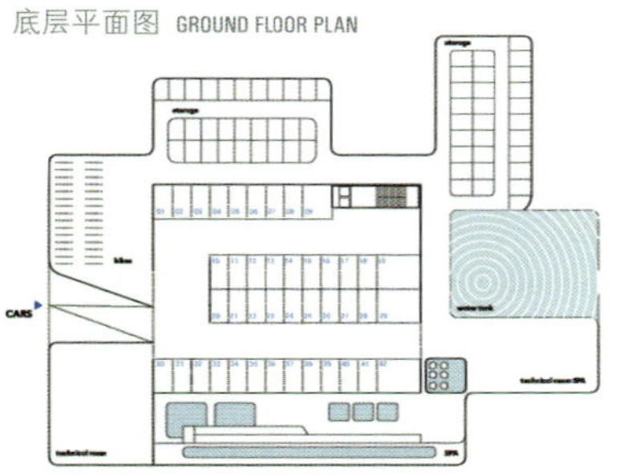

地下层平面图 UNDERGROUND FLOOR PLAN

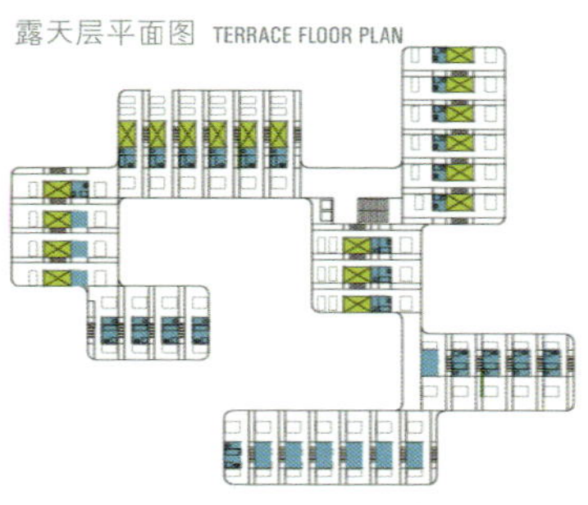

三层平面图 FLOOR PLAN 2

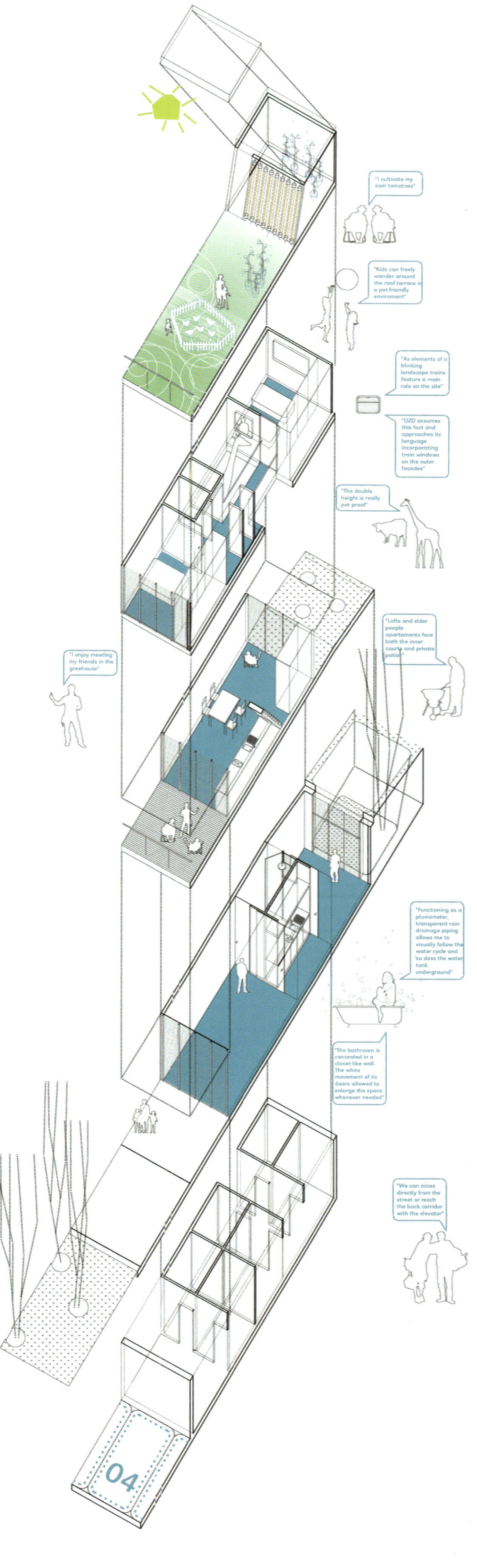

卡塞雷斯 · 西班牙 Cáceres · Spain

(竞标代码) Reactivate la ribera

ToTem arquitectos asociados

(建筑师事务所)
Javier Garcia-Germán · Alia Garcia-Germán (建筑师)
合作 (c) Lola García-Germán · Carlos Moreno · Alberto Porras · Sara Ruiz-Valdepeñas · Jesús Sanabria
Albert Martínez · Encarna Serna

中标 winner

36 种活动

该项目在卡塞雷斯打造了一个新空间，一个有关可持续发展的模型，它将生态系统、生产系统、人口系统、社会系统和娱乐系统融为一体，表明构思一个兼具有机性和强度的新都市环境是有可能的。

36 ACTIONS

It creates a new space for Caceres, a model in terms of sustainability, which makes it compatible with natural, productive, demographic, social and recreational systems, demonstrating that it is possible to think about a new urbanity that is simultaneously organic and intense.

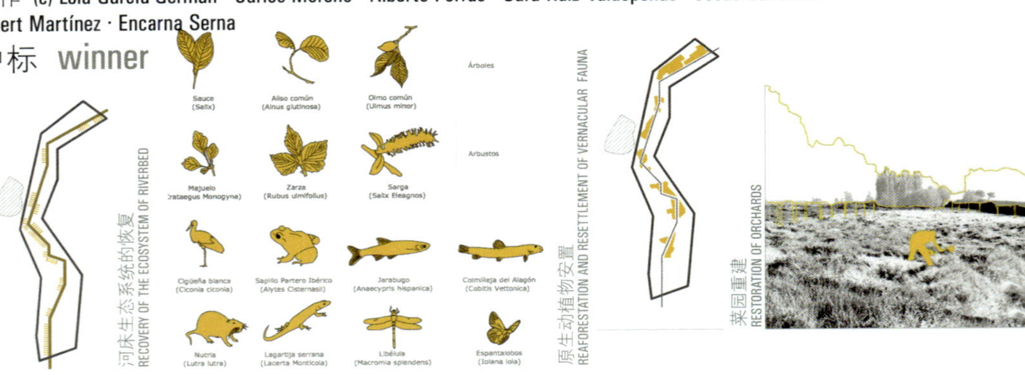

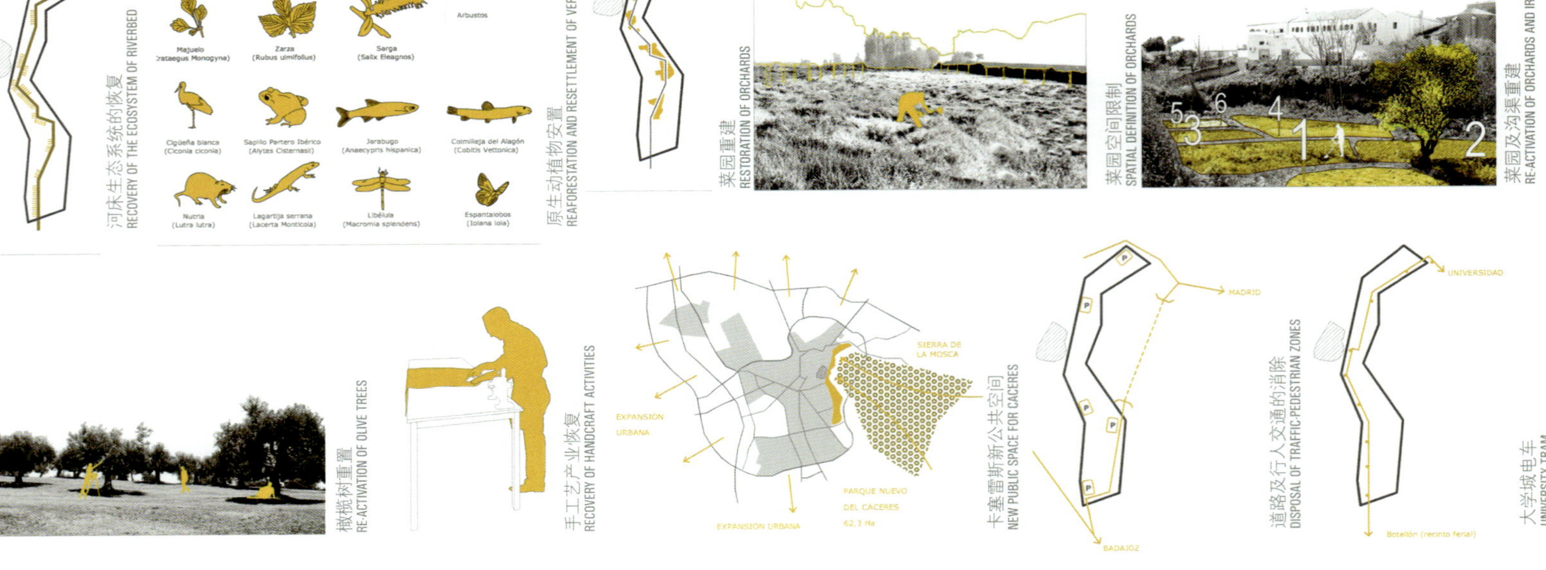

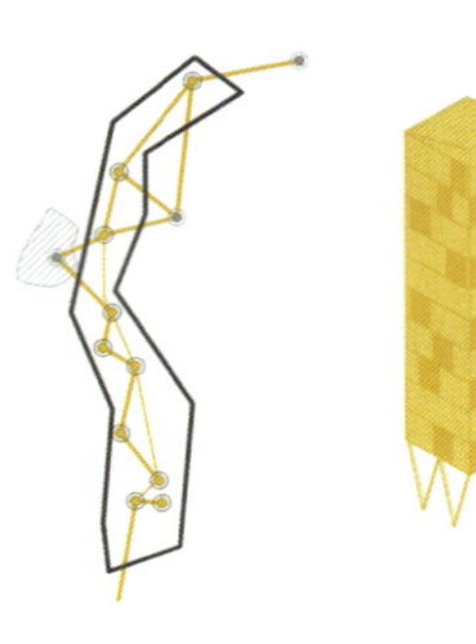

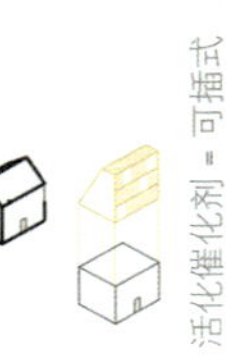

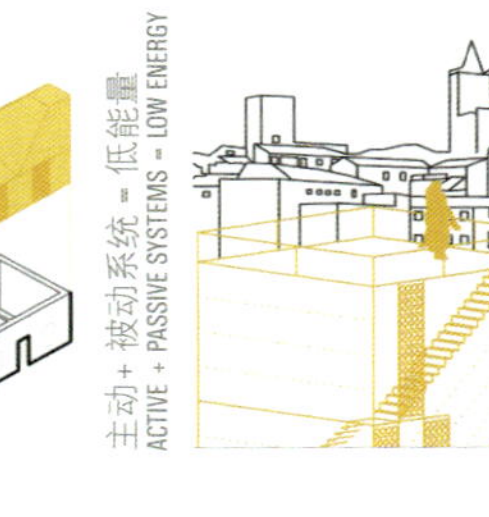

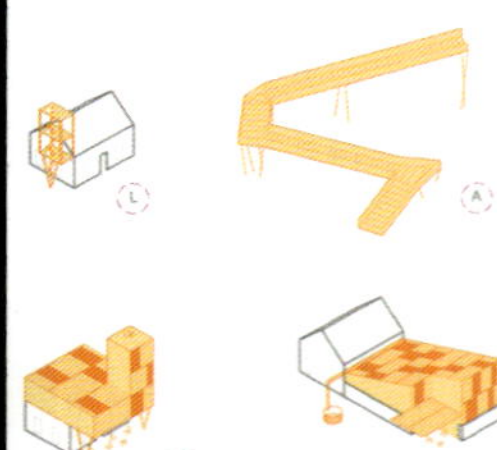

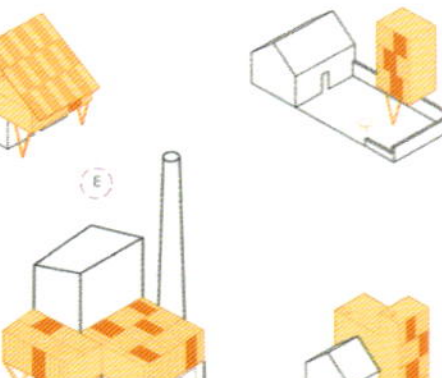
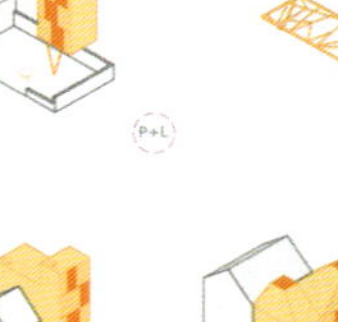

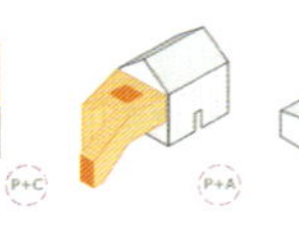
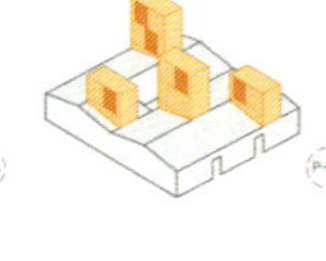

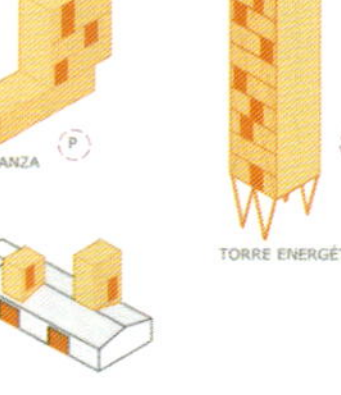

夹式目录: 河床现有建筑的活化剂 CLIP-ON CATALOGUE: RE-ACTIVATORS OF EXISTING BUILDINGS OF THE RIVERBED

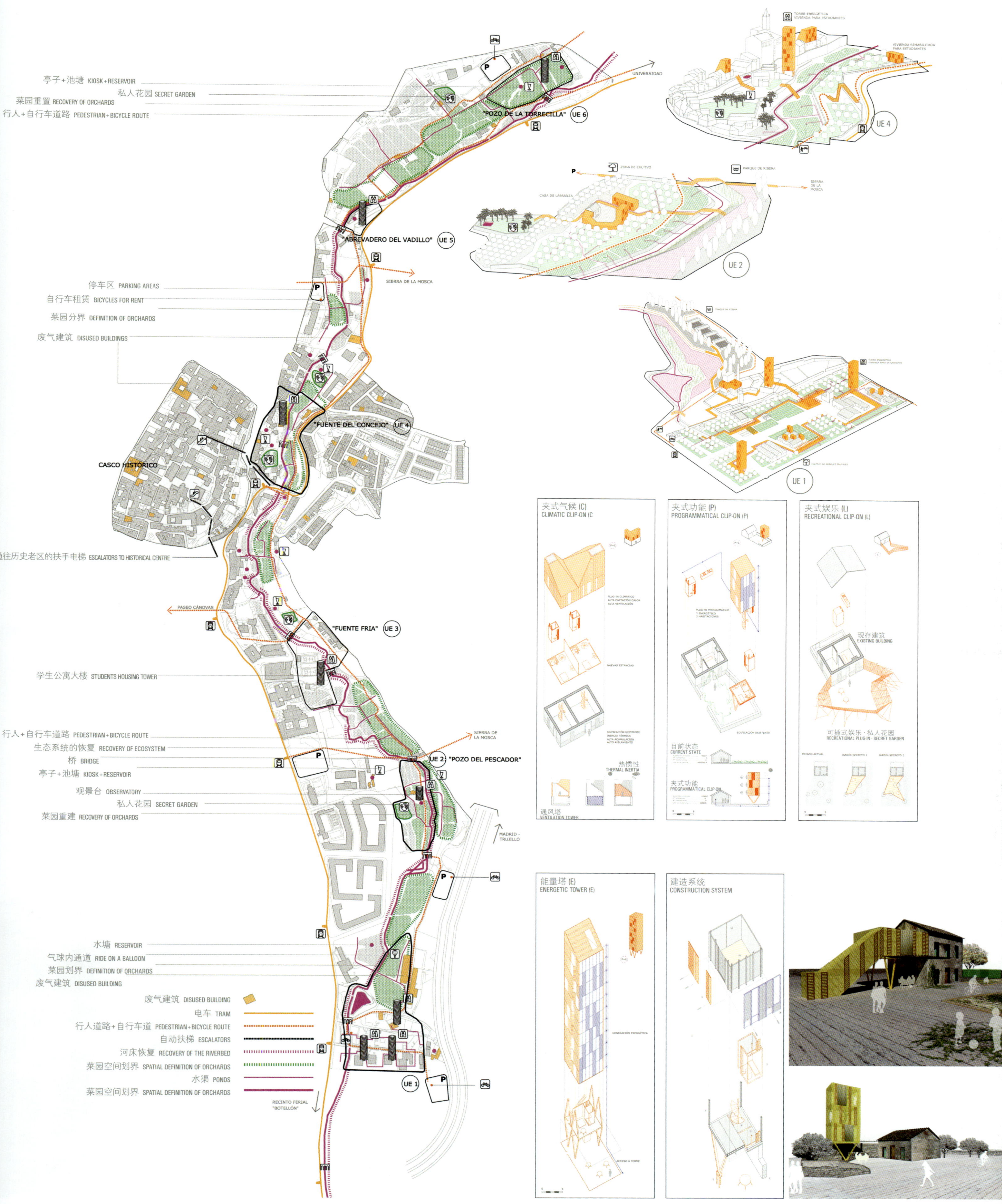

亭子+池塘 KIOSK+RESERVOIR
私人花园 SECRET GARDEN
菜园重置 RECOVERY OF ORCHARDS
行人+自行车道路 PEDESTRIAN+BICYCLE ROUTE
"POZO DE LA TORRECILLA" UE 6
UNIVERSIDAD
"ABREVADERO DEL VADILLO" UE 5
SIERRA DE LA MOSCA
停车区 PARKING AREAS
自行车租赁 BICYCLES FOR RENT
菜园分界 DEFINITION OF ORCHARDS
废气建筑 DISUSED BUILDINGS
"FUENTE DEL CONCEJO" UE 4
CASCO HISTÓRICO
通往历史老区的扶手电梯 ESCALATORS TO HISTORICAL CENTRE
PASEO CÁNOVAS
"FUENTE FRIA" UE 3
学生公寓大楼 STUDENTS HOUSING TOWER
行人+自行车道路 PEDESTRIAN+BICYCLE ROUTE
生态系统的恢复 RECOVERY OF ECOSYSTEM
桥 BRIDGE
亭子+池塘 KIOSK+RESERVOIR
观景台 OBSERVATORY
私人花园 SECRET GARDEN
菜园重建 RECOVERY OF ORCHARDS
SIERRA DE LA MOSCA
UE 2 "POZO DEL PESCADOR"
MADRID - TRUJILLO
水塘 RESERVOIR
气球内通道 RIDE ON A BALLOON
菜园划界 DEFINITION OF ORCHARDS
废气建筑 DISUSED BUILDING
废气建筑 DISUSED BUILDING
电车 TRAM
行人道路+自行车道 PEDESTRIAN+BICYCLE ROUTE
自动扶梯 ESCALATORS
河床恢复 RECOVERY OF THE RIVERBED
菜园空间划界 SPATIAL DEFINITION OF ORCHARDS
水渠 PONDS
菜园空间划界 SPATIAL DEFINITION OF ORCHARDS
RECINTO FERIAL "BOTELLÓN"
UE 1
UE 4
UE 2
UE 1
夹式气候 (C) CLIMATIC CLIP-ON (C
通风塔 VENTILATION TOWER
夹式功能 (P) PROGRAMMATICAL CLIP-ON (P)
目前状态 CURRENT STATE
热惯性 THERMAL INERTIA
夹式功能 PROGRAMMATICAL CLIP-ON
夹式娱乐 (L) RECREATIONAL CLIP-ON (L)
现存建筑 EXISTING BUILDING
可插式娱乐·私人花园 RECREATIONAL PLUG-IN · SECRET GARDEN
能量塔 (E) ENERGETIC TOWER (E)
建造系统 CONSTRUCTION SYSTEM

(竞标代码) Swimming in the green river

Miguel Ángel Rupérez Escribano

(建筑师事务所)

二等奖 runner-up

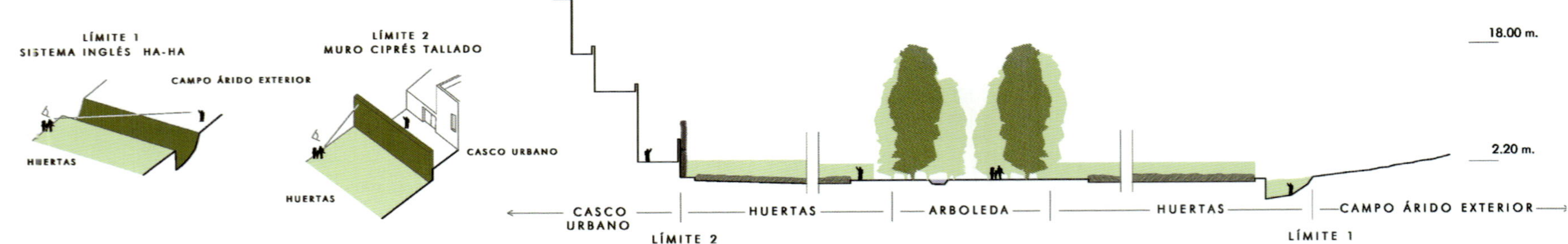

连贯性

过去，"Ribera del Marco"（卡塞雷斯成立和后续发展的重要地区）的主要工作人群是农民、制革工和石油生产商，我们应该在保留其精华的基础上对其主要功能进行升级。它应进行改革并向所有人开放，而记忆和可及性是其修复过程中的两个必要理念。

Ribera处于小树林的阴影下，可任你自由穿越。它将北部新建的住宅区与卡塞雷斯南部的新休闲空间连为一体，将成为城市中又一个公共花园。

住宅区与"Charco del Marco"相邻，"Charco del Marco"现在被称为"Lago del Marco"，是卡塞雷斯南部新兴的休闲繁华地带，实现了该项目的目标——连贯性，既保留了Ribera的完整性，又尽量避免了周围环境的和谐被破坏。

CONTINUITY

The mainly functional use in the past of the "Ribera del Marco" (key location in the founding and subsequent development of Caceres), where the main users were farmers, tanners and oil producers, should be updated without losing its essence. It should change, and be accessible to all. MEMORY and ACCESSIBILITY are the two ideas necessary for its recuperation and rehabilitation.

The Ribera can be traversed under the SHADOW of the grove. Connecting new residential growths in North with the new leisure place in South of Caceres, it can become a new city's public garden.

The dormitory is located next to the "Charco del Marco," now known as the "Lago del Marco," the new focal point for leisure in South Caceres, thus maintaining the CONTINUITY which was the aim of this project: preserving the integrity of the Ribera and trying to avoid perturbing the harmony of the surroundings.

领域范围内假设交织示意图
SKETCH OF INTERVENTIONS CONSIDERING A TERRITORIAL SCALE

Vista de la Ribera

CASCO HISTÓRICO

Vista del Casco

剖面图 A SECTION A

平面图. 单元房间 FLOOR PLAN. HOUSING UNIT

"马克湖" 总平面图 ENVIROMENTAL FLOOR PLAN. "LAGO DEL MARCO"

中午通过庭院斜射反射到起居室的光线
SLANTED LIGHT REFLECTED ON THE LIVING ROOM AT MIDDAY THROUGH THE COURTYARD

南面
SOUTH

西面
WEST

北面
NORTH

没有反射到住宅的光线
LIGHT NOT REFLECTED ON HOUSING

东面
EAST

早上通过庭院斜射入房间的水平光线
HORIZONTAL LIGHT REFLECTED ON THE LIVING ROOM IN THE MORNING THROUGH THE COURTYARD

HA-HA

来自东面的主导微风
DOMINANT EAST BREEZES

视觉连贯而非物质连贯
VISUAL CONTINUITY ALTHOUGH NOT PHYSICAL

6.10
HA-HA
+ 3.00
+ 0.00
- 4.05

剖面图 B SECTION B

(竞标代码) Back to the ribera

Pilar Gutiérrez Pulido · Elena Pérez Palacios

(建筑师)

荣誉提名奖 honourable mention

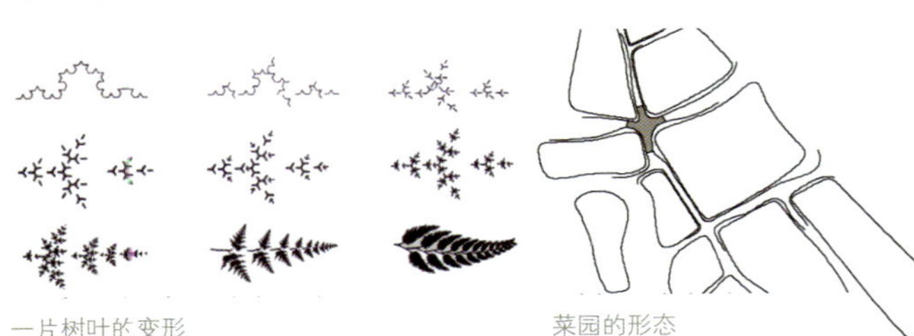

一片树叶的变形
TRANSFORMATION OF A LEAF

菜园的形态
MORPHOLOGY OF THE ORCHARDS

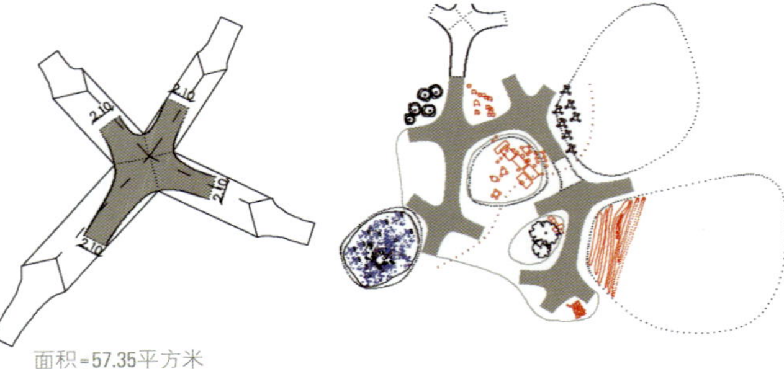

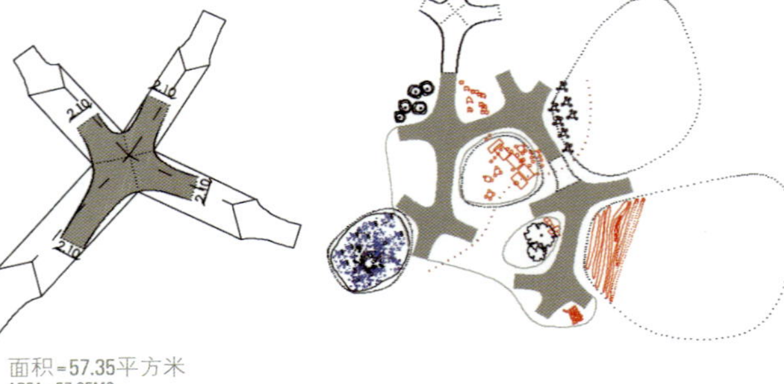

面积=57.35平方米
AREA=57,35M2

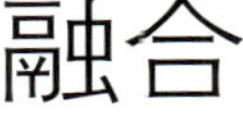

融合

我们打算恢复Ribera的原有生态系统（如集市、花园、工厂……），并增添一项新内容——“慢节奏空间’。“慢节奏空间”的主旨与新的生活方式有关，鼓励人们：

—穿行于卡塞雷斯之间时以步代车，或者选骑自行车。

—支持有利于可持续发展的、健康的社交活动（修建学校花园、回收工作坊、生态集市、步行道……）。Ribera de Marco成了一个汇合了各种不同社交团体的大型公共空间/生态集会地点。

我们选择了一个由四个固定的标准结构组成的系统，这四个结构以自然的方式融合在一起，组成了更大的有多重角度的结构。这些融合的大型结构内部有椭圆形的庭院，有的庭院还带有屋顶。它们大小各异以满足不同的内部活动开展的需要。

COMBINATIONS

Our intervention tries to recover the original Ribera´s ecosystem (such as markets gardens, mills...) and provide a new content: "slow spaces". A backbone of "slow spaces" for the city, related to a new lifestyle, encouraging people to:

-move and connect on foot and by bicycle between parts of Cáceres as an alternative to the car routes.

-take part of sustainable, social and healthy activities (school-gardens, workshops of recycling, ecological market, walks...). Ribera del Marco becomes a great public space / a meeting eco-point that integrates different social groups.

We choose a system of 4 fixed standard pieces which can be combined in a natural way to achieve bigger units with a variety of angles. The bigger-combined structures leave oval patios inside, sometimes covered by a roof. They grow with different sizes adapting the inside program to the necessities.

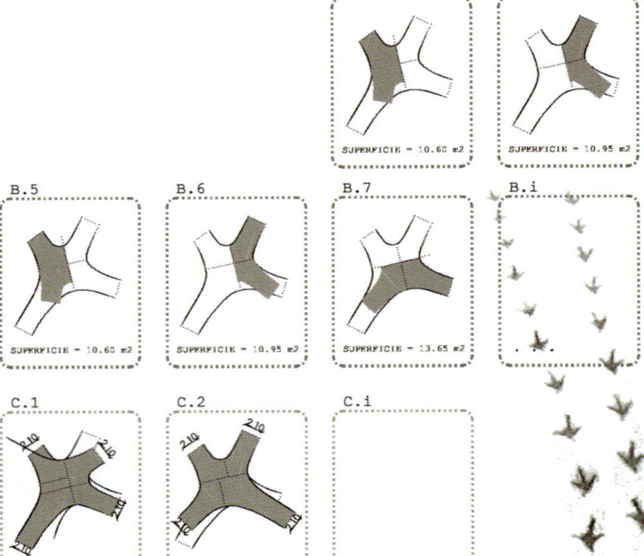

区块的目录 CATALOGUE OF PIECES

区块的分布 DISTRIBUTION OF PIECES

孤立的 ISOLATED

归类的 GROUPPED

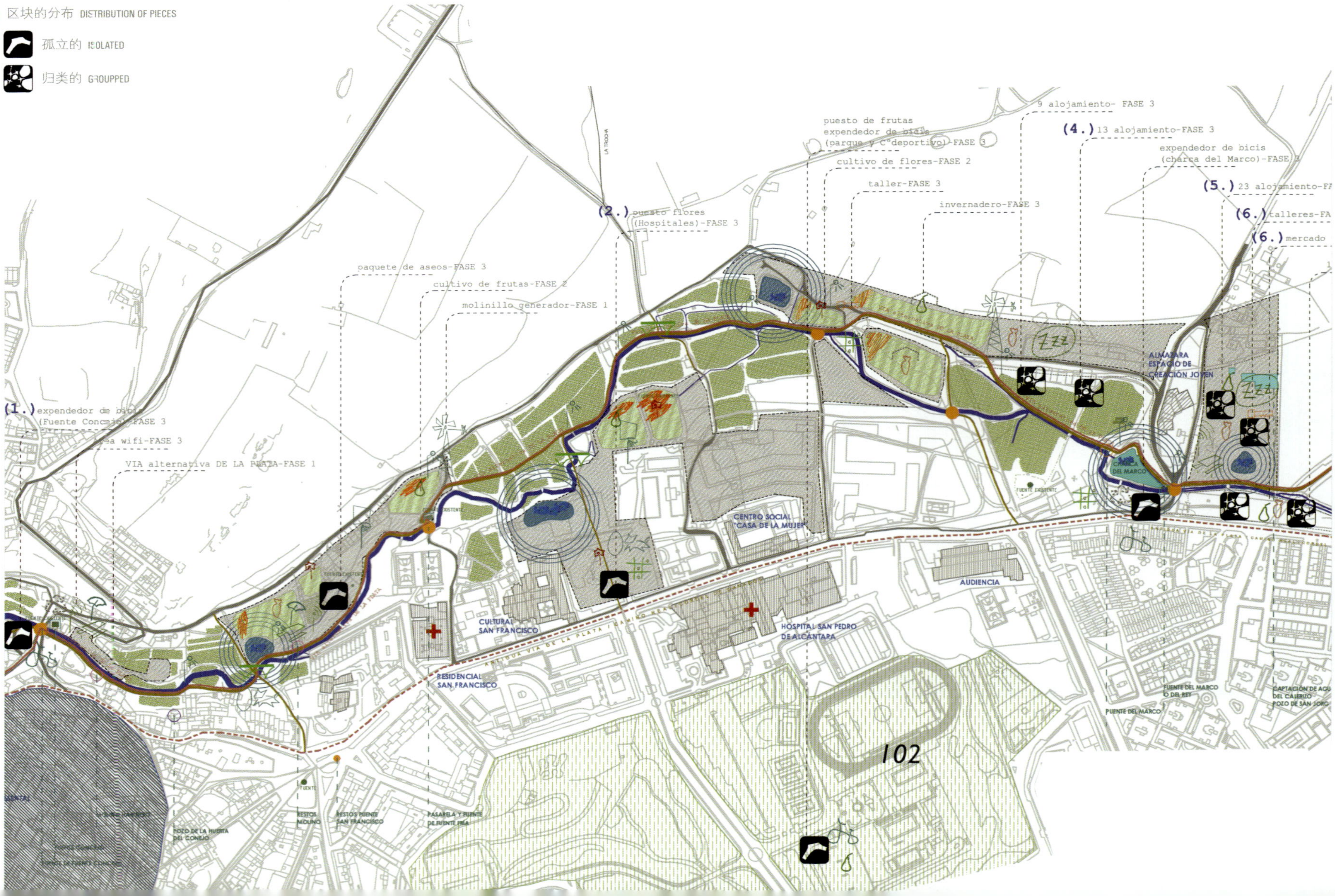

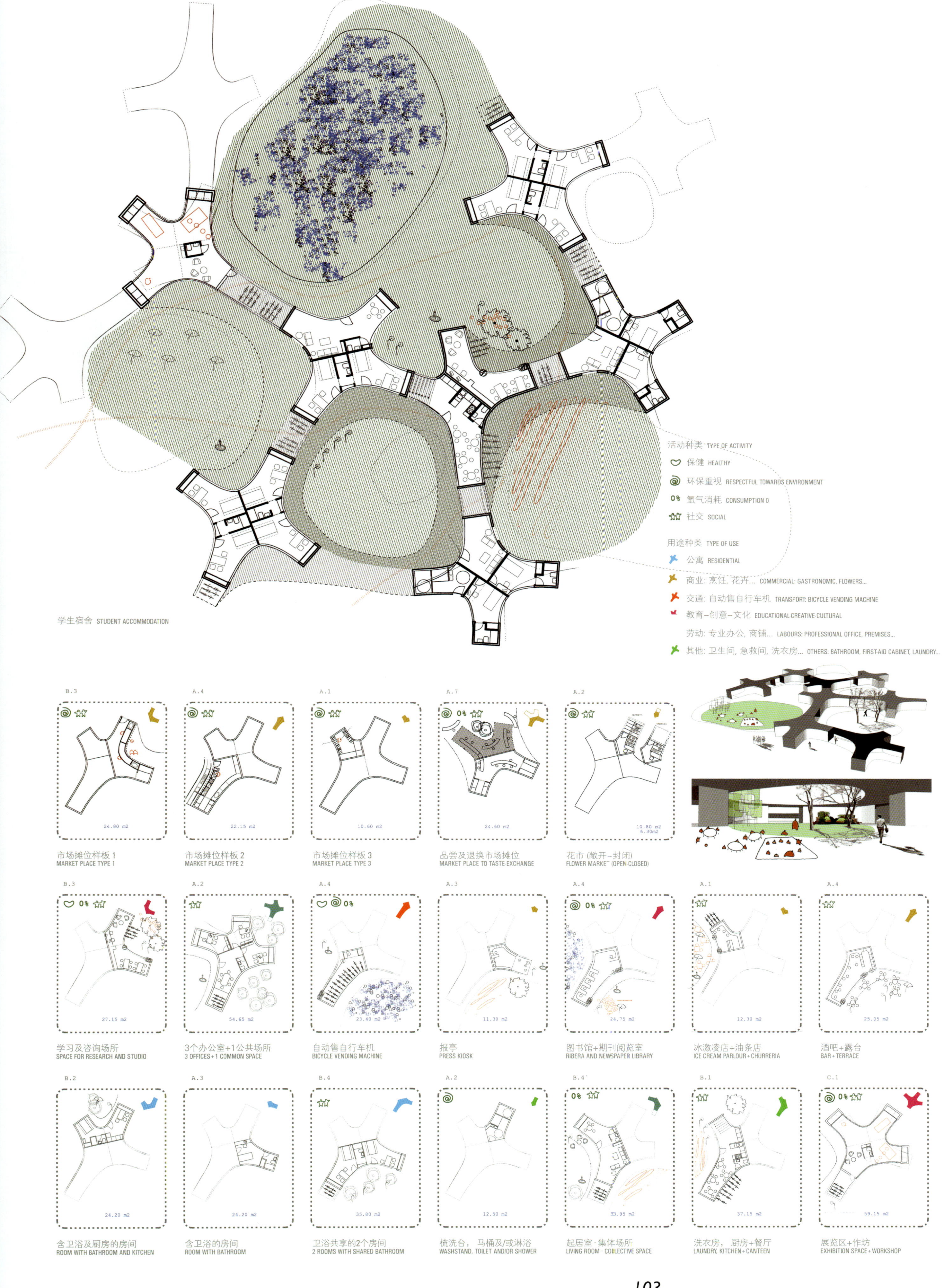

学生宿舍 STUDENT ACCOMMODATION
活动种类 TYPE OF ACTIVITY
保健 HEALTHY
环保重视 RESPECTFUL TOWARDS ENVIRONMENT
0% 氧气消耗 CONSUMPTION 0
社交 SOCIAL
用途种类 TYPE OF USE
公寓 RESIDENTIAL
商业：烹饪，花卉… COMMERCIAL: GASTRONOMIC, FLOWERS...
交通：自动售自行车机 TRANSPORT: BICYCLE VENDING MACHINE
教育–创意–文化 EDUCATIONAL CREATIVE CULTURAL
劳动：专业办公，商铺… LABOURS: PROFESSIONAL OFFICE, PREMISES...
其他：卫生间，急救间，洗衣房… OTHERS: BATHROOM, FIRST-AID CABINET, LAUNDRY...
B.3
24.80 m2
市场摊位样板 1
MARKET PLACE TYPE 1
A.4
22.15 m2
市场摊位样板 2
MARKET PLACE TYPE 2
A.1
10.60 m2
市场摊位样板 3
MARKET PLACE TYPE 3
A.7
0%
24.60 m2
品尝及退换市场摊位
MARKET PLACE TO TASTE-EXCHANGE
A.2
10.80 m2
6.30m2
花市（敞开–封闭）
FLOWER MARKET (OPEN-CLOSED)
B.3
0%
27.15 m2
学习及咨询场所
SPACE FOR RESEARCH AND STUDIO
A.2
54.65 m2
3个办公室+1公共场所
3 OFFICES + 1 COMMON SPACE
A.4
0%
23.40 m2
自动售自行车机
BICYCLE VENDING MACHINE
A.3
11.30 m2
报亭
PRESS KIOSK
A.4
0%
24.75 m2
图书馆+期刊阅览室
RIBERA AND NEWSPAPER LIBRARY
A.1
12.30 m2
冰激凌店+油条店
ICE CREAM PARLOUR + CHURRERIA
A.4
25.05 m2
酒吧+露台
BAR + TERRACE
B.2
24.20 m2
含卫浴及厨房的房间
ROOM WITH BATHROOM AND KITCHEN
A.3
24.20 m2
含卫浴的房间
ROOM WITH BATHROOM
B.4
35.80 m2
卫浴共享的2个房间
2 ROOMS WITH SHARED BATHROOM
A.2
12.50 m2
梳洗台，马桶及/或淋浴
WASHSTAND, TOILET AND/OR SHOWER
B.4
0%
33.95 m2
起居室·集体场所
LIVING ROOM · COLLECTIVE SPACE
B.1
37.15 m2
洗衣房，厨房+餐厅
LAUNDRY, KITCHEN + CANTEEN
C.1
0%
59.15 m2
展览区+作坊
EXHIBITION SPACE + WORKSHOP

Isaac Martí Alabart · Sergi Lois Alcázar · Maria Cifuentes Rius
(建筑师)

荣誉提名奖 honourable mention

水: 沿着河床的纵向联系 WATER: LONGITUDINAL RELATIONSHIPS FOLLOWING THE RIVERFRONT

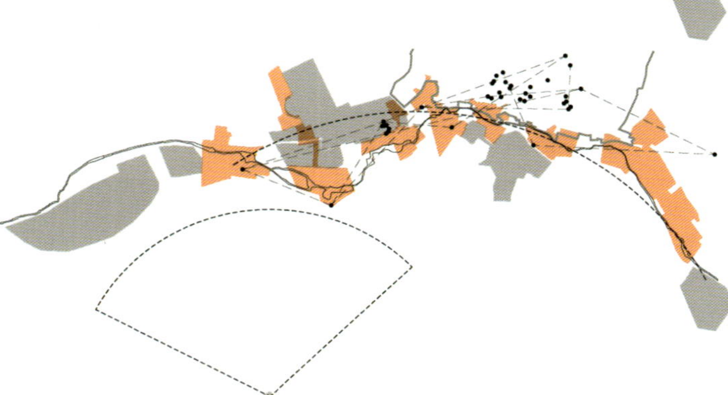

土: 通过河床的横向联系 LAND: CROSS RELATIONSHIPS CROSSING THE RIVERFRONT

空气: 周边与河床建的视觉联系 AIR: VISUAL RELATIONSHIPS BETWEEN THE SURROUNDING AND THE RIVERFRONT

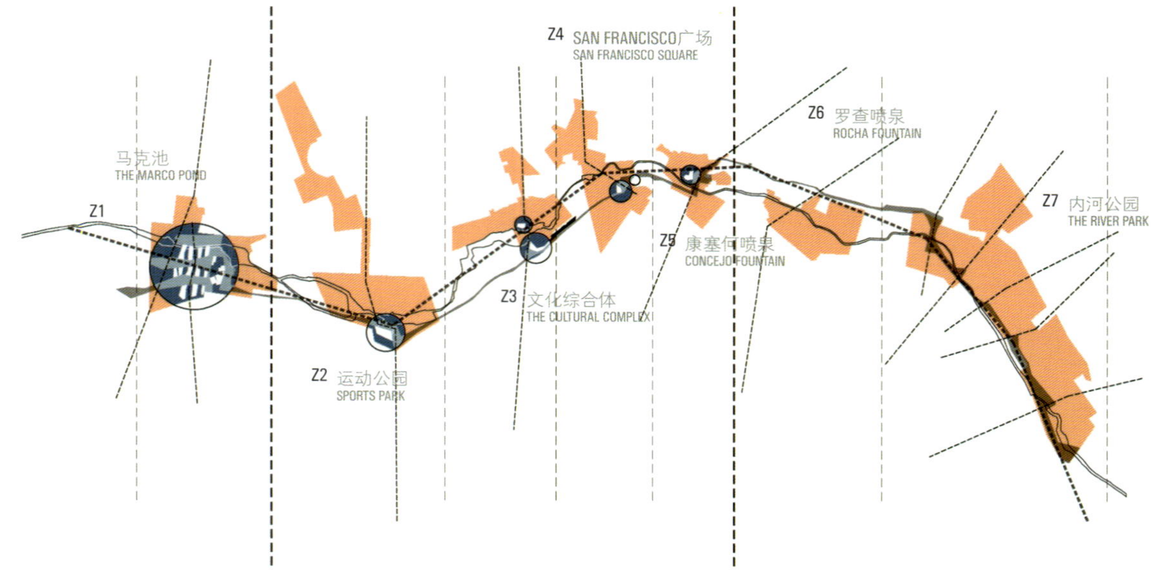

三种大体关系

河床呈纵向延伸，因此有利于将河床改造成为城市和自然之间的断口。该项目意在完善沿岸路径并为城市打造一系列空间。每个区域均位于城市和乡村的交叉点，都有一个交叉路口，起着桥梁作用，连接两岸，并将都市结构与自然景观连为一体。我们将在大厦之间打造自然景观，让其成为周边风景的一部分，而且这里还可成为鹳的栖息地。该项目在设计时，设想河床沿岸其他大厦将一一拔地而起，到那时，城市容纳了自然，自然融入了城市，彼此结为一体。

THREE GLOBAL RELATIONSHIPS

The Riverbed has a longitudinal dimension. This condition has made the riverbed turn into a fracture between city and nature. The intention is to enhance the routes along the bank and create a sequence of spaces for the city. Each area promotes a cross connection. They are bridges that lead from one side to another of the bank and stitch the urban fabric with the landscape. Each area locates in a meeting point between the city and the countryside. We establish a network of visual relationships between the towers that are part of the landscape. This is the habitat of storks which cross the sky. The project envisages the emergence of other towers in the riverbed so that the city can colonize the nature.

Teruel · Spain 特鲁埃尔 · 西班牙

(竞标代码) oioio — Genius Locci

Berta Barrio Uria · Josep Peraire Selva

(建筑师)

中标 winner

MUELA DE CASCO HISTÓRICO
MUELA NORTE
MUELA DEL ENSANCHE
ÁMBITO DE PROYECTO
LA VEGA
MUELA DEL PINAR
TURIA

背景 CONTEXT

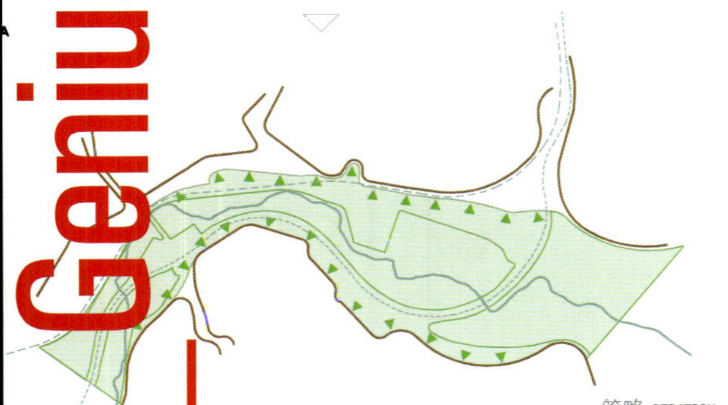

策略 STRATEGY

草场目前占用率 CURRENT OCCUPATION OF THE MEADOW

自然占用率方案 NATURAL OCCUPATION PROPOSAL

传统的白杨树种植系统 TRADITIONAL SYSTEM TO CULTIVATE BLACK POPLARS

白杨树田 BLACK POPLARS FIELD

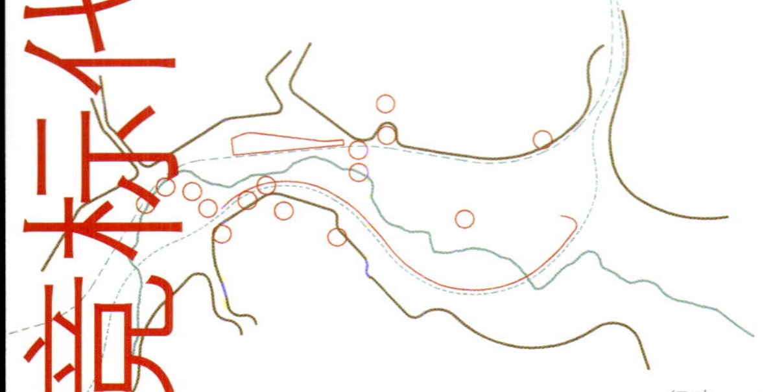

行动 ACTIONS

行动所需空间 SPACES FOR INTERVENTION

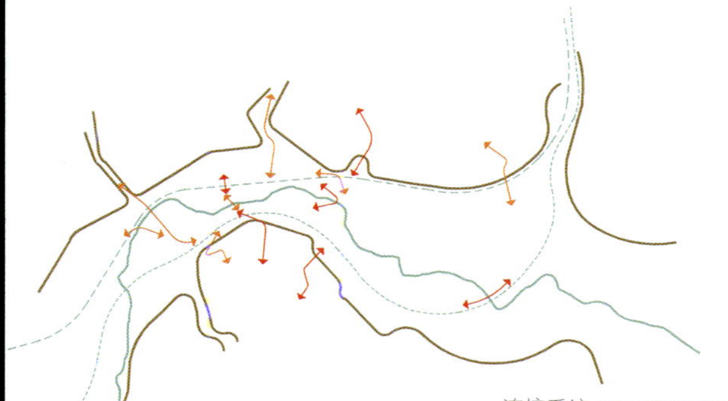

连接系统 CONECTION SYSTEM

现有连接加固 CONSOLIDATION OF EXISTING CONNECTION

新的连接 NEW CONNECTION

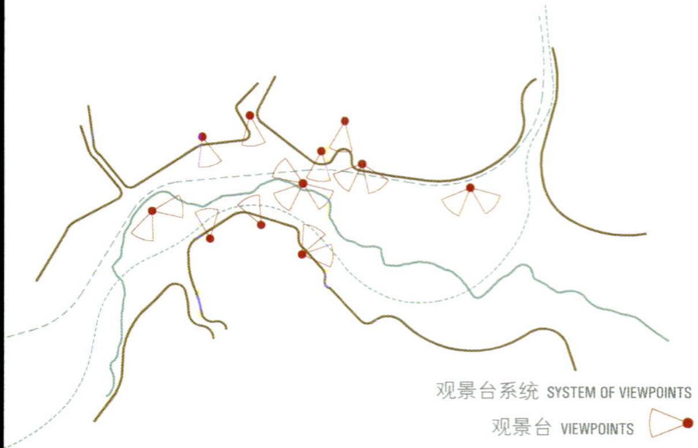

观景台系统 SYSTEM OF VIEWPOINTS

观景台 VIEWPOINTS

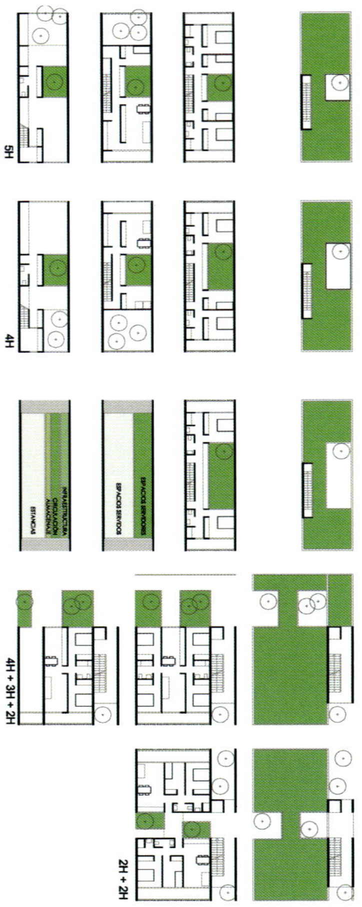

5H
4H
4H + 3H + 2H
2H + 2H

底层平面图 GROUND FLOOR PLAN

二层平面图 FLOOR PLAN 1

三至四层平面图 FLOOR PLAN 2-3

屋顶平面图 ROOF PLAN

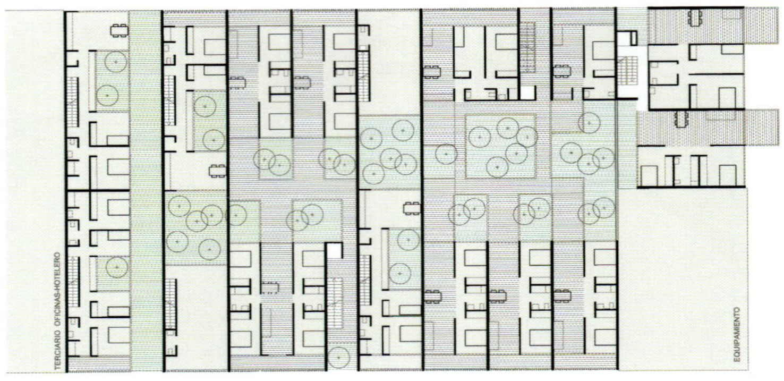

TERCIARIO OFICINAS-HOTELERO
EQUIPAMENTO

公寓单元样式 EXAMPLE OF RESIDENTIAL UNITS

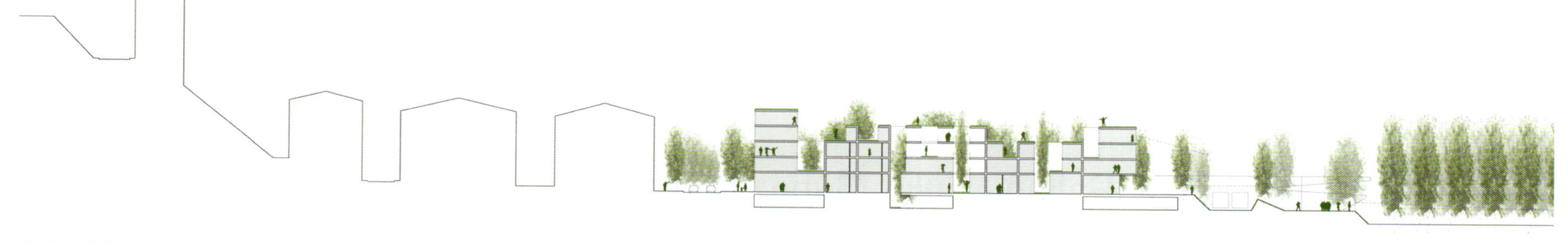

草场的空间连续 SPATIAL SEQUENCE OF THE MEADOW

不同的关联

都市的压力造成了与草场环境格格不入的外表和功能，如基础设施和后勤保障、工业项目和住宅工程。如今，草场被弄得支离破粹，与特鲁埃尔之间的关联弱化了，也不再属于特鲁埃尔了。该项目旨在通过以下方式使草场融入城市的社会、经济和文化生活：

1. 空间复原和功能复兴，加入公共活动（种子带来新的活力）。

2. 完善其与城市之间的可达性。

该项目通过长期以来因地制宜的具体策划提出了一个可灵活调整的计划。

草场的活力来自于它的生产活动以及自然特征，基于此我们能将草场打造成一个重要的生活场所和城市的未来空间，因此我们不能仅仅将其视为一个静态而刻板的景观。

DIFFERENT CONNECTIONS

The urban pressure has generated the appearance of elements and uses that are not proper in the meadow, as major infrastructures and logistics, industrial and residential programs. Today the meadow is fragmented, weak and isolated from Teruel. The aim is to integrate the meadow into the social, economic and cultural life of the city by:
1. The recovery of space and the revitalization of its own uses, injecting public activities (seeds generating new dynamics).
2. The generation of good accessibility from the city.
The project proposes a gobal and flexible order through detailed interventions developed over the time and adaptable to the conditions of each moment.

The meadow must not be understood as a static and frozen landscape, since its energy lies precisely on its productive activity and on its natural features that transform it into a critical space for life and the future of the city.

Teruel fusion (竞标代码)

Xabier Barrutieta (建筑师)
Team Layercake:Manuel Galián · Julio Muñoz · Héctor Torres
Jaime Traspaderne · Begoña Arrate

二等奖 runner-up

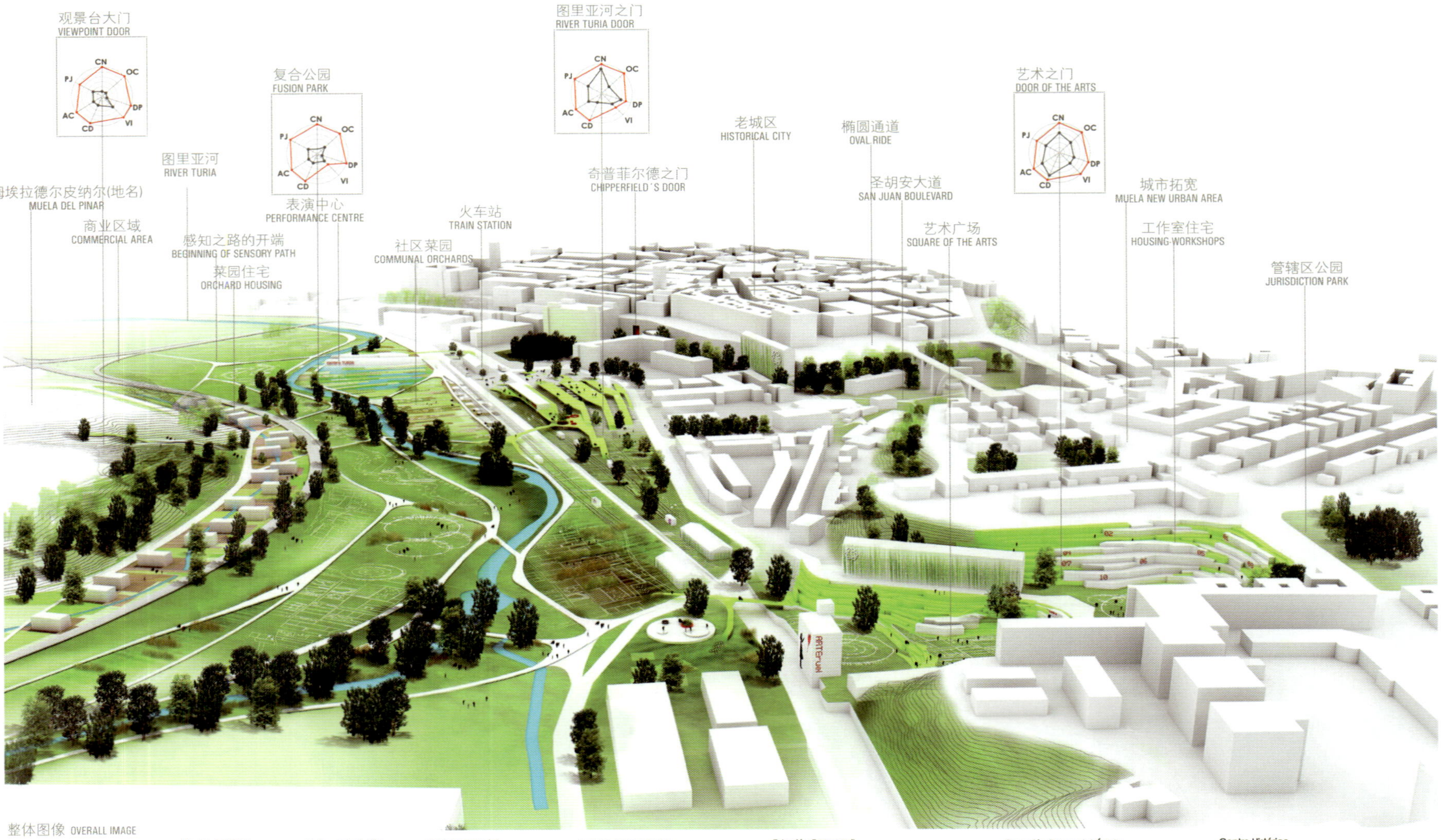

观景台大门 VIEWPOINT DOOR
复合公园 FUSION PARK
图里亚河之门 RIVER TURIA DOOR
艺术之门 DOOR OF THE ARTS
图里亚河 RIVER TURIA
姆埃拉德尔皮纳尔(地名) MUELA DEL PINAR
商业区域 COMMERCIAL AREA
感知之路的开端 BEGINNING OF SENSORY PATH
菜园住宅 ORCHARD HOUSING
表演中心 PERFORMANCE CENTRE
社区菜园 COMMUNAL ORCHARDS
火车站 TRAIN STATION
奇普菲尔德之门 CHIPPERFIELD´S DOOR
老城区 HISTORICAL CITY
椭圆通道 OVAL RIDE
圣胡安大道 SAN JUAN BOULEVARD
艺术广场 SQUARE OF THE ARTS
城市拓宽 MUELA NEW URBAN AREA
工作室住宅 HOUSING-WORKSHOPS
管辖区公园 JURISDICTION PARK

整体图像 OVERALL IMAGE

Muela del Pinar / Bulevar de fusión / Nodo comercial / Huertos comunitarios / Estación Ferrocarril / Conexión Paseo del Óvalo / Centro Histórico

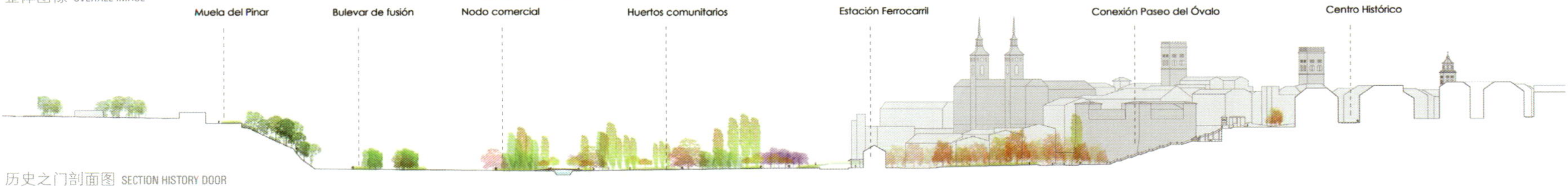

历史之门剖面图 SECTION HISTORY DOOR

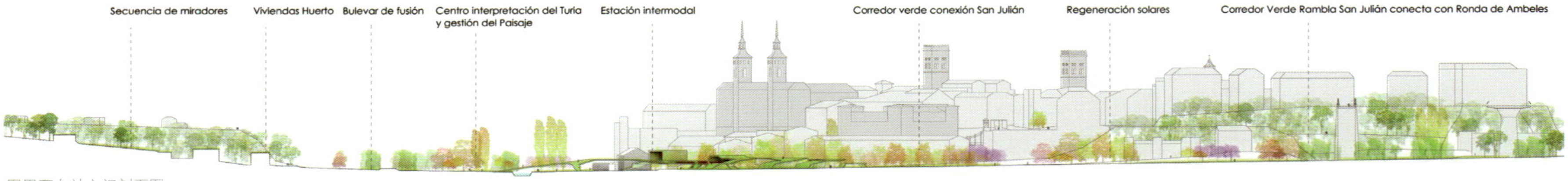

Secuencia de miradores / Viviendas Huerto / Bulevar de fusión / Centro interpretación del Turia y gestión del Paisaje / Estación intermodal / Corredor verde conexión San Julián / Regeneración solares / Corredor Verde Rambla San Julián conecta con Ronda de Ambeles

图里亚车站之门剖面图 SECTION TURIA´S STATION DOOR

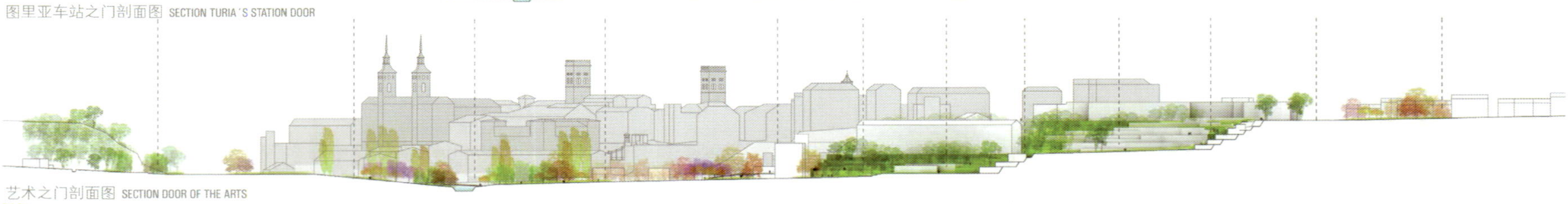

艺术之门剖面图 SECTION DOOR OF THE ARTS

该建筑的每扇门都综合了不同的都市元素。它们通过“历史之门”、“图里亚车站大门”、“艺术之门”和“米拉多大门”打造了一系列连接图里亚低地和特鲁埃尔高城的磁极。低地将成为一个景色优美、生机盎然的景观，集休闲、运动和生态科技为一体，并将循环产业图标和铁路设施打造成富有艺术性和景观性的结构。

Each gate is a singular combination of different urban components. They shape a series of new magnetic poles that connect the Turia lowlands and the high city of Teruel through the Gate of History, the Turia Station Gate, the Gate of the Arts and the Mirador Gate. The lowlands will be a dynamic and sensorial landscape that will host leisure, sports, and ecotechnology nodes, and foresees recycling industrial icons and railway facilities as art and landscape objects.

低程度景观介入
LOW DEGREE OF LANDSCAPE INTERVENTION
中程度景观介入
MEDIUM DEGREE OF LANDSCAPE INTERVENTION
高程度景观介入
HIGH DEGREE OF LANDSCAPE INTERVENTION
视觉趣味路线
ROUTE WITH VISUAL INTEREST
内部路线
INTERIOR ROUTES
图里亚小树林
GROVE OF THE TURIA
地形路线
TOPOGRAPHICAL ROUTES
图里亚水上公园
AQUATIC PARK OF THE TURIA
芳香花卉植被
VEGETATION MADE OF AROMATIC PLANTS
视觉趣味点
INTERESTING VIEWPOINTS
观景台
VIEWPOINT SPACE
咨询处
INFORMATION POINT
表演中心
PERFORMANCE CENTRE
社区菜园
COMMUNAL ORCHARDS
可回收车厢
RECYCLED WAGONS
植物园及农作物花园
BOTANNICAL AND AGRICULTURAL GARDENS
垂钓放生区
FISHING AREA. NO DEATH
防洪区
AREAS PREPARED FOR FLOODS
露天舞台
OUTDOOR SCENARIO
野餐，烧烤+儿童游戏
PICNICS, BARBEQUE + CHILDREN'S GAMES
足球+排球
FOOTBALL + VOLLEYBALL
景观整合
LANDSCAPE INTEGRATION
现有工业建筑
EXISTING INDUSTRIAL BUILDINGS
小型服务及设备
SMALL SERVICES AND FACILITIES
感觉 SENSATIONS

(竞标代码) New Zagreb "the place to be"

萨格勒布 · 克罗地亚 Zagreb · Croatia

Geert De Groote · Pieter D'haeseleer

(建筑师)

中标 winner

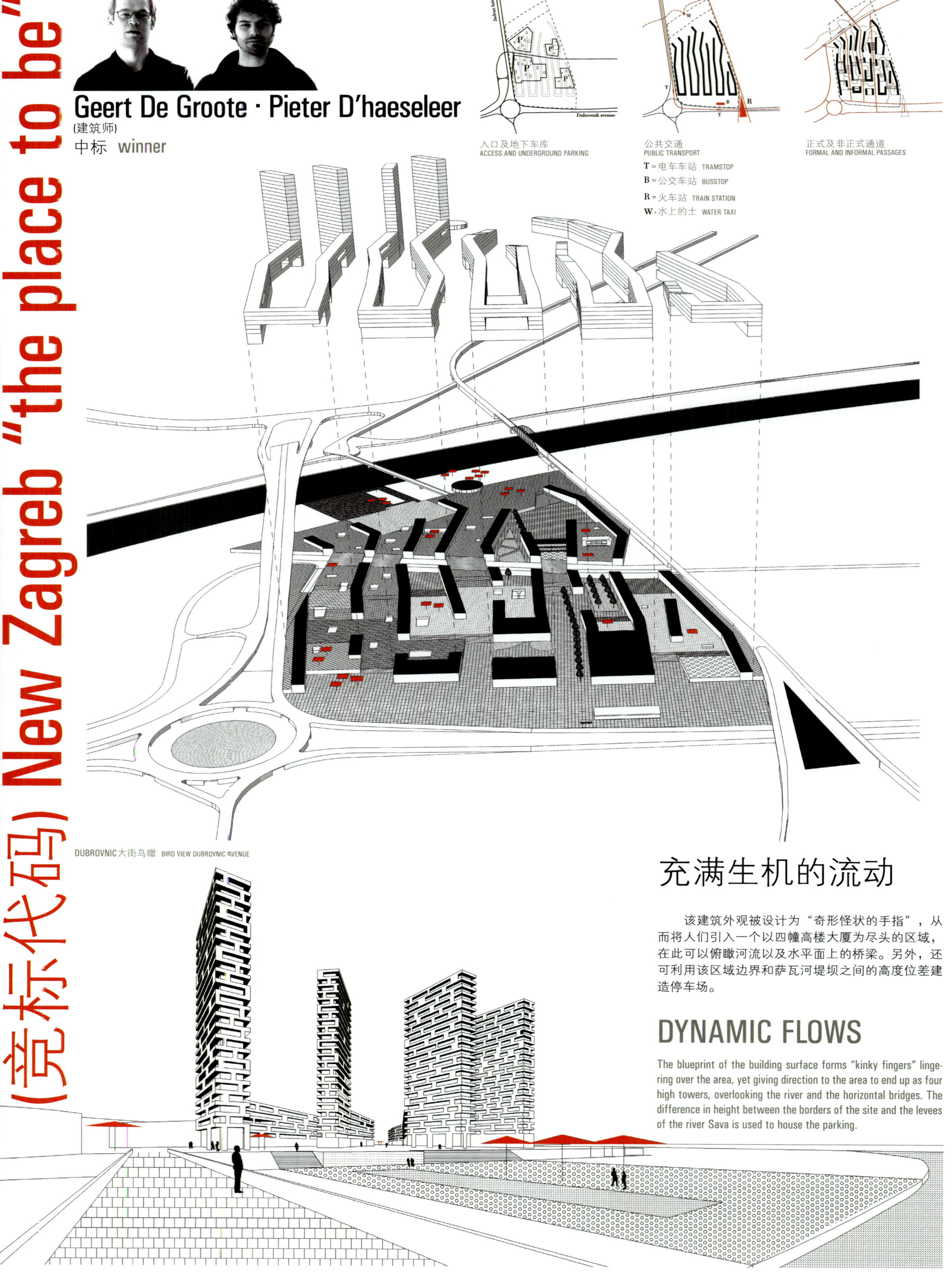

DUBROVNIC大街鸟瞰 BIRD VIEW DUBROVNIC AVENUE

充满生机的流动

该建筑外观被设计为“奇形怪状的手指”，从而将人们引入一个以四幢高楼大厦为尽头的区域，在此可以俯瞰河流以及水平面上的桥梁。另外，还可利用该区域边界和萨瓦河堤坝之间的高度位差建造停车场。

DYNAMIC FLOWS

The blueprint of the building surface forms "kinky fingers" lingering over the area, yet giving direction to the area to end up as four high towers, overlooking the river and the horizontal bridges. The difference in height between the borders of the site and the levees of the river Sava is used to house the parking.

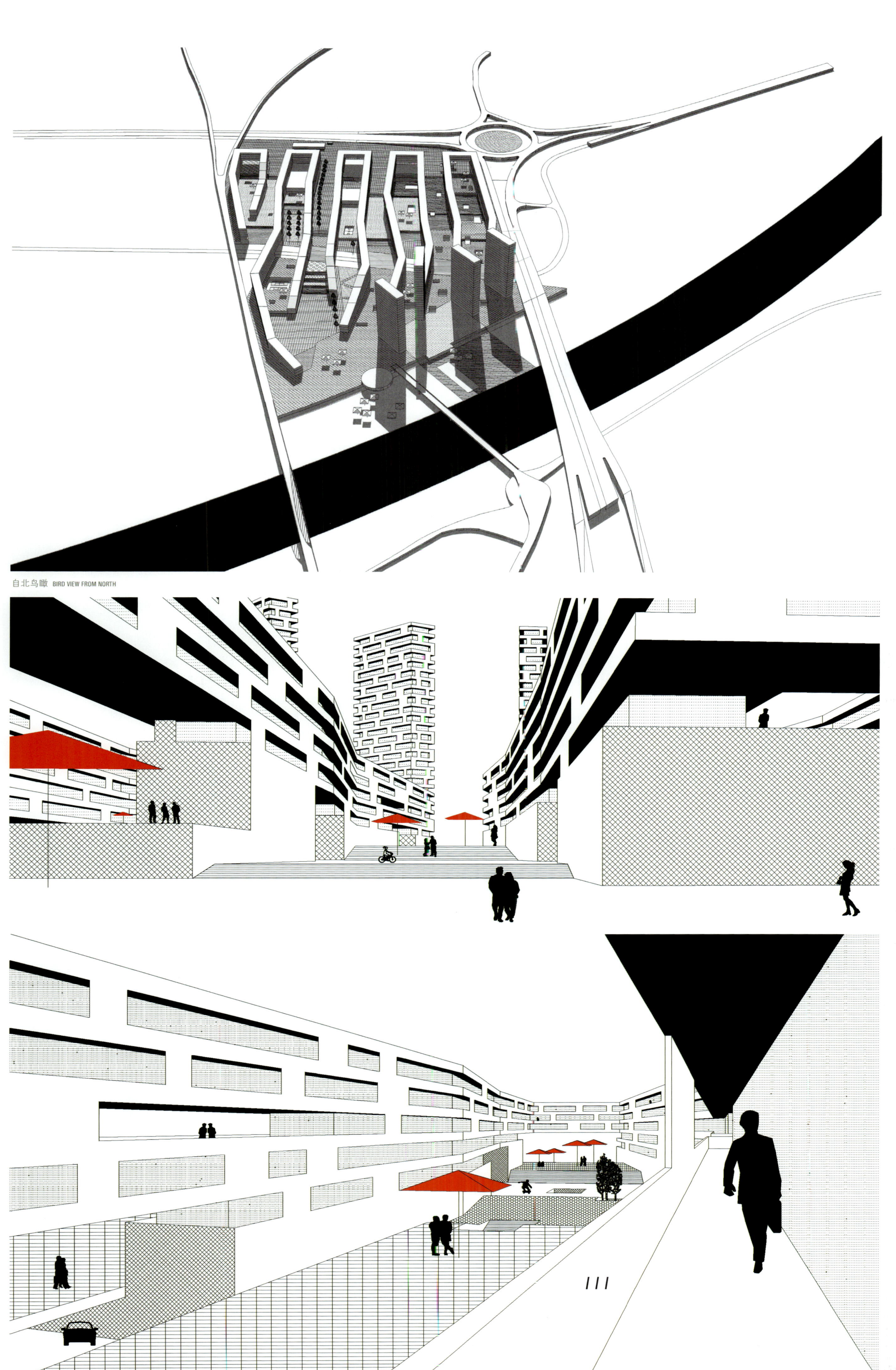

自北鸟瞰 BIRD VIEW FROM NORTH

(竞标代码) Land Ramification

Nicolas Jomain · Boriana Tchonkova
(建筑师)

二等奖 runner-up

分支

该项目计划利用自然划分的空间在一个典型的后工业化郊区结构中创造新的互动，旨在修建一个能带动周边地区多元化发展的基础设施。

BRANCHES

The work tries to create new interactions in a typical post-industrial suburban tissue, through the appropriation of space by nature. The objective is to introduce an infrastructure which could be the support of diversity for the surrounding area.

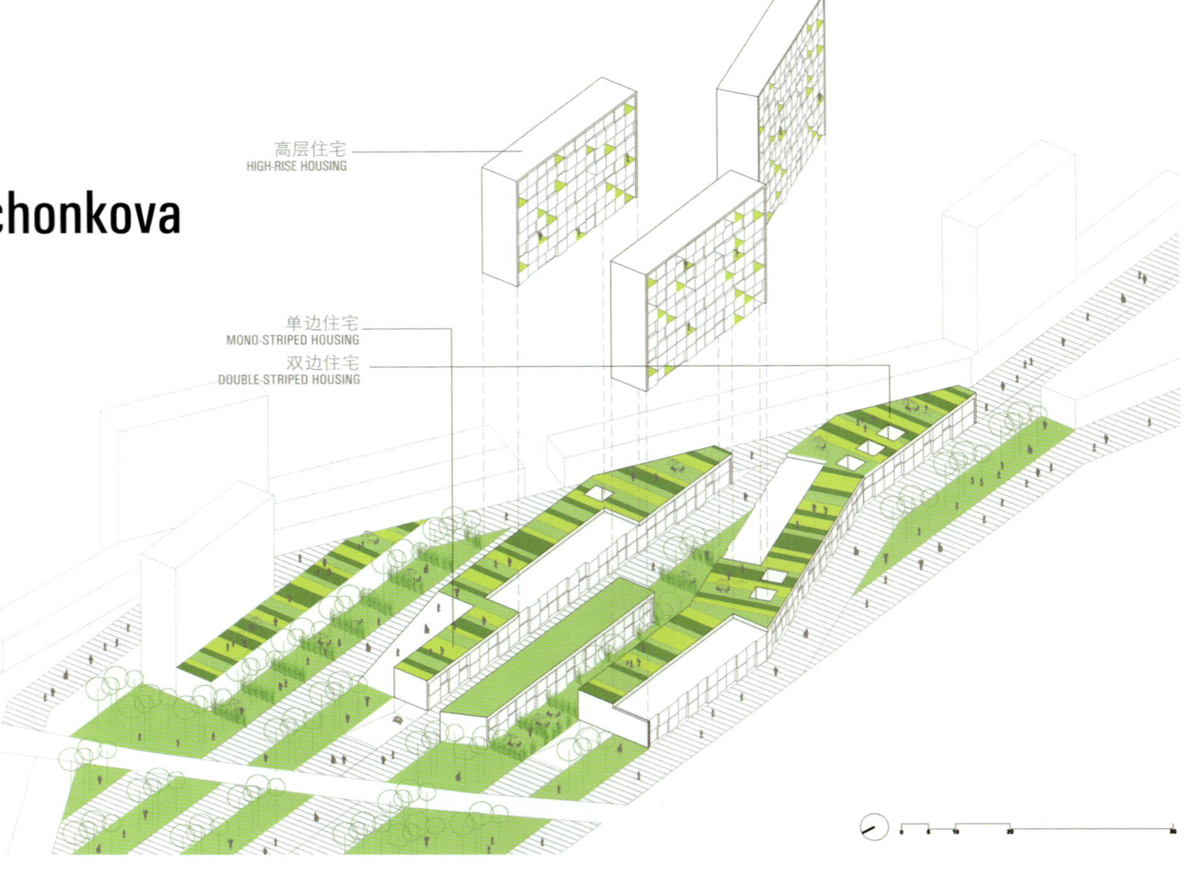

PLAN DIRECTOR MASTERPLAN

高层住宅 HIGH-RISE HOUSING

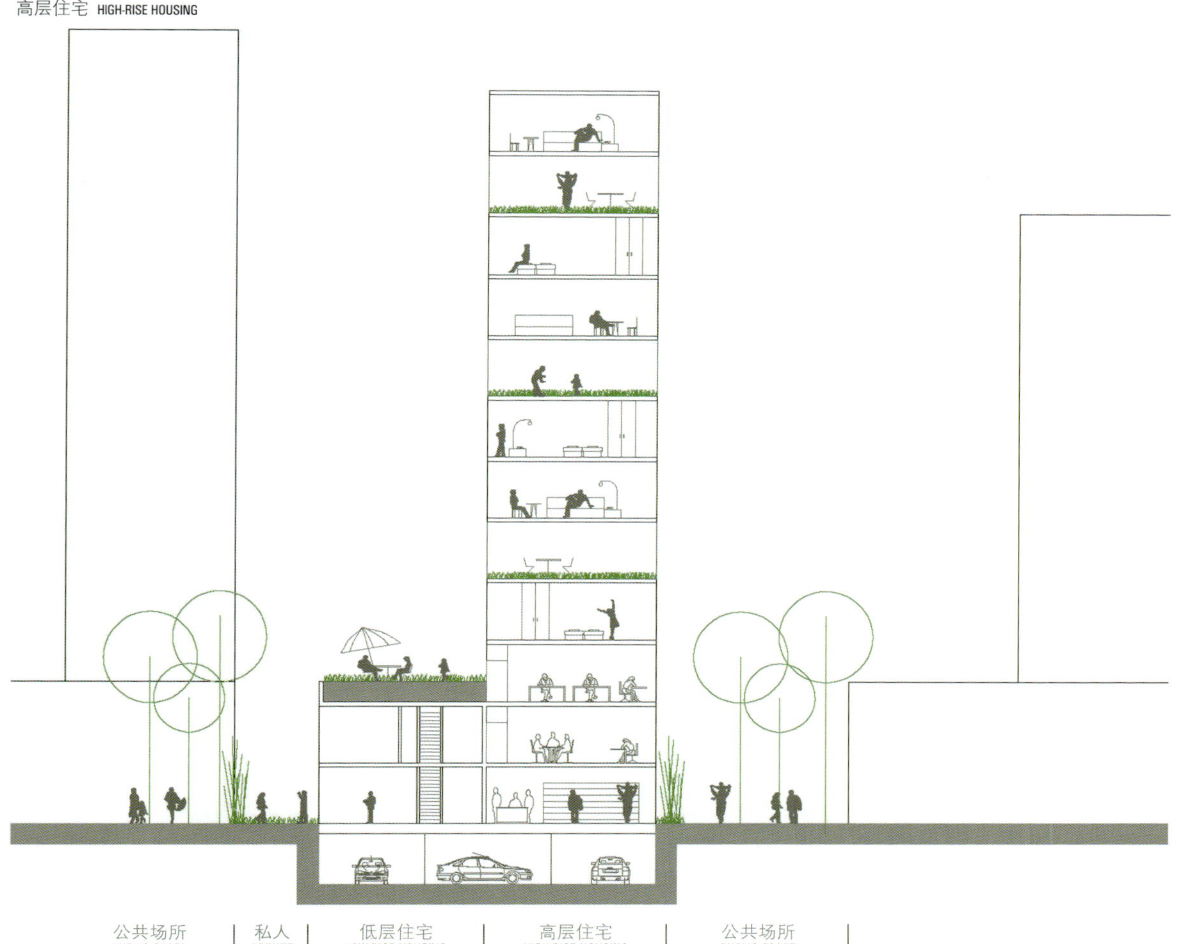

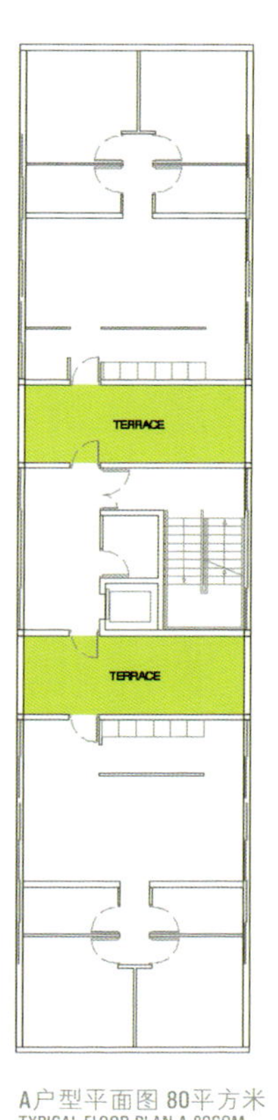

A户型平面图 80平方米
TYPICAL FLOOR PLAN A 80SQM

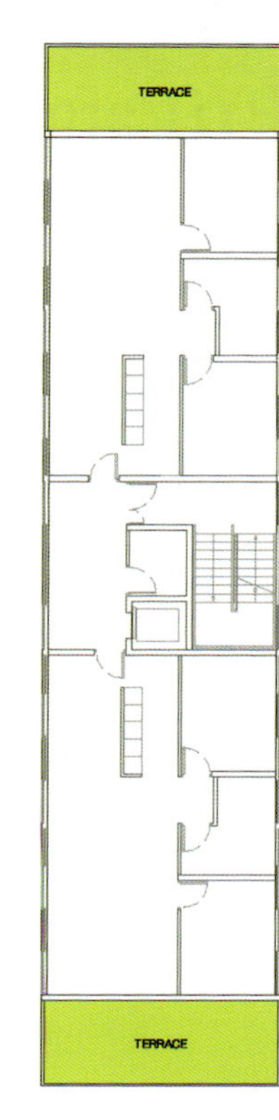

B户型平面图 80平方米
TYPICAL FLOOR PLAN B 80SQM

低层住宅 LOW-RISE HOUSING

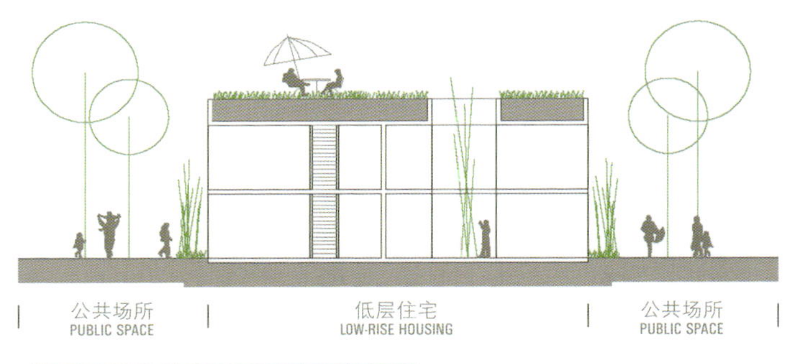

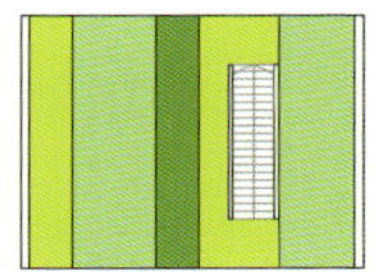

花园顶部
ROOF GARDEN

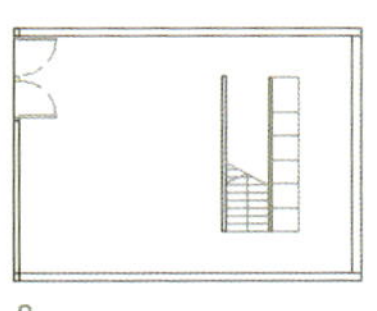

A. 单边住宅 90平方米
A. MONO-STRIPED HOUSING 90SQM

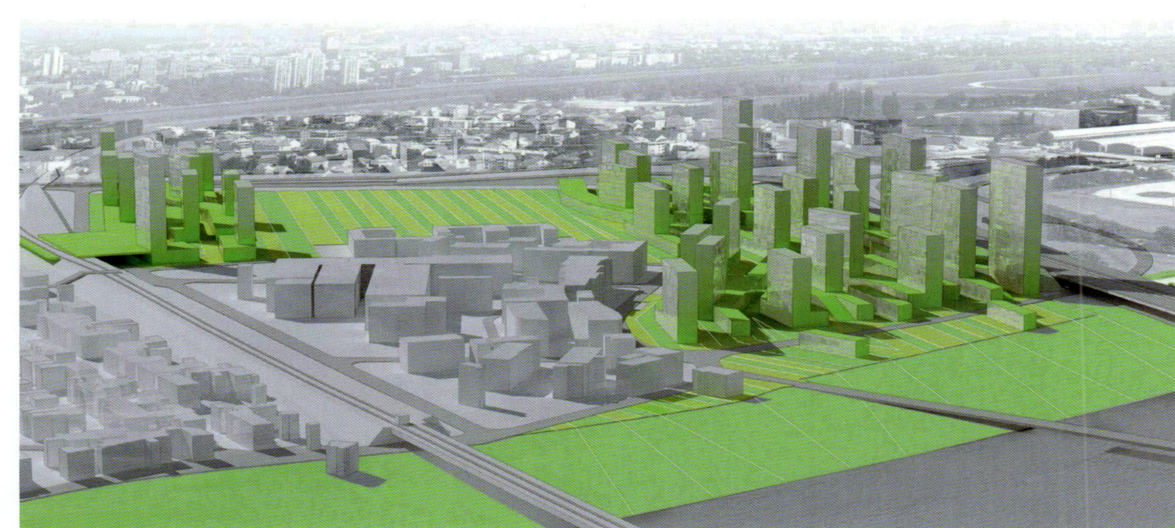

公共场所
PUBLIC SPACE
低层住宅
LOW-RISE HOUSING
公共场所
PUBLIC SPACE

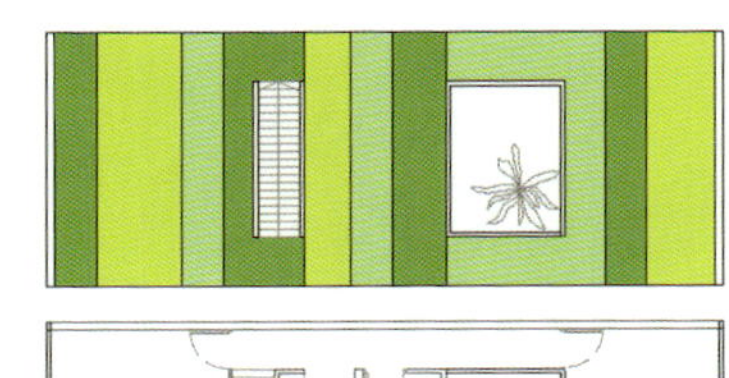

花园顶部
ROOF GARDEN

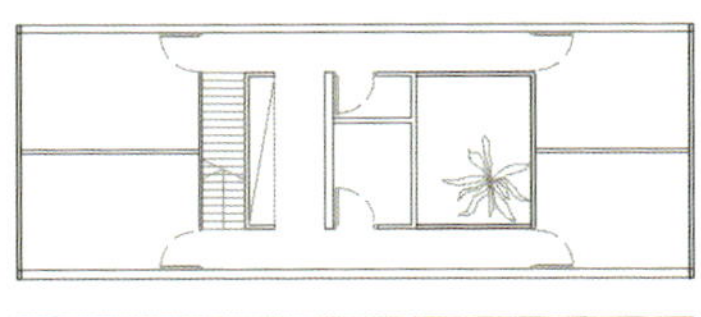

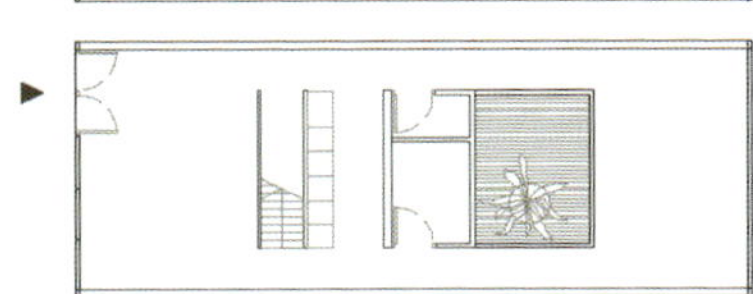

B. 双边住宅 160平方米
B. DOUBLE-STRIPED HOUSING 160SQM

(竞标代码) zigzagerb

PYO arquitectos

(建筑师事务所)

Ophélie Herranz Lespagnol ·Paul Galindo Pastre (建筑师)

参标 proposal

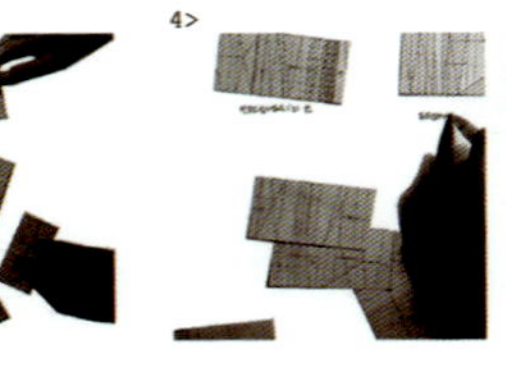
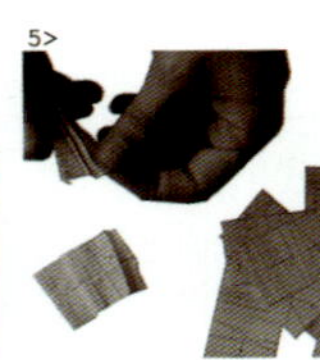

城市进程
URBAN PROCEDURES

该建筑可看作一个可变的结构，一个按照临时安排的顺序进行开发的蓝本。ZIGZAGREB理念中的结构包括多个相互交织的结构体，每一个结构体都由多个同等高度（7层）的条状建筑构成。

The project has to be understood as a variable structure, a growing prototype which is developed through a temporary sequence. ZIGZAGREB concept consists of a structure of interwoven blocks, each one made of multiple strip-buildings with the same height (7 floors).

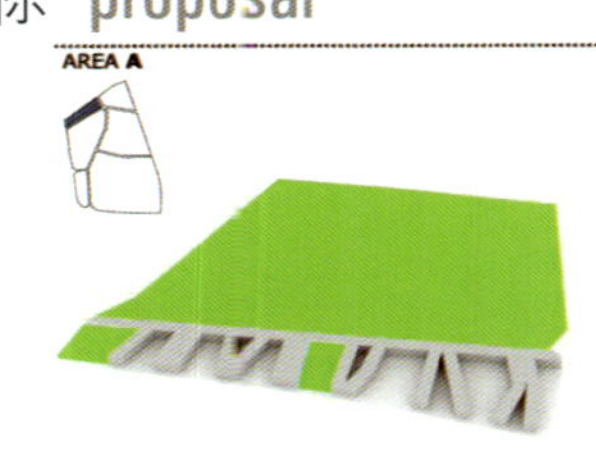

低密度主要公共区域
MAINLY PUBLIC WITH LOW DENSITY

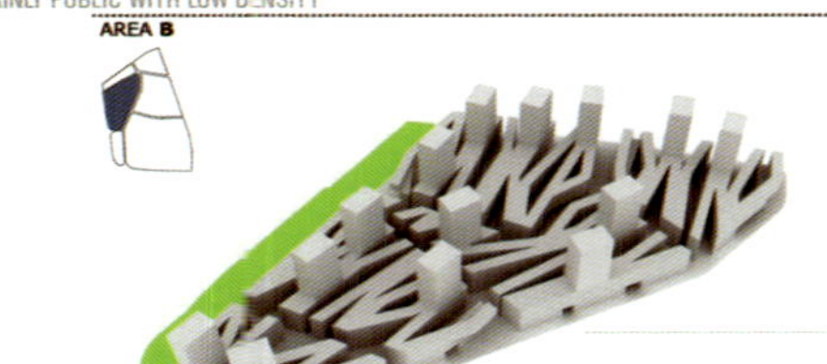

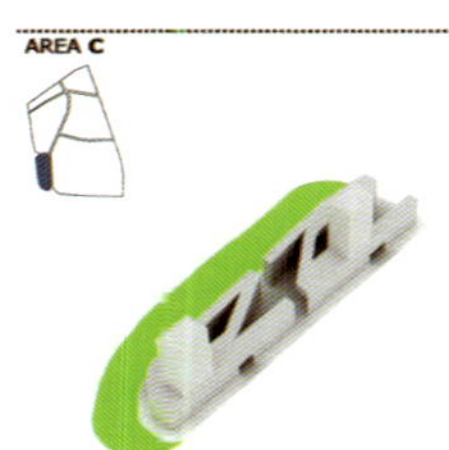

住宅，商铺，医院集中的极度高密度区域
EXTREMELY DENSE WHERE DWELLINGS, BUSINESS AND HOSPITAL FACILITIES ARE MIXED

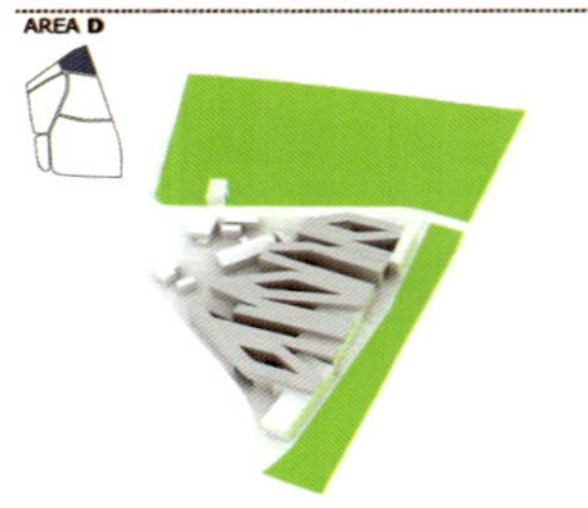

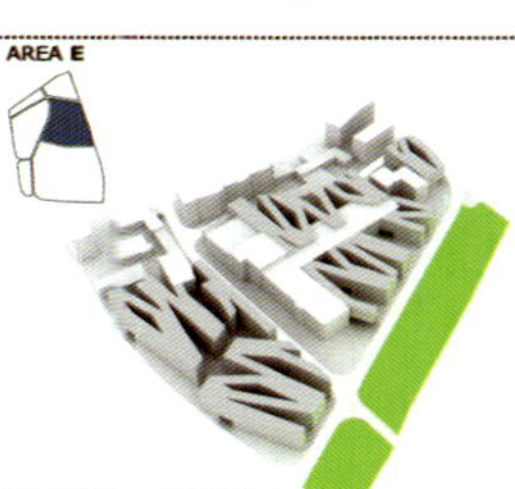

期望在现有环境下重新建立
IT EXPECTS TO REACTIVATE AN EXISTING BUILT SITUATION

拥有一个如同磁场般的多式联运车站
IT CONTAINS THE NEW INTERMODAL STATION WHICH WORKS AS A MAGNETIC FIELD

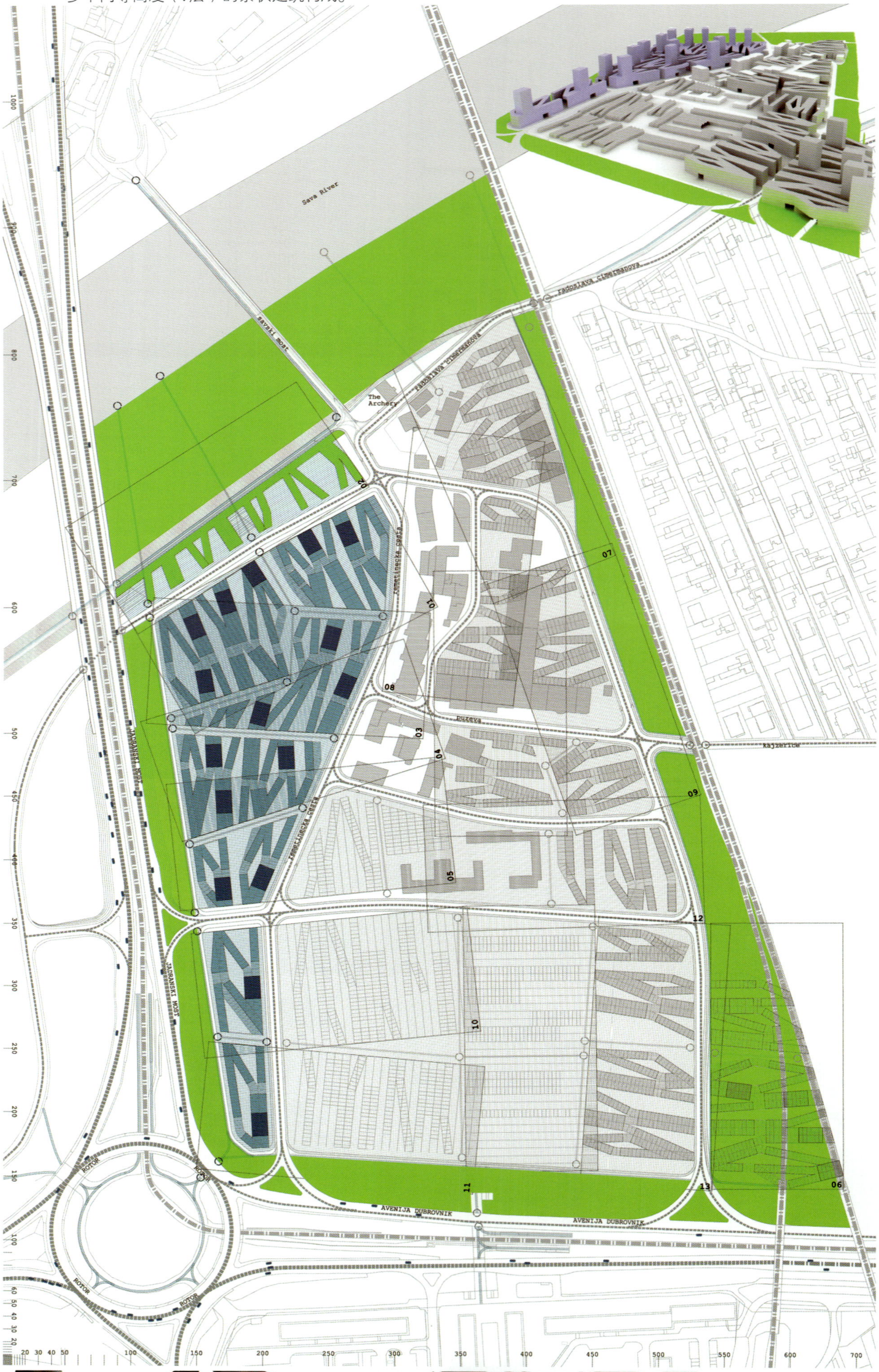

纹理构造
TEXTURAL MECHANISM

城市平面图 URBAN PLAN

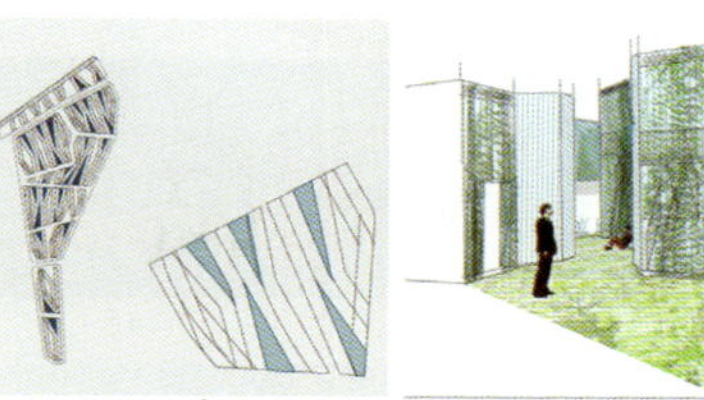
绿地裂缝 GREEN FISSURES

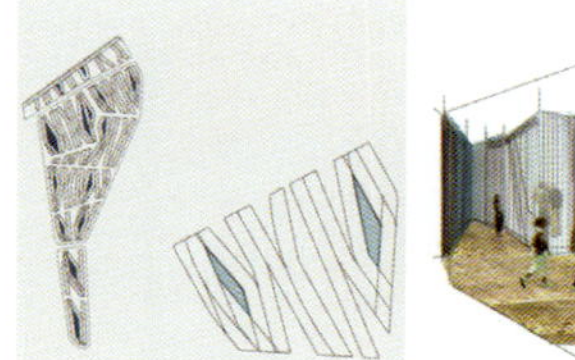

树叶状的庭院 LEAF-SHAPED COURTYARDS

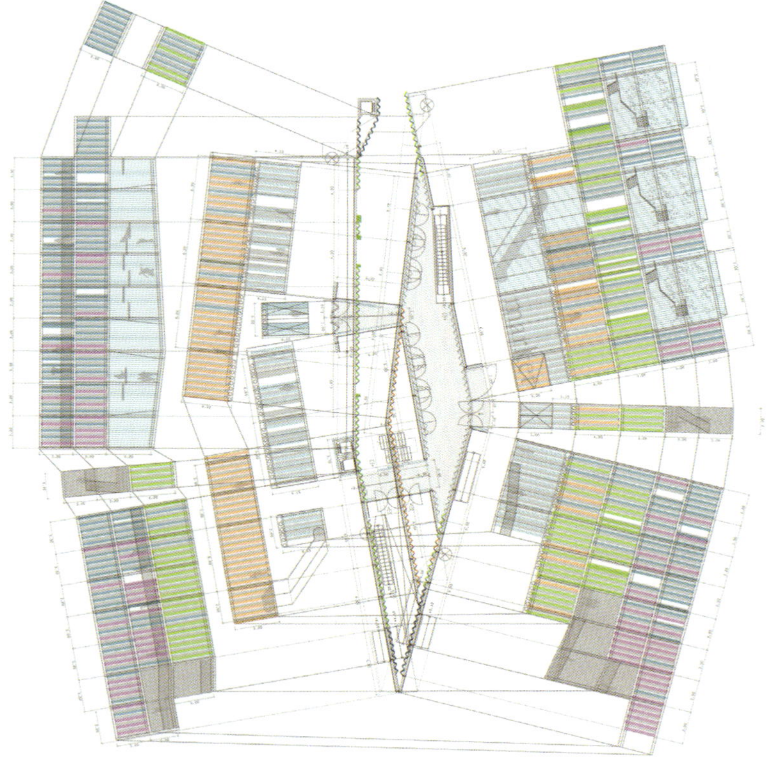
树叶状的庭院·平面图及立面图展开 LEAF-SHAPED COURTYARDS · PLAN AND UNFOLDED ELEVATION

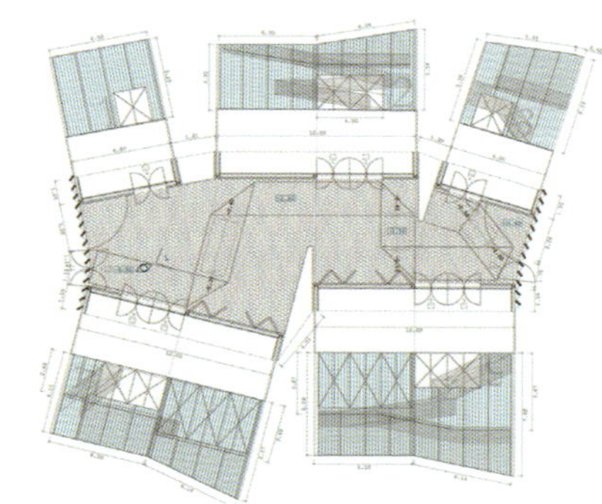
入口广场·平面图及立面图展开 ACCESS PLAZA · PLAN AND UNFOLDED ELEVATION

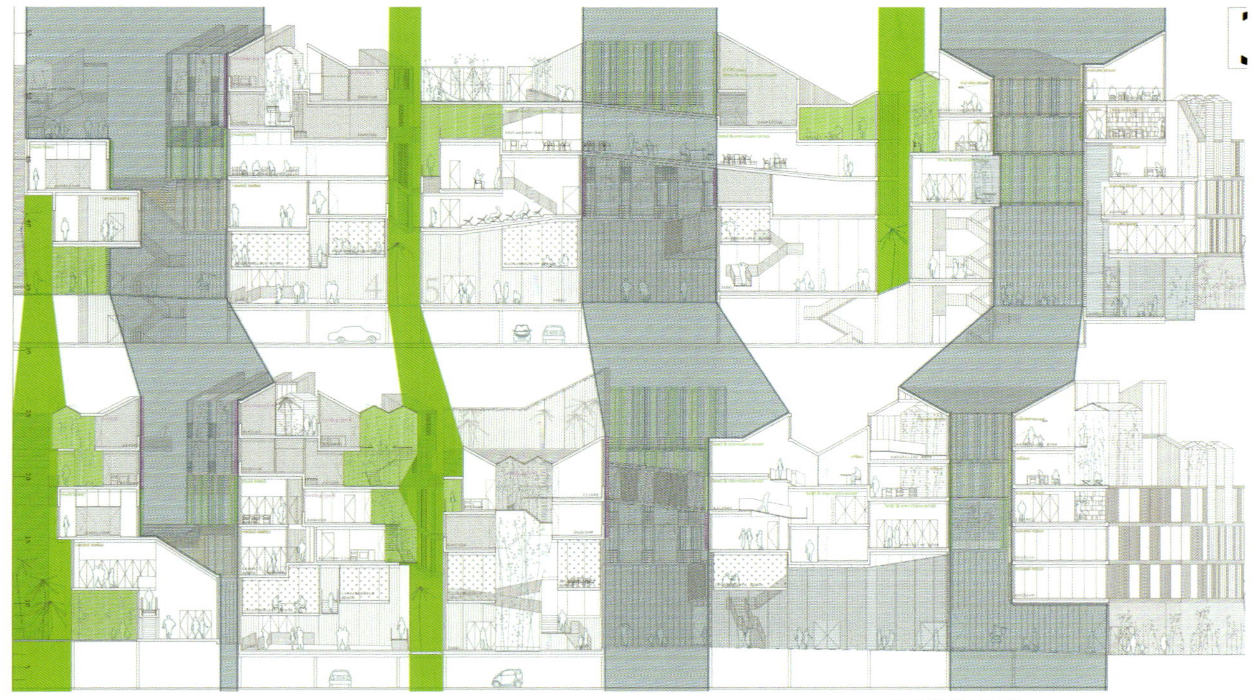
连续样板剖面图 CONSECUTIVE PROTOTYPICAL SECTIONS

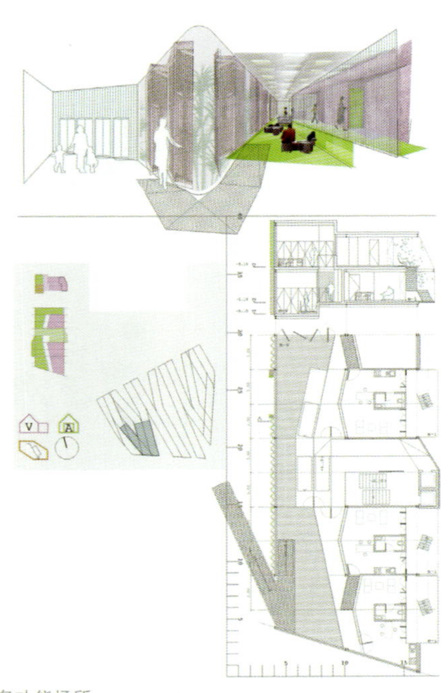
多功能场所 MIXTED-USE SPACE

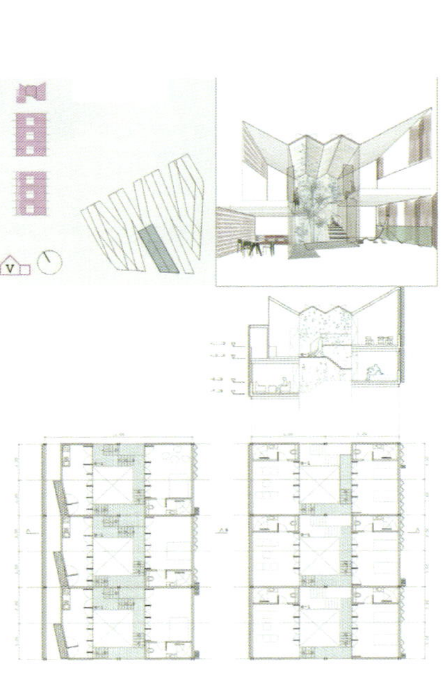
O户型平面图 DWELLINGS TYPE O

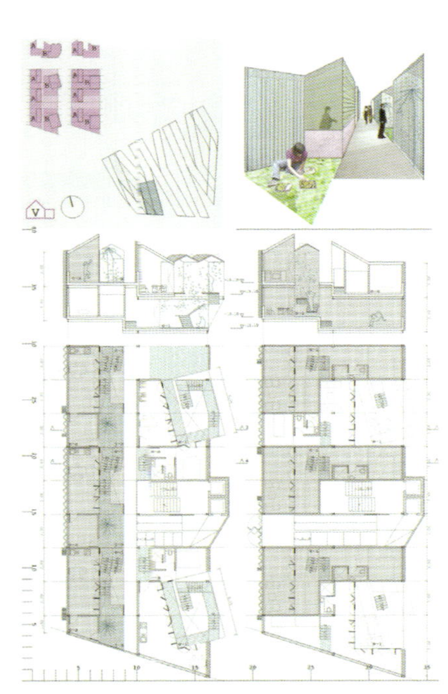
X户型平面图 DWELLINGS TYPE X

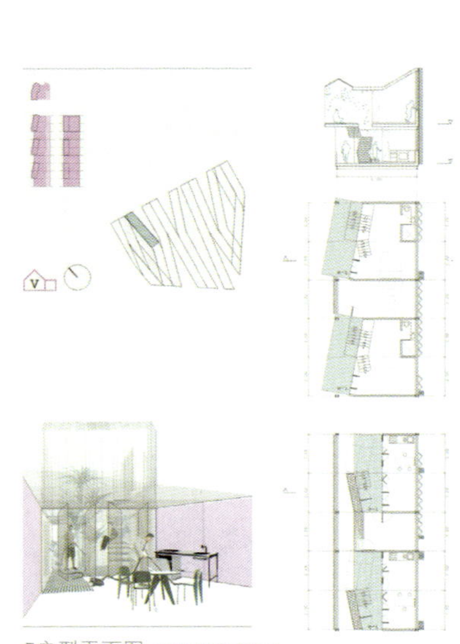
Z户型平面图 DWELLINGS TYPE Z

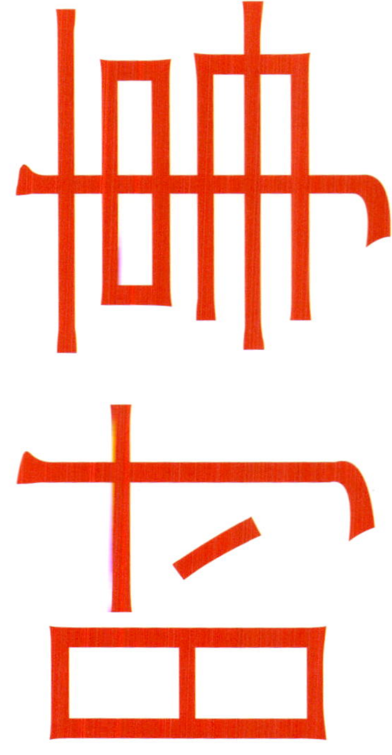

西尔斯赫卡藤中心 · 慕尼黑
Hirschgarten Center · Germany

Allmann Sattler Wappner Architekten 刚刚开始鹿苑中心的初步设计，其中心位于慕尼黑最大的啤酒花园中（在德语中：hirsch=鹿，garten=苑）。他们在2008年的竞标大赛中获胜。

Allmann Sattler Wappner Architekten just started the preliminary design of the Hirschgarten Centre, located in the largest beer garden in Munich (in German: hirsch=deer, and garten=garden). They won the competition in 2008.

一等奖 FIRST PRIZE
ALLMANN SATTLER WAPPNER ARCHITEKTEN
(建筑师事务所)
Amandus Sattler · Markus Allmann + Ludwig Wappner
团队 team: Sabrina Schuppener · Raphael Huber
Thorsten Overberg · Eva Walczyk · Maximilian Köth
Felix Nölke · Severin Oswald · Sebastian Exner
Elise Matter.

地块位置 SITE PLAN

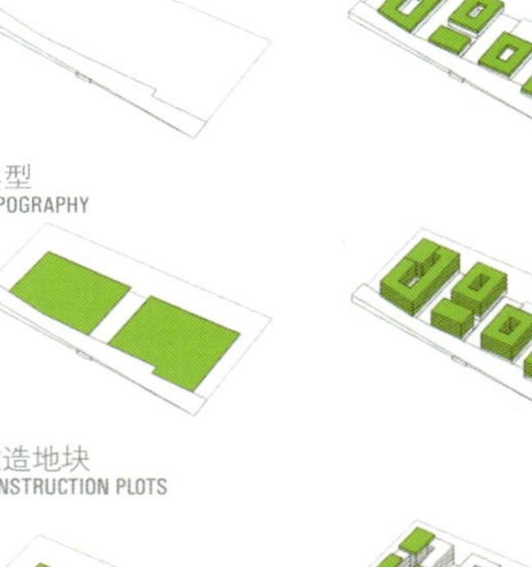

类型 TOPOGRAPHY

样板田 TYPE OF FARM

建造地块 CONSTRUCTION PLOTS

叠加 STACKING

发展层 DEVELOPMENT PLAN

社区边缘 URBAN EDGE

足迹 FOOTPRINT

制高点 HIGH POINTS

基础层 HOST PLAN

总构架 OVERALL STRUCTURE

剖面图 AA SECTION AA

帝国战争博物馆北馆外部空间设计·曼彻斯特
Exterior spaces of the Imperial War Museum North · UK

Topotek I 设计团队在RIBA国际设计大赛中脱颖而出，从而被选为负责帝国战争博物馆北馆（IWMN）博物馆外部空间开发的团队。根据评选小组的意见，设计方案最佳地诠释了一个综合体，表现了冥想和创意的理念，并将与2002年竣工的丹尼尔·里伯斯金设计的建筑相辅相成。

The Imperial War Museum North (IWMN) has selected **Topotek I** as the winner design team to develop the external spaces of the museum following an RIBA international design competition. The proposed scheme was the one which, in the opinion of the selection panel, best addressed a complex brief asking for zones of contemplation, creativity and play which would complement the Daniel Libeskind building opened in 2002.

新艺术中心·韦斯特考克
New Arts Centre · Ireland

韦斯特考克的新艺术大楼正在为开工筹集最后2%的资金。爱尔兰皇家建筑院（RIAI）代表大赛发起者韦斯特考克艺术中心（WCAC），组办了一项分为两阶段的公开建筑设计大赛，搜寻质量上乘、特色鲜明的公用设施设计，并希望该设施能使韦斯特考克人民在Skibbereen接触并了解当地甚至是全球性的艺术。该赛事吸引了216位参赛选手。Donaghy + Dimond、O'Donnell + Tuomey、Clancy Moore和Matter Design / Office dA建筑师事务所入选了决赛候选人名单。

The new Arts Building for West Cork is looking for the final 2% funding needed to start construction. The RIAI (Royal Institute of the Architects of Ireland) administered an open, two-stage competition on behalf of the competition promoters, West Cork Arts Centre (WCAC) for the architectural design of a high-quality, purpose-built facility that will enable the people of West Cork to have access to and to engage with local and global arts at Skibbereen. The competition attracted 216 entries. Along with the winning firm, **Donaghy + Dimond Architects**, O'Donnell + Tuomey Architects, Clancy Moore Architects and Matter Design / Office dA comprised the shortlisted finalists.

通过画廊剖面图 SECTION THROUGH GALLERY

北立面 NORTH ELEVATION

一等奖 FIRST PRIZE

Donaghy + Dimond Architects（建筑师事务所）
预算师 quantity surveyor: David Langdon PKS
结构工程 structural engineer: Casey O'Rourke Associates
设备工程 services engineer: Buro Happold Consultants Ltd.
防火顾问 fire consultants: Michael Slattery and Associates

可持续性创新研发中心·巴达霍斯
Sustainability Centre I+D+i · Spain

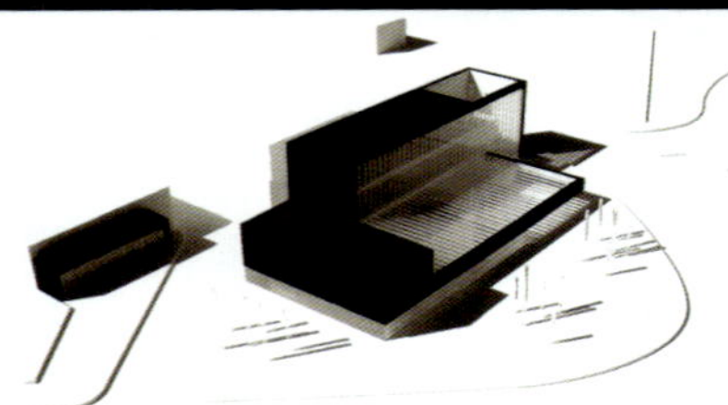

底层平面图 GROUND FLOOR PLAN

二层平面图 FIRST FLOOR PLAN

三层平面图 SECOND FLOOR PLAN

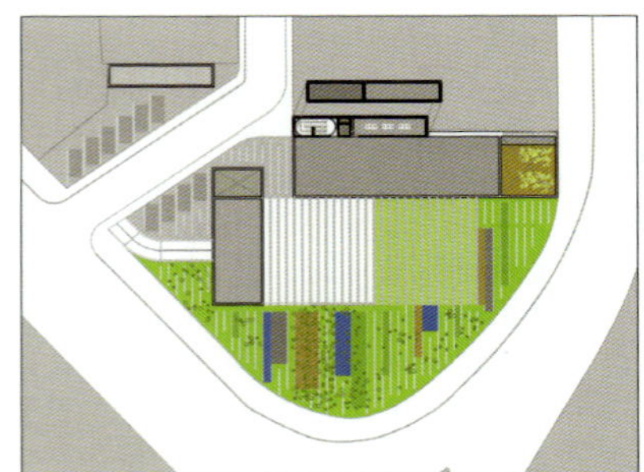
顶层平面图 ROOF PLAN

一等奖 FIRST PRIZE
DANIEL JIMÉNEZ+JAIME OLIVERA ARQUITECTOS (建筑师事务所)
合作 (c) Juan Yruela · Joao Durao · Aurora Fernández
Carlos Olivera · Nacho Jiménez · Fernando García

底层平面图 GROUND FLOOR PLAN

二层平面图 FIRST FLOOR PLAN

三层平面图 SECOND FLOOR PLAN

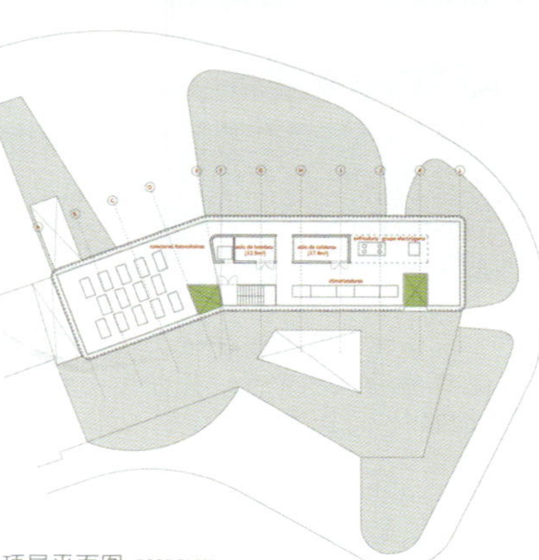
顶层平面图 ROOF PLAN

入围 FINALIST
Juan Elvira Peña · Enrique Krahe Marina
Carlos Rubio Manso · Joaquín Escribano Mediero
Oscar Martínez Valero · José Pablo González Valiente (建筑师)
合作 (c) Joaquín Longhi · Aranzazu Montero · Jesús Isla

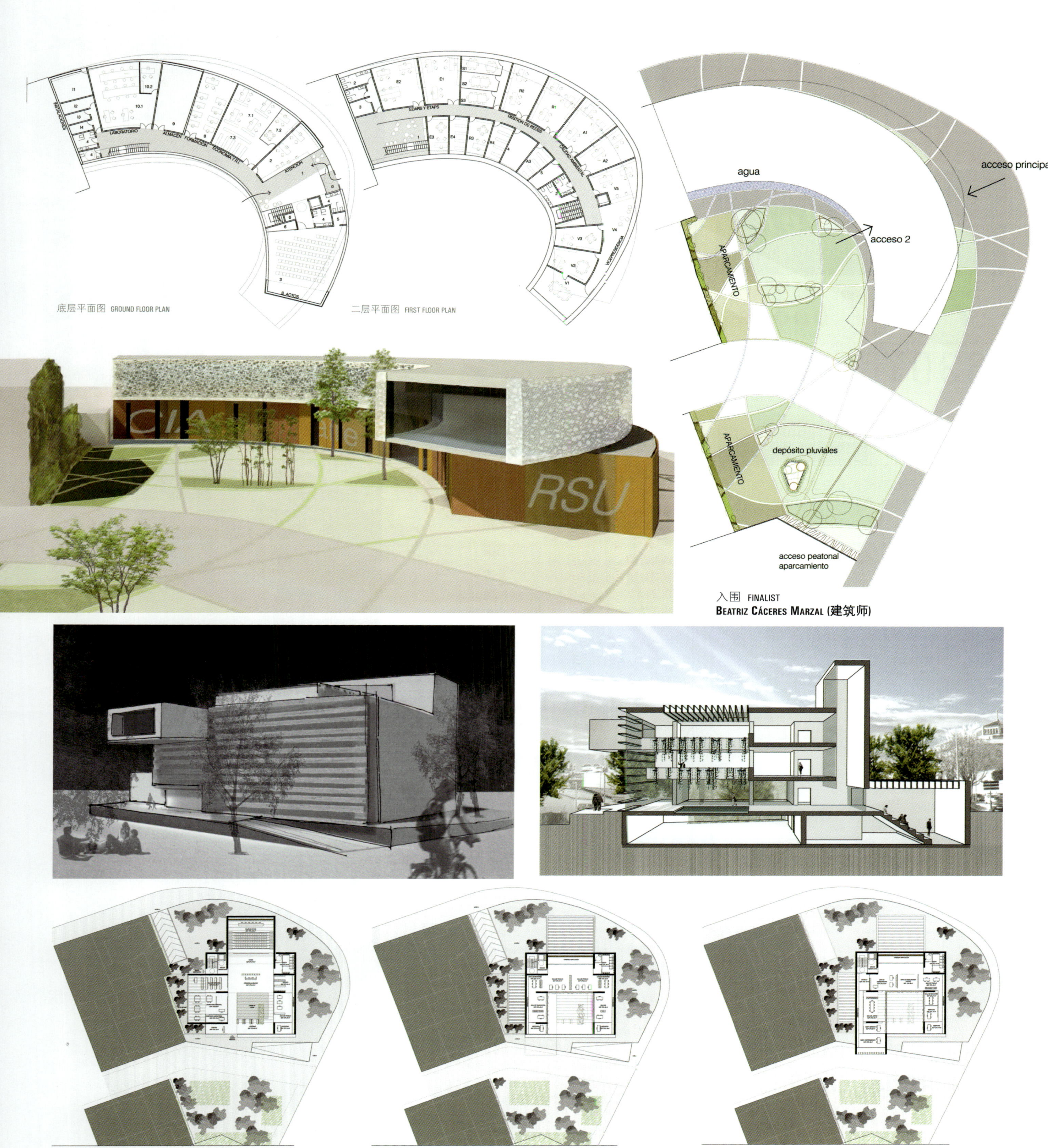

底层平面图 GROUND FLOOR PLAN

二层平面图 FIRST FLOOR PLAN

入围 FINALIST

Beatriz Cáceres Marzal (建筑师)

底层平面图 GROUND FLOOR PLAN

二层平面图 FIRST FLOOR PLAN

三层平面图 SECOND FLOOR PLAN

入围 FINALIST

Julián Prieto Fernández · Thilo Gumbsch (建筑师)

图书预购

pre-order

7本　7 books

630¥ 含税及邮资 taxes and shipping costs included

570¥ 学生优惠价 special prize for students 含税及邮资 taxes and shipping costs included

我想从第几辑开始购买未来建筑系列图书
I wish to subscribe to ***future*** arquitecturas starting from series number..

预购信息 *pre-order information*

姓名 NAME ....

身份证号码 ID NUMBER ....

单位 COMPANY ....

地址 ADDRESS ....

....

邮编 ZIP CODE ....

电话 PHONE ....

电邮 E-MAIL ....

腾讯 QQ ....

更多信息请联系 For more information contact:

未来建筑中国联络办公室 *future* arquitecturas s.l. China
杭州市文晖路303号交通大厦11楼
邮编：310014

china@arqfuture.com

QQ 860464402

日期 DATE　　　　签名 SIGNATURE

www.arqfuture.com

*2011*年有效 valid only for year *2011*

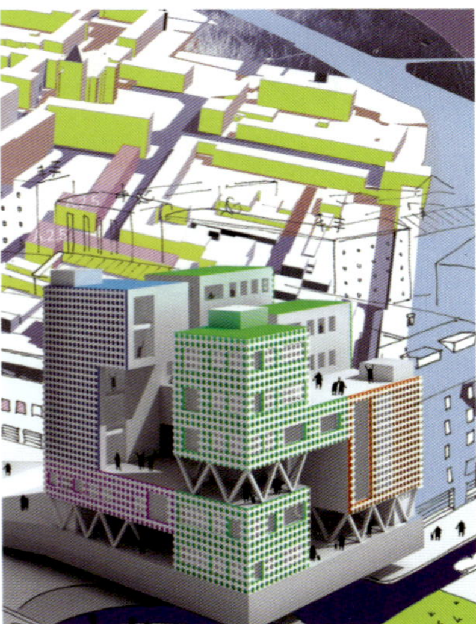

主编 · directors and publishers
Gerardo Mingo Pinacho · Gerardo Mingo Martínez (西)
联合主办单位 · co-sponsors
浙江大学建筑工程学院 · ZUCCEA
浙江大学建筑设计研究院 · ADRZU
西班牙未来建筑 · future arquitecturas s.l.
联合出版 · co-publisher
浙江大学出版社 Zhejiang University Press
图形 · layout
Carlos de Navas Paredes (西)
执行编辑 · managing editor
Gabriela Vélez Trueba (西) · gabriela@arqfuture.com
中国地区公司合伙人 · corporate partner in China
赵磊 Zhao Lei · leizhao@arqfuture.com
美洲地区公司合伙人 · corporate partner in America
Santiago Vélez (厄) · svelez@arqfuture.com
行政人员 · administration
Belén Carballedo (西) · belen@arqfuture.com
Ignacio Rodríguez (西) · future@arqfuture.com
销售部 · distribution department
曾江福 Zeng Jiangfu
手机 cell phone: 13564489269
电话 telephone: 20 65877188
广告 · advertising
china@arqfuture.com

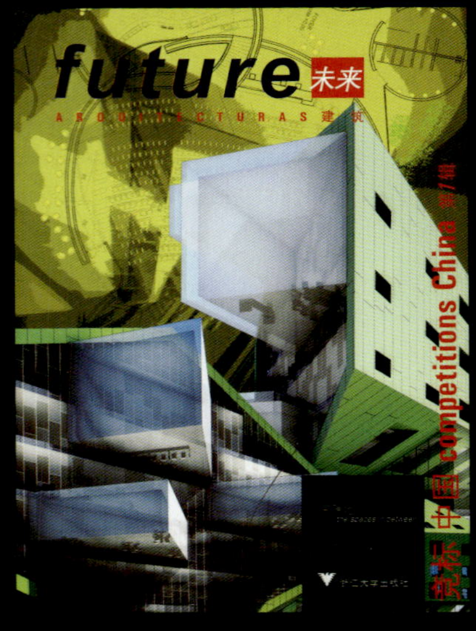